Free Radical Mechanisms of Tissue Injury

Editors

Mary Treinen Moslen, Ph.D.
Department of Pathology
University of Texas Medical Branch
Galveston, Texas

Charles V. Smith, Ph.D.
Department of Pediatrics
Baylor College of Medicine
Houston, Texas

CRC Press
Boca Raton   Ann Arbor   London   Tokyo

Library of Congress Cataloging-in-Publication Data

Free radical mechanisms of tissue injury / editors, Mary Treinen
    Moslen, Charles V. Smith.
        p. cm.
    Includes bibliographical references and index.
    ISBN 0-8493-5161-8
    1. Free radicals (Chemistry)—Pathophysiology. 2. Active oxygen-
-Pathophysiology. I. Moslen, Mary Treinen. II. Smith. Charles V.
    [DNLM: 1. Free Radicals. 2. Oxygen—adverse effects. 3. Oxygen-
-metabolism. 4. Pathology. QV 312 F853]
    RB170.F72 1992
    616.07'1—dc20
    DNLM/DLC
    for Library of Congress   91-29598
                                                                         CIP

Direct all inquiries to CRC Press, Inc., 2000 Corporate Blvd., N.W., Boca Raton, Florida
33431.

© 1992 by CRC Press, Inc.

International Standard Book Number 0-8493-5161-8

Library of Congress Card Number 91-29598
Printed in the United States of America

3 4 5 6 7 8 9 0

Printed on acid-free paper

PREFACE

Considerable skepticism greeted the initial hypotheses that free radical-mediated transformations made a significant contribution to tissue injury and disease. Now the pendulum has swung toward a general acceptance that radicals or reactive oxygen species are involved in some way in many human diseases. This area of research is thus maturing from searches for radicals in association with tissue injury toward a testing of more refined hypotheses regarding specific contributions of radicals and reactive oxygen species to disease processes.

An impetus for this book was a symposium on free radical mechanisms of tissue injury organized by the editors for the 199th national meeting of the American Chemical Society in Boston. The purpose of the symposium was to present an overview of the involvement of radicals and reactive oxygen species in tissue injury. The high quality of the individual presentations and the unique focus on multiple tissues suggested that a book organized along the same lines would be a useful contribution to the field. One goal of the book was to attract the attention of chemists who have not applied their expertise to problems in human disease. The chapters that comprise this book are updated and expanded versions of the symposium presentations.

The editors sincerely thank the authors of the chapters for their timely submission of critical articles and inclusion of new observations. We acknowledge the research and guidance of the scientists whose pioneering efforts stimulated our interest in this area. In particular, we note the late Edward S. Reynolds who introduced MTM to the mechanisms of liver injury by the prototypic free radical toxin, carbon tetrachloride. His observations about $°CCl_3$ binding to lipids of target membranes are illustrated in Figure 2 of the Chapter by CVS. We thank Renée Taub of CRC Press for her careful editing.

We hope that this book will contribute to this field of research and its maturation toward a more critical, specific, and chemically sound science.

Mary Treinen Moslen
Charles V. Smith

THE EDITORS

Mary Treinen Moslen, Ph.D., is an Associate Professor of Pathology and Director of the Graduate Program in Pathology at the University of Texas Medical Branch, Galveston, Texas.

She received a B.S. in Chemistry from Vassar College in Poughkeepsie, New York in 1967 and an M.S. in Scientific Communication from Boston University in 1973. Her career goals shifted from writing about other people's research to doing research while she received research training in the cell injury laboratory of the late Dr. Edward S. Reynolds. She received a Ph.D. in Environmental Toxicology, from the University of Texas Medical Branch, in Galveston, in 1983.

Dr. Moslen is the author of more than 50 papers and has received grants from NIEHS, NIH and the John Sealy Foundation. She is a member of the Society of Toxicology, American Association of Pathologists, American Society for Pharmacology and Experimental Therapeutics. Other memberships are the Editorial Boards of *Toxicology* and of *Toxicology and Applied Pharmacology* and the NIH Toxicology Study Section. Research interests of Dr. Moslen are hepatic pathophysiology after injury by chemical toxins or hydrostatic pressure and the mechanisms of cell protection.

Charles Vincent Smith, Ph.D., is a Research Associate Professor of Pediatrics at Baylor College of Medicine in Houston, Texas. Dr. Smith received his B.A. degree in Chemistry from Rice University in 1970 and his Ph.D. in Chemistry in 1974. He worked as a research chemist for Shell Development Company until 1978, when he moved to Baylor. Dr. Smith is a member of the American Chemical Society, the American Society for Pharmacology and Experimental Therapeutics, the Society of Toxicology, and the Society for Pediatric Research. He is Chair-elect of the Chemical Pathology and Toxicology subdivision of the American Chemical Society. Dr. Smith's major research interests are the chemical mechanisms involved in acute cell and tissue injury, particularly the effects of radicals and reactive oxygen species in biological systems.

CONTRIBUTORS

Bernadette R. Corbett, M.S.
Department of Food Science and
  Human Nutrition
University of Florida
Gainesville, Florida

Michael D. Corbett, Ph.D.
Department of Food Science and
  Human Nutrition
University of Florida
Gainesville, Florida

John W. Eaton, Ph.D.
Department of Laboratory Medicine
  and Pathology
University of Minnesota Medical
  School
Minneapolis, Minnesota

Stephen J. Elliott, M.D.
Department of Pediatrics
Baylor College of Medicine
Houston, Texas

Margaret E. Haberland, Ph.D.
Department of Medicine
UCLA School of Medicine
Los Angeles, California

Yvonne M. W. Janssen
Department of Pathology
University of Vermont
Burlington, Vermont

James P. Kehrer, Ph.D.
Department of Pharmacology and
  Toxicology
College of Pharmacy
University of Texas at Austin
Austin, Texas

Bernhard H. Lauterburg, M.D.
Department of Clinical
  Pharmacology
University of Berne
Berne, Switzerland

Joanne P. Marsh, M.S.
Department of Pathology
University of Vermont
Burlington, Vermont

Muniraj Manohar, Ph.D.
Department of Pathology
University of Vermont
Burlington, Vermont

Mary Treinen Moslen, Ph.D.
Department of Pathology
University of Texas Medical Branch
Galveston, Texas

Brooke T. Mossman, Ph.D.
Department of Pathology
University of Vermont
Burlington, Vermont

Michael E. Murphy, Ph.D.
Department of Pharmacology
University of Vermont
Burlington, Vermont

John S. Penn, Ph.D.
Arkansas Center for Eye Research
University of Arkansas for Medical
  Sciences
Little Rock, Arkansas

Brigitte de Quay, Ph.D.
Department of Clinical
  Pharmacology
University of Berne
Berne, Switzerland

Sayed M. H. Sadrzadeh, Ph.D.
Department of Pathology
New England Deaconess Hospital
Harvard Medical School
Boston, Massachusetts

William P. Schilling, Ph.D.
Department of Molecular
  Physiology and Biophysics
Baylor College of Medicine
Houston, Texas

Scott B. Shappell, M.D., Ph.D.
Department of Medicine
Baylor College of Medicine
Houston, Texas

Susan Shull, Ph.D.
Department of Biochemistry
University of Vermont
Burlington, Vermont

Charles V. Smith, Ph.D.
Department of Pediatrics
Baylor College of Medicine
Houston, Texas

Addison A. Taylor, M.D., Ph.D.
Department of Medicine
Baylor College of Medicine
Houston, Texas

TABLE OF CONTENTS

Chapter 1

FREE RADICAL MECHANISMS OF TISSUE INJURY

Charles V. Smith

TABLE OF CONTENTS

I. REQUIREMENTS FOR FREE RADICAL-MEDIATED INJURY

To study free radical mechanisms of tissue injury, one must first have a tissue injury to study. This requirement is often either totally or partially overlooked. Not all biological responses to perturbations are detrimental; therefore, functional homeostatic responses need to be distinguished from injury. A number of parameters have been employed as markers of injury, but seldom is a definition of the term presented. In the present context, tissue injury will be defined as any alteration that detracts from the viability or essential function of an organism. Brief presentations of the criteria employed for evaluation of tissue damage and, in some cases, the methods employed for the measurement of these parameters can be most helpful in the discussion of mechanisms of injury.

The second requirement for the study of free radical mechanisms of tissue injury is the participation of free radicals, for which there are two definitions. One is easily recognized by most chemists, namely, any atom or molecule containing one or more unpaired electrons (radical) that has escaped the solvent cage or enzyme active site in which it was generated (free). A second meaning of the term appears with disturbing frequency in biological literature and seems to cover just about anything that is apparently bad, is poorly understood, and is hopefully important to reviewers of manuscripts and proposals. In the long run, the chemical definition should prove to be more useful.

A subset of related terms often employed in biological publications includes oxyradicals, oxygen free radicals, or various permutations or perturbations of the buzz words "oxygen" and "radical." The term "reactive oxygen species" would be more appropriate for many of these studies because singlet oxygen, hydrogen peroxide, and peroxide, hydroperoxide, and epoxide metabolites of endogenous lipids and xenobiotics contain chemically reactive oxygen-containing functional groups, but are not radicals and do not necessarily alter tissue molecules through radical reactions (Table 1). Conversely, one of the most studied free radical intermediates, the trichloromethyl radical, contains no oxygen. Although reactive oxygen species and free radicals are not identical, the close association and frequent confusion of the two concepts in the existing biological literature must be recognized. Comparisons and contrasts between toxins such as carbon tetrachloride and certain hydrazines that appear to act through organic radical reactive intermediates and the agents such as diquat, paraquat, and nitrofurantoin, that act through the generation of reactive oxygen species are outlined in Table 2. Also outlined in Table 2 are the characteristics of acute toxins that alter biological molecules through reactions that are not free radical in nature.

II. EVIDENCE FOR FREE RADICALS

Although radical-mediated processes occur normally in many biological

TABLE 1
Reactive Oxygen Species

Species	Target	Product
O_3	R–CH=CH-R	RCHO
$^1O_2$	LH	LOOH
$^3O_2$	L$\cdot$	LOO$\cdot$
$O_2^{\bar{\cdot}}$	LH	No reaction
$O_2^{\bar{\cdot}}$	FERRITIN	Fe^{2+}
$O_2^{\bar{\cdot}}$	SOD	H_2O_2
H_2O_2	GSH	GSSG
H_2O_2	Fe^{2+}	$\cdot$OH, FeO$\cdot$, FeOH
$\cdot$OH	Anything	Many things
HOH	Epoxides	Diols

systems,[1-3] the evidence that free radical intermediates contribute to a wide variety of diseases and examples of tissue damage is extensive.[1,4-6] A number of lines of evidence have been developed that support the involvement of free radical processes in tissue injury. Electron spin resonance (ESR) spectroscopy has been used to detect radical intermediates in biological systems in association with tissue damage.[7,8] However, significant limitations to this approach are readily recognizable. The steady-state concentrations of radical species reflect not only the rates of their formation, but the measured concentrations also are determined by the rates with which the radicals react with surrounding molecules. The more reactive radical species will be less likely to accumulate, and, therefore, should be more difficult to observe directly. Further, the more reactive radicals would be expected to be potentially more damaging than radical intermediates that are less reactive and more readily observed by ESR.

Radical intermediates that are too reactive to be studied directly can be observed with the use of spin-trapping agents such as α-phenyl-*t*-butylnitrone (PBN) or 5,5-dimethyl-1-pyrroline-*N*-oxide (DMPO). The radical species formed initially are trapped as the corresponding nitroxyl radicals, which are more stable chemically and are easier to observe by ESR.[7] Although these radical traps have been used *in vivo*, the quantitative relationship between radical formation and trap adduct formation remains to be elucidated.[8] Additional complications to the interpretation of experiments with spin traps arise from the need to understand the metabolism of the spin-trapping agents and the respective radical adducts. The cell-mediated disappearance and the chemical degradation of radical adducts of DMPO have been reported to be very rapid.[9]

Much of the biological evidence for the participation of radicals and reactive oxygen species in tissue damage is based on observations of protection against or potentiation of the extent of injury caused by a given challenge. Protection is achieved experimentally by administration of radical traps or "antioxidants," while the potentiation of injury is accomplished by depletion

TABLE 2
Chemical Nature of Reactive Metabolites

	Class A	Class B	Class C	Class D
Parent drugs	Furosemide Dimethylnitrosamine	Acetaminophen Bromobenzene Thiophene 2-Furamide	CCl$_4$ Halothane hydrazines	Diquat Paraquat Nitrofurantoin Hyperoxia Ozone ?(ischemia/reflow) Reactive oxygen species
Metabolite	Hard electrophile	Soft electrophile	Organic radical	Reactive oxygen species
Substrate for GSH Transferase	No	Yes	No	+/−
Parameters				
Protein alkylation	++++	++++	++++	0
Lipid Alkylation	0	0	++++	0
GSH depletion	0	++++	0	++
Alkane expiration	0	++++	++++	++
LOOH and LOH	0	0	++++	++
GSSG formation	0	−	0	++++

TABLE 3

Pharmacological Modifications of Reactive Metabolite-Mediated Toxicities

Metabolite	Class A Hard electrophile	Class B Soft electrophile	Class C Organic radical	Class D Reactive oxygen species
Effect on tissue damage of				
GSH depletion	0	+ + + +	0	?
GSH precursors	0	− − − −	0	?
Se deficiency	0/ −	−	0	+ + + +
Vitamin E deficiency	−	+ / −	0	+ + + +
BSO	(0?)	+ + + +	?	(+ + ?)
BCNU	(0?)	0	0	+ + + +
Desferrioxamine	(0?)	0	− −	− − −

of endogenous substances that are capable of acting as antioxidants or radical traps.[10] The pharmacologic evidence obtained from these studies is extremely useful, particularly since this experimental approach is readily applicable to studies in relevant live animal experimental models.

Nevertheless, the expectation that a given chemical or experimental manipulation will exert the single intended effect is assumed more frequently than it is evaluated critically. For example, the hepatotoxicity of the herbicide diquat is dramatically potentiated in animals that have been fed a diet deficient in selenium and in whom significantly lower glutathione peroxidase activities are observed.[11] Similarly, diquat toxicity is potentiated by pretreatment of the animals with 1,3-bis(2-chloroethyl)-1-nitrosourea (BCNU), which inhibits the enzyme glutathione reductase.[12] The apparently critical role of the glutathione peroxidase/glutathione reductase system in protecting against diquat hepatotoxicity is consistent with the involvement of hydrogen peroxide or substrate lipid hydroperoxides in the toxicity of diquat. Other interpretations are possible — one being that the observed potentiation of injury is nonspecific. However, the hepatic damage caused by acetaminophen, which is also hepatotoxic in these same animal models but apparently through different mechanisms (Table 2), is not exacerbated by selenium deficiency[13] or by pretreatment with BCNU,[14] thus supporting the mechanistic interpretation for the involvement of substrates for glutathione peroxidase/glutathione reductase in the damage done by diquat. The effects of these and other related pharmacological manipulations on selected hepatotoxic agents are outlined in Table 3, which further illustrates the differences in the mechanisms involved in the toxicities of reactive electrophiles, organic radicals, and reactive oxygen species (ROS).

The third type of evidence for radical-mediated tissue injury is the identification and quantitation of products characteristic of the radical-mediated alteration of biological molecules and the correlation of these chemical alterations with biological determinations of tissue injury. The coupling of quantitative chemical determinations with studies of biological outcome is arguably the most definitive type of evidence for the participation of radical-mediated reactions in tissue damage. Intuitively, a radical or other reactive intermediate would be expected to initiate tissue injury through chemical alterations of biological molecules. Further, the nature and extent of these alterations would be expected to determine the pathophysiological sequelae. It should be noted that the initiation of tissue injury by chemically reactive intermediates is distinguished from toxicities caused by exaggerated pharmacological responses, which are due to excessive concentrations of drug and lead to receptor-mediated cellular responses that initiate cell dysfunction or death.

In most examples of tissue injury a number of chemical alterations can be identified, a complicating factor that is particularly true of toxicities mediated by radicals or ROS. Not all of the observed changes will contribute to injury. Some may be only slightly related to the biologically significant processes, while other chemical alterations may be *results from* rather than *causes of* the injury. In addition, the mechanisms responsible for initiation of injury may be different from the mechanisms responsible for the biological propagation of the lesion. However, examination of chemical evidence for the involvement of radical processes associated with tissue injury in conjunction with pharmacologic manipulations of the extent of injury can be very informative. In some cases, it is possible to increase or decrease one or more of the chemical alterations dramatically while only slightly affecting the biological parameters of injury (or vice versa). A lack of correlation between the chemical alterations and the biological outcomes indicates the need for a more careful consideration of other mechanisms for the initiation of damage.

III. TYPES OF RADICAL REACTIONS

Free radicals are not noted for the specificity with which they react.[1,4] This lack of specificity presents a major obstacle to the study of radical-mediated processes occurring in biological systems, which are chemically very complex even without pathological disruption. However, three major types of radical interaction with biological molecules have been identified, studied, and related to tissue injury. Much of what is known about radical-mediated pathology can be considered from the perspective of these three types of reactions and the transformations to be expected, given the available tissue target molecules. These reactions are radical addition, electron transfer, and atom abstraction (Figure 1).

A. RADICAL ADDITION REACTIONS

Early studies of radical addition reactions were based on measurements of the covalent attachment to tissue lipids of metabolites of radiolabeled xenobiotics, such as carbon tetrachloride.[4,15] Measurements of covalent binding to lipids have been used to provide estimates of the extent to which radical metabolites interact with tissue molecules because the target structures, carbon–carbon double bonds found primarily in unsaturated fatty acids and cho-

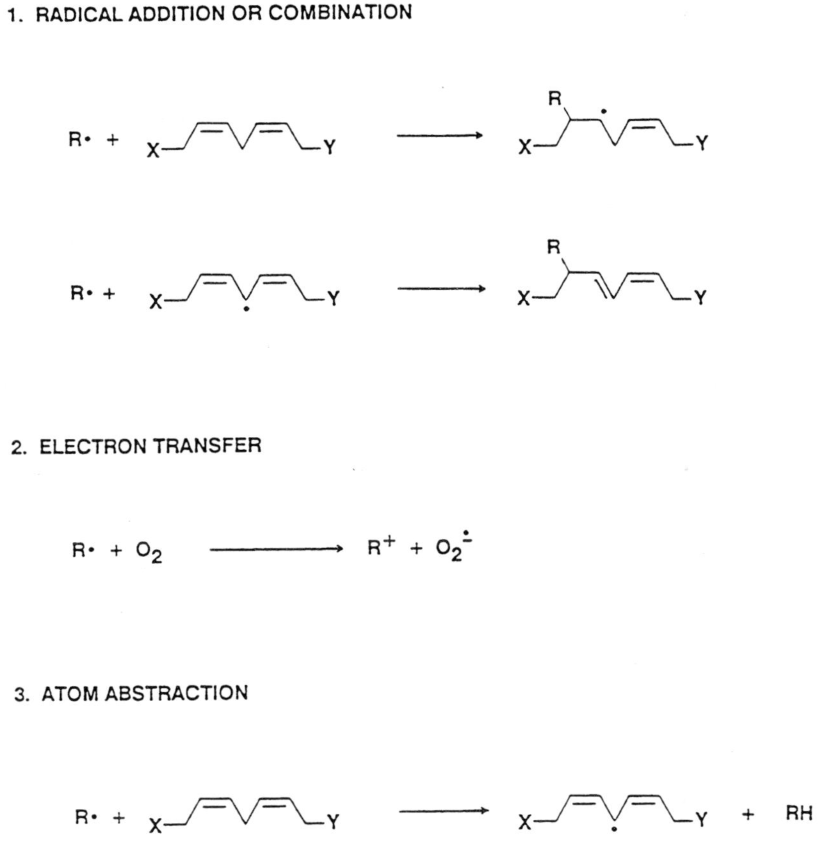

1. RADICAL ADDITION OR COMBINATION

2. ELECTRON TRANSFER

$$R\cdot + O_2 \longrightarrow R^+ + O_2^{\cdot-}$$

3. ATOM ABSTRACTION

FIGURE 1. Three characteristic reactions of free radicals in biological systems.

lesterol, are located preferentially in tissue lipids (Figure 2). Most quantitative estimates of binding to tissue lipids have been obtained by administering radiolabeled drug, isolating tissue lipid, repeated evaporation of solvent and added unlabeled agent to remove parent compound and radioactive metabolites not bound to tissue lipids, and quantitation by liquid scintillation counting.[15] A major limitation of this technique is the requirement that the toxin and its nonlipid-bound metabolites be volatile enough for removal by evaporation. In studies of a selected series of nonvolatile hepatotoxins, we found that transesterification of the extracted lipid and chromatographic separation offered a reasonable experimental approach to the study of binding to lipids by a broader range of substances.[16]

A second limitation of the more commonly employed technique of extraction and evaporation was uncovered in this study.[16] Although the parent drug was removed readily by evaporation in control studies, we observed substantial binding to hepatic lipids by [^{14}C]-dimethylnitrosamine *in vivo*,

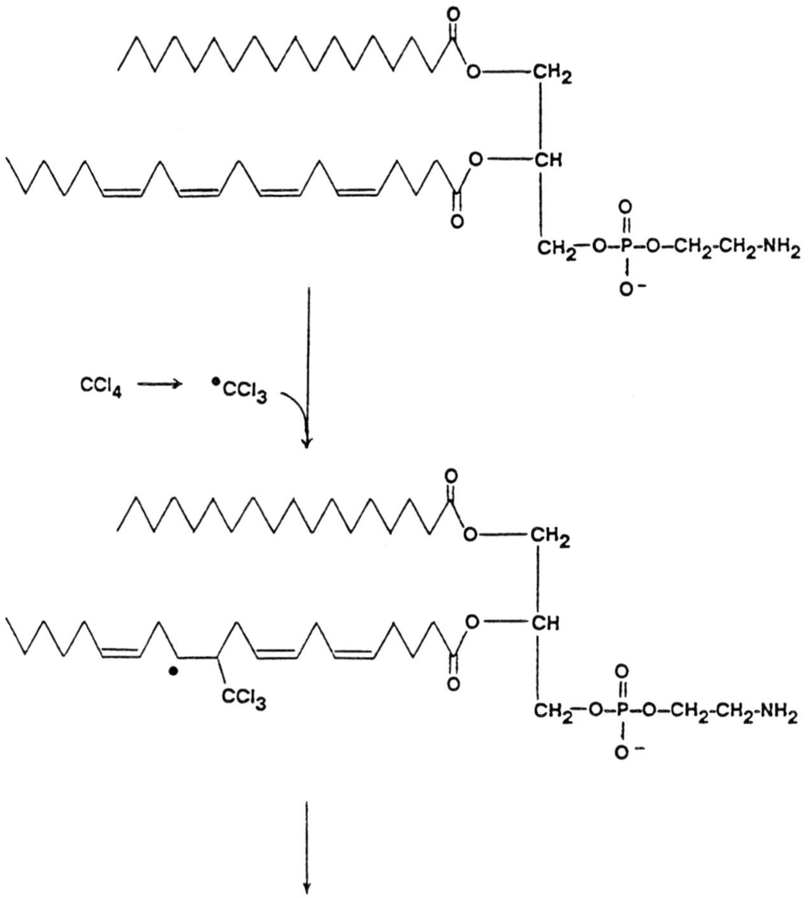

FIGURE 2. Covalent binding to lipids of metabolites of carbon tetrachloride. The covalent binding of metabolites of carbon tetrachloride to tissue lipids appears to result from addition of the metabolically generated trichloromethyl radical to carbon–carbon double bonds as are found in unsaturated fatty acids.

despite the strong evidence that the reactive metabolite of dimethylnitrosamine more closely resembles a methyl carbonium ion than a free radical.[17] Transesterification of the lipid caused the radioactivity from the labeled dimethylnitrosamine to be partitioned to the aqueous phase of a hexane/water extraction. We interpreted these results to indicate that the binding of the reactive metabolites of [14C]-dimethylnitrosamine had occurred to the polar functional groups of the tissue lipids rather than to the lipophilic side chains (Figure 3).

In contrast, essentially all of the radiolabel in lipid extracts of rats treated with [14C]-carbon tetrachloride remained associated with the lipophilic components (fatty acid methyl esters and cholesterol) following transesterification, as would be expected for products derived from radical addition reactions

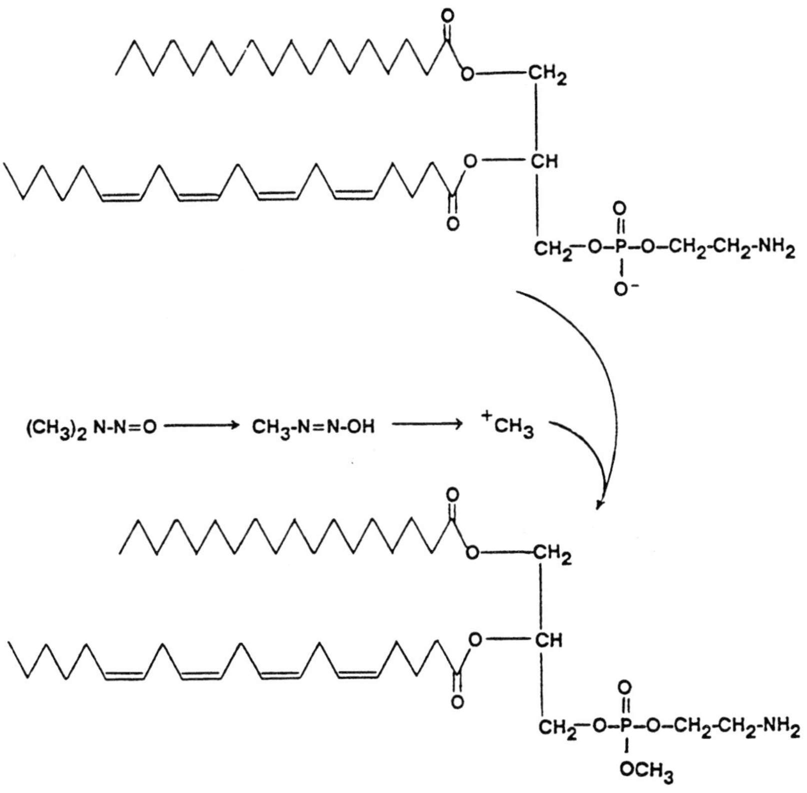

FIGURE 3. Covalent binding to lipids of metabolites of dimethylnitrosamine. The covalent binding of metabolites of dimethylnitrosamine to tissue lipids appears *not* to involve radical addition reactions, but is best explained in terms of alkylation of nucleophilic sites on lipid molecules by highly electrophilic metabolites, represented here as the methyl carbonium ion.[16]

(Figure 2). Mass spectral evidence for metabolism of carbon tetrachloride providing trichloromethyl adducts to unsaturated fatty acids *in vitro*[18] (Figure 2) and to cholesterol *in vivo*[19] has been reported. Studies providing quantitative data based on methods that are as chemically specific as mass spectrometry are needed.

Metabolites of [¹⁴C]-carbon tetrachloride bind to tissue protein as well as to lipid.[4,16] Tissue proteins would appear to be less susceptible to radical-mediated covalent bond formation, because of the relative absence of isolated carbon–carbon double bonds in proteins, but the extent of the binding to protein observed is similar to the binding to lipid.[16] Phosgene[20] and dichloromethyl carbene[21] have also been reported as metabolites of carbon tetrachloride and these electrophilic intermediates could account for the binding to nucleophilic sites on tissue proteins through nonradical reactions. Recent studies by Osawa et al.,[22,23] however, indicate that much, if not all, of the

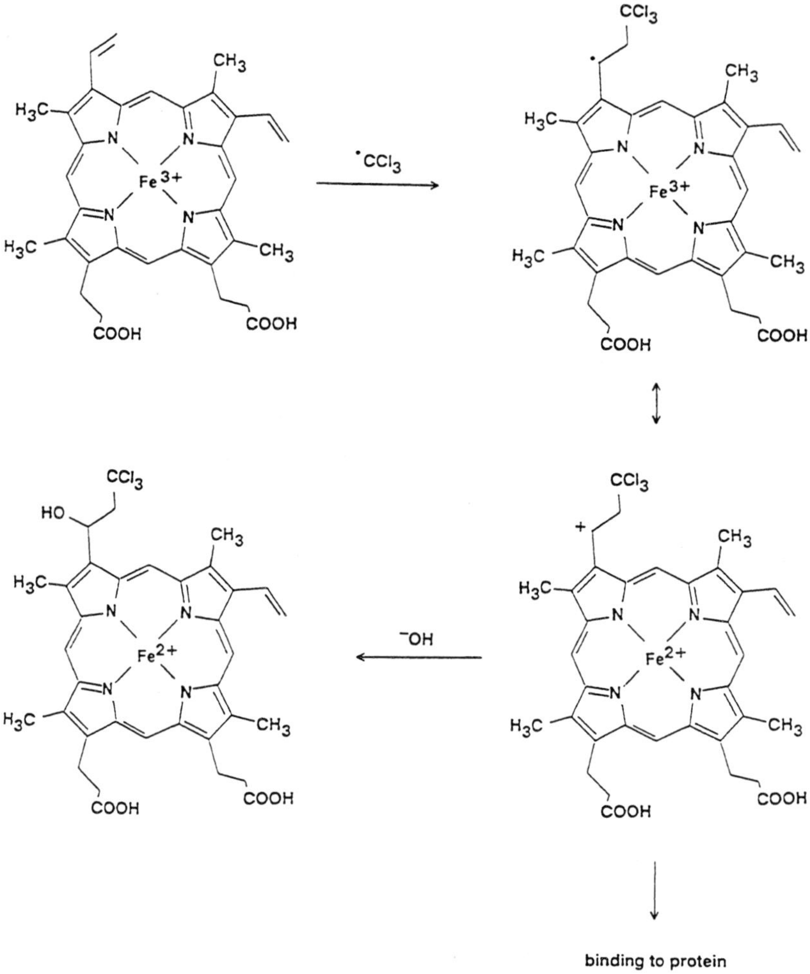

FIGURE 4. Radical-mediated covalent binding of carbon tetrachloride metabolites to protein. Osawa et al.[22,23] have suggested that the covalent bonding of carbon tetrachloride metabolites to tissue protein proceeds through the heme-derived carbonium ion produced by radical addition of trichloromethyl radical to the I-vinyl moiety.

protein binding can be attributed to radical addition of the trichloromethyl radical to the beta carbon of the I-vinyl group of heme (Figure 4), leading to formation of a heme-protein covalent crosslink through electrophilic intermediates such as the carbonium ion shown in Figure 4.

B. ELECTRON TRANSFER REACTIONS

Radical metabolites also can interact with tissue molecules by electron transfer reactions in which the reactive intermediate accepts an electron,

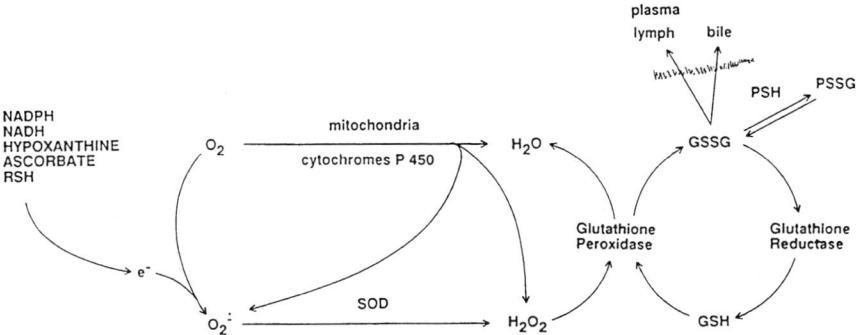

FIGURE 5. Determinants of the steady state concentrations of glutathione disulfide (GSSG).

causing a net reduction of the intermediate at the expense of concomitant oxidation of the tissue molecule. Conversely, oxidation of the reactive intermediate by electron transfer yields the net reduction of the tissue molecule (Figure 1, reaction 2). Electron transfer mechanisms have received considerable attention from investigators studying mechanisms of biological damage, particularly in the generation and actions of ROS. The biological reduction of oxygen to water and carbon dioxide is carefully controlled in an apparent attempt to minimize the release of partially reduced oxygen species in free or unbound form. However, the capabilities and capacities of these control mechanisms are not entirely foolproof and numerous studies of the mechanisms of tissue injury attribute the damage to reactions of these partially reduced, chemically reactive, oxygen intermediates.

Transfer of a single electron to dioxygen yields the superoxide anion radical (Figure 5) and the involvement of superoxide in many different forms of tissue injury has been proposed.[24-27] Increased production of superoxide is observed in response to exposure to elevated concentrations of oxygen or in response to increased electron flow (reductive stress?) during normoxia. The increased superoxide production during hyperoxic exposure has been attributed to saturation of the mitochondrial electron flow regulation or to leakage from microsomal cytochromes P-450.[28] Another potential source of increased production of superoxide is from endogenous metabolites that accumulate in reduced form during ischemia and release this excess reductive capacity suddenly during reflow, an effect potentially augmented in some tissues by conversion of xanthine dehydrogenase to xanthine oxidase.[26,29,30] Exogenous compounds can stimulate superoxide production by mechanisms that have been studied intensively over the last several years.[25,31,32]

The ubiquity and high activities of superoxide dismutases in most tissues limit the observable reactions attributable to free superoxide, and increases in superoxide generation frequently are documented through the effects of its dismutation product, hydrogen peroxide.[31,33] Intracellular hydrogen peroxide concentrations are kept low by catalase and by glutathione peroxidase, the latter system producing glutathione disulfide (GSSG) in the process of reduction of hydrogen peroxide (Figure 5). Cellular concentrations of GSSG are normally kept far lower than the concentrations of glutathione (GSH) by glutathione reductase-mediated reduction of GSSG back to GSH and by the ATP-dependent export of GSSG from some cells. Development of a number of highly sensitive methods for the measurement of GSSG in cells and extracellular fluids has facilitated characterization of the factors affecting formation and clearance of GSSG. Consequently, GSSG measurement has become a useful index of exposure of cells and tissues to increased production of ROS.[31,33-36]

The critical position of glutathione peroxidase and glutathione reductase in the metabolism of hydrogen peroxide and the number of sensitive methods available for the measurement of GSSG and GSH in biological systems support the utility of measuring GSSG as a marker for increased production of reactive oxygen species. However, the connections between shifts in thiol/disulfide status and biological dysfunction need careful and specific examination.

GSH/GSSG ratios in rat liver of around 300/1 have been reported[37] and it is reasonable to speculate that large shifts in this ratio from normal homeostasis could be deleterious to tissues. Oxidant stress is a term used to categorize either a decrease in the GSH/GSSG ratio or an increase in GSSG production. However, oxidant stress is not a term that is intrinsically well defined, and in most publications no definitive meaning can be identified. One suggested definition of the term is a shift in the pro-oxidant/antioxidant status of a system.[38] We have suggested that a more useful definition of oxidant stress would be any measurable shift in one or more redox couples to more electron-deficient (oxidized) equilibria or steady states. The most frequently measured parameters of oxidant stresses are tissue concentrations or rates of export of GSH and GSSG, but even massive shifts in GSH/GSSG ratios are not always accompanied by observable tissue injury, particularly *in vivo*.[12,39] Increased production of GSSG is reasonable evidence for increased production of reactive oxygen species, particularly the nonradical hydrogen peroxide, but a change in the GSH/GSSG ratio due to a decreased concentration of GSH can be produced by mechanisms involving neither radicals nor ROS.[10,14,40]

Large decreases in GSH/GSSG ratios, effected by stimulation of GSSG production or by depletion of GSH, can be produced without the initiation of observable tissue injury.[12,31,40] These studies are useful in providing quantitative criteria for the interpretation of data obtained from other studies in which tissue injury and evidence of oxidant stresses are observed and the mechanisms responsible for the tissue damage are being sought. For example, ischemia and reflow in the perfused rat liver are accompanied by measurable increases in the production of GSSG.[29] However, the amount of GSSG produced as a result of ischemia and reflow is orders of magnitude lower than the GSSG produced by doses of diquat that do not cause observable hepatic

$$\text{PSH} + \text{GSSG} \quad \Longleftrightarrow \quad \text{PSSG} + \text{GSH}$$

$$K_{ox} = \frac{[\text{PSSG}]\,[\text{GSH}]}{[\text{PSH}]\,[\text{GSSG}]}$$

$$\frac{[\text{PSH}]}{[\text{PSH}] + [\text{PSSG}]} = \frac{[\text{GSH}]/[\text{GSSG}]}{K_{ox} + [\text{GSH}]/[\text{GSSG}]}$$

FIGURE 6. Protein thiol *S*-thiolation by thiol—disulfide exchange with GSSG. At equilibrium, the PSH/PSSG ratios are determined by the values of K_{ox} for a given PSH and the GSH/GSSG ratios, but are independent of the absolute concentrations of GSH.[37]

damage. The quantitative criteria produced by the studies with diquat, therefore, indicate that little, if any, of the injury caused by ischemia and reflow should be attributed to the observed oxidant stress response.

The cell injury putatively caused by shifts in GSH/GSSG ratios presumably would be mediated by thiol/disulfide exchange reactions with critical protein thiol groups (PSH). Evidence supporting the hypothesis that protein thiol *S*-thiolation by GSSG contributes to oxidant-induced cell death has been reported from studies in isolated cell models,[41] but comparable changes *in vivo* were not observed in related studies.[42] At chemical equilibrium, the fraction of a given protein thiol that would become converted to the corresponding disulfide, PSSG, by reaction with GSSG (Figure 5) is determined by the ratio of GSSG to GSH and by the equilibrium constant of the redox pair, K_{ox}, but would be independent of the absolute concentrations of GSH and GSSG (Figure 6).[37] For a protein thiol with a $K_{ox} = 1$, a 30-fold decrease in the GSH/GSSG ratio, from 300 to 10, would decrease the cellular content of the reduced form of the protein to 90% of its original status. Even for an oxidant shift of this magnitude (30-fold), it could be difficult to demonstrate a 10% decrease in the activity of a thiol-dependent enzyme and even more difficult to establish a change of this nature as a cause of cell death or dysfunction. The equilibrium content of protein–S–S–protein disulfides formed through thiol-disulfide exchange reactions with GSH and GSSG is a function of the absolute concentrations of GSH and GSSG as well as the relative concentrations (Figure 7), unlike the formation of PSSG.[37]

It is important to recognize that changes in tissue or cellular concentrations of GSH or GSSG will not necessarily effect *in vivo* the thiol status of a protein predicted by thermodynamic equilibria. If the interaction of the potential

$$\underset{SH}{\overset{SH}{P}} + GSSG \rightleftharpoons \underset{S}{\overset{S}{P}} + 2\,GSH$$

$$K_{ox} = \frac{[PSS]\,[GSH]^2}{[P(SH)_2]\,[GSSG]}$$

$$\frac{[P(SH)_2]}{[P(SH)_2] + [PSS]} = \frac{[GSH]^2/[GSSG]}{K_{ox} + [GSH]^2/[GSSG]}$$

FIGURE 7. The formation of intramolecular protein disulfides through thiol—disulfide exchange with GSH and GSSG. At thermodynamic equilibrium, the formation of intramolecular protein disulfides is a function of the absolute concentrations of GSH, as well as the GSH/GSSG ratios.

reactants is inhibited by cellular compartmentation, protein folding, or related phenomena, the protein thiol status will not change in parallel to changes in GSH and GSSG. Conversely, irreversible denaturation of a protein by disulfide formation would lead to an underestimation of the effects on such proteins by measurements of GSH and GSSG.[37]

Some tissues also maintain low intracellular concentrations of GSSG, in part, through its active export (Figure 5).[33] The excretion of intracellular GSSG into bile by parenchymal hepatocytes has been studied extensively and measurement of GSSG in sequential bile samples is a useful means of following changes in intrahepatic oxidant responses.[12,31,33,39,40] Substances that stimulate the intrahepatic production of ROS, such as diquat, paraquat, and nitrofurantoin, at doses that do not necessarily result in hepatic necrosis, cause rapid and marked increases in the biliary excretion of GSSG.[31] In male Fischer 344 rats, a 0.05 mmol/kg (ip) dose of diquat causes a greater than tenfold increase in biliary excretion of GSSG, but causes no necrosis.[39] Administration of 0.24 mmol/kg of ferrous sulfate 15 min prior to this dose of diquat results in massive hepatic damage, but no difference in the GSSG production in response to diquat is observed. Conversely, pretreatment with the iron chelator desferrioxamine offers significant protection against the hepatic injury caused by a hepatotoxic dose of diquat (0.1 mmol/kg), but causes no change in the biliary GSSG response to this higher dose of diquat.

The chemical determinations of GSSG coupled with the pharmacological manipulations of the hepatic injury with ferrous sulfate and desferrioxamine indicate that the necrosis caused in these rats by diquat is not readily attributed to shifts in thiol-disulfide status, whatever the uncertainties regarding compartmentation or kinetics. The apparent separation of shifts in thiol-disulfide status from the initiation of injury in this model does not disprove the relevance of thiol-disulfide status to oxidant-initiated cell injury, but the application of similar approaches and quantitative comparisons to other models of tissue injury will be helpful in evaluating the potential contributions of oxidant stress responses to the observed damage.

C. ATOM ABSTRACTION REACTIONS

The third general reaction of free radicals in biological matrices is atom abstraction (Figure 1, reaction 3). Atom abstraction reactions probably are involved in the radical-mediated alteration of other tissue structures such as DNA,[43] proteins,[44] and cholesterol,[45] but the peroxidation of polyunsaturated fatty acids has received more attention. The classical example of this reaction is the abstraction of a *bis*-allylic methylenic H atom from homoconjugated polyunsaturated fatty acids (Figure 8). The product of this abstraction reaction is the corresponding pentadienyl radical. Molecular oxygen, which is itself a triplet diradical, react with carbon-centered radicals such as this in a step that has essentially no energy of activation and thus proceeds at rates approaching diffusion control.[46] This transformation to the corresponding fatty acid peroxyl radical constitutes the generic process called lipid peroxidation, namely the introduction of a peroxyl (O–O) bond into a lipid molecule. The abstraction of a hydrogen atom from another lipid molecule by the lipid peroxyl radical yields the fatty acid hydroperoxide and another lipid radical in a propagation cycle that will continue to consume fatty acid substrate and to produce fatty acid hydroperoxide until the substrate is depleted or, more commonly, the radical chain is interrupted in a termination step. Hydrogen atom donation by vitamin E to the lipid peroxyl radical is the primary route through which the radical chains are terminated in biological systems.[47] The tocopheroxyl radical is relatively ineffective in propagating the chain and might be reduced back to tocopherol by ascorbate or by GSH-dependent systems.

Radical abstraction-mediated alterations of target tissue molecules in association with tissue injury have been most extensively characterized for the hepatic necrosis caused by carbon tetrachloride.[5,48,49] Ironically, an early study of the association of lipid peroxidation as a potential mechanism for the toxicity of carbon tetrachloride led investigators to conclude that peroxidation was not likely to be involved in the initiation of injury, because hepatotoxic doses did not increase the observable lipid peroxidation.[50] This study was based upon the quantitation of lipid peroxidation by measurement of thiobarbituric acid-reactive substances (TBARS). The subsequent application of other methods for the measurement of the products of the peroxidation of biological lipids has shown, beyond a reasonable doubt, that lipids are peroxidatively altered in reactions initiated by metabolites of carbon tetrachloride.[5,48,49]

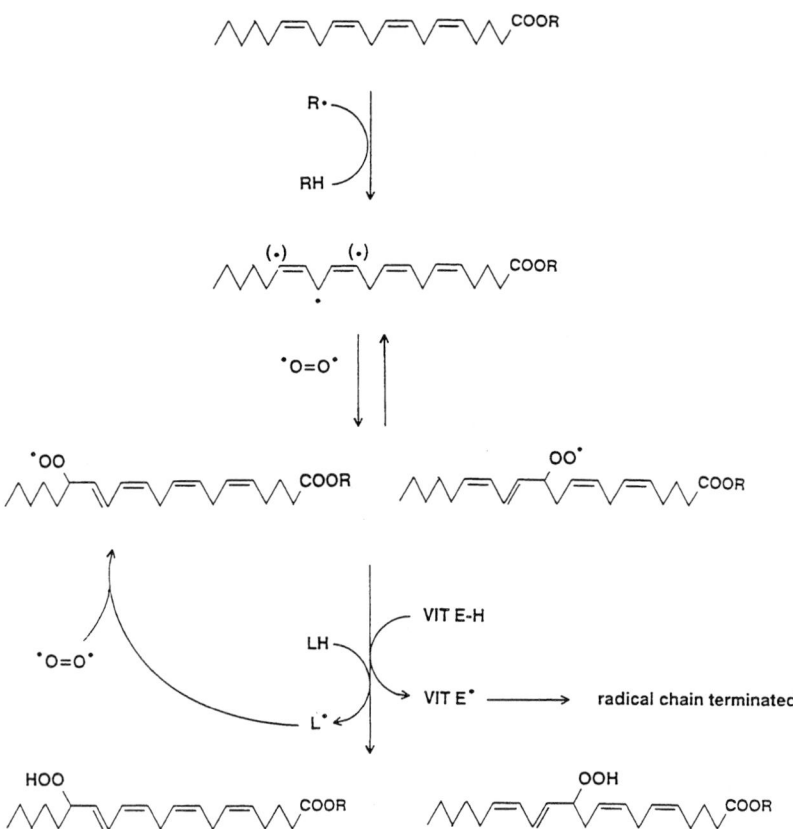

FIGURE 8. Radical chain peroxidation of polyunsaturated fatty acids. Abstraction of a bis-
allylic hydrogen atom (in this example from the 13-position of arachidonate) produces the
pentadienyl radical shown in its major resonance forms. Reaction of the pentadienyl radical with
oxygen preferentially preserves the conjugated diene and yields the 11- or 15-peroxyl radicals.
The peroxyl radicals give the respective hydroperoxides by hydrogen atom abstraction. If the
hydrogen atom is derived from another lipid molecule (LH), the radical thus produced (L˙) can
propagate the radical chain peroxidation. Hydrogen atom donations by vitamin E or related
antioxidants also provide the corresponding hydroperoxides, but do not actively propagate the
radical chain reaction sequence.

IV. MEASUREMENT OF LIPID PEROXIDATION

The thiobarbituric acid (TBA) test is the most frequently used method of
analysis used in studies of lipid peroxidation. Operationally, the assay methodology is simple and the test is very sensitive. However, the chemical
transformations required to produce effective precursors for the malondialdehyde moiety are relatively complex and estimates of the efficiency of conversion of specific fatty acid hydroperoxides into the TBA-derived chromophore range from less than 0.1% to as much as 10%.[51-54]

The great majority of the studies of the involvement of lipid peroxidation in tissue injury have employed the TBA test or other chemically nonspecific methods of analysis. Interpretations of the data thus obtained usually have failed to take into consideration the differences in the chemical properties of the wide variety of primary and secondary products that can be formed through the peroxidation of biological lipids.[52,53,55-60] The numerous fundamentally different methods for measuring products of lipid peroxidation have been developed in large part because of the numerous chemically different classes of products formed. Despite the early and clear example offered by the early studies on carbon tetrachloride,[5,50] however, little attention has been paid to the specific factors influencing correlations with and competition among the distinct types of product. Increases in the one parameter measured usually are interpreted as demonstrations of increased lipid peroxidation.

The report by Riely et al.[61] of increased expiration of ethane by mice treated with carbon tetrachloride was greeted with considerable enthusiasm because it was shown to give a positive response *in vivo* and the nature of the air sampling technique offered the chance to examine experimental animals noninvasively and, therefore, repeatedly, and also made a broader range of studies in humans more approachable. However, we have found that although rates of expiration of ethane and pentane increase dramatically in mice treated with hepatotoxic doses of acetaminophen, the contents of the hydroperoxy or hydroxy derivatives of polyunsaturated fatty acids in hepatic lipids of these animals are not measurably increased.[48,49,59] One possible explanation for this apparent contradiction is suggested by the reaction scheme presented in Figure 9. The formation of the 1-pentyl radical and, eventually, of pentane from the beta scission of the precursor ω-6 hydroperoxide will decrease the amount of the hydroxy derivative that can be produced from a given amount of hydroperoxide. Increased production of pentane, and, by analogy, of ethane from the ω-3 fatty acids, thus may reflect either an increase in the amount of lipid peroxidized or in a change in the disposition between beta scission and reduction. In carbon tetrachloride-treated mice we observe marked increases in the expiration of ethane and pentane and in the hydroxy fatty acid content of hepatic lipids, but the corresponding hydroperoxy derivatives are not detectable.[49,59]

Reactive chelates of iron or copper can stimulate the beta scission of hydroperoxides, which suggests the availability of these species may play a major role in determining the rates of expiration of ethane and pentane. In the animal model of diquat-induced hepatic necrosis discussed above, we found that the rates of ethane and pentane expiration were significantly enhanced in response to diquat in the animals pretreated with ferrous sulfate, and were diminished in the animals pretreated with desferrioxamine, in parallel with the observed changes in hepatotoxicity. The concordance of alkane expiration rates and hepatic injury stand in marked contrast with the lack of effects of the pretreatments on diquat-induced GSSG production.[39]

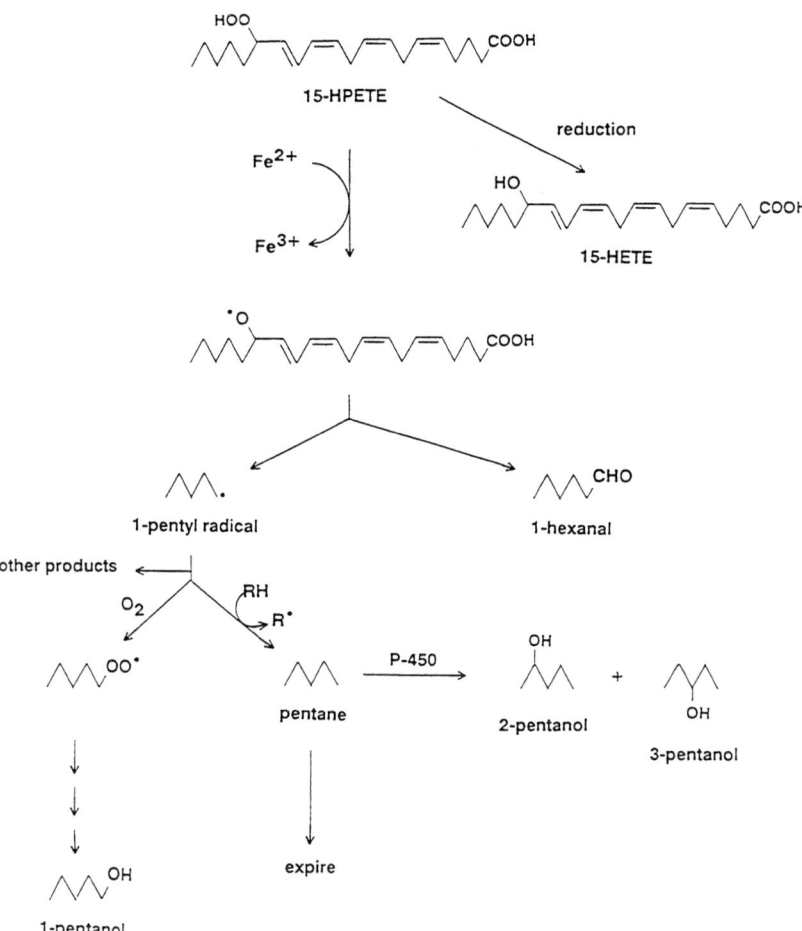

FIGURE 9. Mechanistic considerations in relating pentane expiration to lipid peroxidation. The processes involved in the conversion of 15-hydroperoxyeicosatetraenoic acid (15-HPETE) to pentane are outlined and the several product-determining branch points in this process are illustrated.

Intracellular iron content is regulated tightly and much of the stored iron is deposited in ferritin in the ferric form. The reduction and release of iron from ferritin is stimulated by superoxide and by the monoreduced cation radical of diquat.[62] The nonphysiological release of iron in chemically active forms could be a big problem for cells either through the stimulation of lipid peroxidation or through the destruction of physiologically essential cyclooxygenase and lipoxygenase intermediates.

An additional complication in the use of ethane and pentane for "measuring" lipid peroxidation is that the yields are low even *in vitro* with added transition metal catalyst and these yields can be inhibited by over 99% by the presence of molecular oxygen. We found that the ferrous sulfate-mediated decomposition of synthetic 15-hydroperoxyeicosatetraenoic acid (Figure 9) gave pentane in 2.5% yield if oxygen was removed from the system.[63] If the buffer and atmosphere were saturated with oxygen, the yield was decreased to only 0.01% of the hydroperoxide decomposed.[63] In the oxygenated system, 1-pentanol was produced and accounted for 8.5% of the starting material, which we interpret as the facile trapping of the 1-pentyl radical by molecular oxygen with subsequent conversion to the primary alcohol. An additional complication of the use of pentane in the measurement of lipid peroxidation *in vivo* is that pentane is metabolized;[64] however, the products of metabolism of pentane are the 2- and 3-pentanols,[65] whereas the product of trapping the beta-scission intermediate is 1-pentanol (Figure 9). Oxidative metabolism is less of a concern in the measurement of ethane,[64] but the factors influencing beta scission versus reduction and the trapping of the ethyl radical by oxygen versus H-atom abstraction should be analogous.

Other analytical methods used in the study of the peroxidation of biological lipids have potential complications that are comparable to the factors discussed above in relation to the measurement of ethane and pentane. The apparent contradictions often noted in different studies may arise in large part from insufficient consideration of the chemical and physiological subtleties that are capable of influencing the experimental parameters.[66] The more recent recognition of the potent biological properties of the products of lipid peroxidation and the surprising chemical specificities in their biological activities have brought increased attention to the need for chemical specificity in the analytical methodologies employed in such studies.[60]

Recent advances in bioanalytical capabilities, particularly in mass spectrometry, have made increasing numbers of these studies possible. However, it is essential that these elegant bioanalytical methods be applied to relevant biological problems. In addition, the well-studied principles of the chemistry of free radicals and ROS need to be considered more thoroughly in the interpretation of the results of the relevant biological studies. The requirement for the integration of fundamentally different disciplines underscores the need for active collaboration between investigators with different areas of primary expertise.

ACKNOWLEDGMENT

The support of NIH grant GM44263 is gratefully acknowledged.

REFERENCES

1. **Pryor, W. A.**, The role of free radical reactions in biological systems, in *Free Radicals in Biology*, Pryor, W. A., Ed., Academic Press, San Diego, 1976, 1.
2. **Hongslo, J. K., Bjornstad, C., Schwarze, P. E., and Holme, J. A.**, Inhibition of replicative DNA synthesis by paracetamol in V79 Chinese hamster cells, *Toxicol. In Vitro*, 3, 13, 1989.
3. **Holme, J. A., Bjorge, C., Brogger, A., Schwarze, P. E., Mann, G., Thelander, L., and Hongslo, J. K.**, Acetaminophen-induced sister chromatid exchanges and chromosomal aberrations are caused by a selective inhibition of ribonucleotide reductase, *Toxicologist*, 10, A378, 1990.
4. **Reynolds, E. S. and Moslen, M. T.**, Free radical damage in liver, in *Free Radicals in Biology*, Pryor, W. A., Ed., Academic Press, San Diego, 1980, 49.
5. **Recknagel, R. O., Glende, E. A., Jr., Dolak, J. A., and Waller, R. L.**, Mechanisms of carbon tetrachloride toxicity, *Pharm. Ther.*, 43, 139, 1989.
6. **Preuss-Schwartz, D., Nimesheim, A., and Marnett, L. J.**, Peroxyl radical- and cytochrome P-450-dependent metabolic activation of (+)-7,8,-dihydroxy-7,8-dihydrobenzo[a]pyrene in mouse skin *in vitro* and *in vivo*, *Cancer Res.*, 49, 1732, 1989.
7. **McCay, P. B.**, Application of ESR spectroscopy in toxicology, *Arch. Toxicol.*, 60, 133, 1987.
8. **Flaherty, J. T. and Weisfeldt, M. L.**, Reperfusion injury, *Free Radical Biol. Med.*, 5, 409, 1988.
9. **Samuni, A. and Swartz, H. M.**, The cellular-induced decay of DMPO spin adducts of $\cdot$OH and $\cdot O_2^-$, *Free Radical Biol. Med.*, 6, 179, 1989.
10. **Smith, C. V. and Mitchell, J. R.**, Pharmacological aspects of glutathione in drug metabolism, in *Coenzymes and Cofactors*, Dolphin, D., Poulsen, R., and Avramovic, O., Eds., John Wiley & Sons, New York, 1989, 1.
11. **Burke, R. F., Lawrence, R. A., and Lane, J. M.**, Liver necrosis and lipid peroxidation in the rat as a result of paraquat and diquat administration. Effect of selenium deficiency, *J. Clin. Invest.*, 65, 1024, 1980.
12. **Smith, C. V.**, Effect of BCNU pretreatment on diquat-induced oxidant stress and hepatotoxicity, *Biochem. Biophys. Res. Commun.*, 144, 415, 1987.
13. **Burk, R. F. and Lane, J. M.**, Ethane production and liver necrosis in rats after administration of drugs and other chemicals, *Toxicol. Appl. Pharmacol.*, 50, 467, 1979.
14. **Smith, C. V. and Mitchell, J. R.**, Acetaminophen hepatotoxicity *in vivo* is not accompanied by oxidant stress, *Biochem. Biophys. Res. Commun.*, 133, 329, 1985.
15. **Castro, J. A. and Gomez, M. I. D.**, Studies on the irreversible binding of ^{14}C-CCl$_4$ to microsomal lipids in rats under varying experimental conditions, *Toxicol. Appl. Pharmacol.*, 23, 541, 1972.
16. **Smith, C. V., Hughes, H., and Mitchell, J. R.**, Free radicals *in vivo:* covalent binding to lipids, *Mol. Pharmacol.*, 26, 112, 1984.
17. **Margison, G. P. and O'Connor, P. G.**, Nucleic acid modification by N-nitroso compounds, in *Chemical Carcinogens and DNA*, Grover, R. L., Ed., CRC Press, Boca Raton, FL, 1979, 111.
18. **Trudell, J. R., Bosterling, B., and Trevor, A. J.**, Reductive metabolism of carbon tetrachloride by human cytochromes P-450 reconstituted in phospholipid vesicles: mass spectral identification of trichloromethyl radical bound to dioleoyl phosphatidylcholine, *Proc. Natl. Acad. Sci. U.S.A.*, 79, 2678, 1982.
19. **Ansari, G. A. S., Moslen, M. T., and Reynolds, E. S.**, Evidence for *in vivo* covalent binding of $\cdot$CCl$_3$ derived from CCl$_4$ to cholesterol of rat liver, *Biochem. Pharmacol.*, 31, 3509, 1982.
20. **Pohl, L. R., Schulick, R. D., Highet, R. J., and George, J. W.**, Reductive-oxygenation mechanism of metabolism of carbon tetrachloride to phosgene by cytochrome P-450, *Mol. Pharmacol.*, 25, 318, 1984.

21. **Pohl, L. R. and George, J. W.**, Identification of dichloromethyl carbene as a metabolite of carbon tetrachloride, *Biochem. Biophys. Res. Commun.*, 117, 367, 1983.
22. **Osawa, Y., Highet, R. J., Murphy, C. M., Cotter, R. J., and Pohl, L. R.**, Formation of heme-derived products by the reaction of ferrous deoxymyoglobin with BrCCl$_3$, *J. Am. Chem. Soc.*, 111, 4462, 1989.
23. **Osawa, Y. and Pohl, L. R.**, Covalent bonding of the prosthetic heme to protein: a potential mechanism for the suicide inactivation or activation of hemoproteins, *Chem. Res. Toxicol.*, 2, 131, 1989.
24. **McCord, J. M. and Fridovich, I.**, Superoxide dismutase: the first twenty years, *Free Radical Biol. Med.*, 5, 363, 1988.
25. **Thor, H., Smith, M. T., Hartzell, P., Bellomo, G., Jewell, S. A., and Orrenius, S.**, The metabolism of menadione (2-methyl-1,4-naphthoquinone) by isolated hepatocytes. A study of the implications of oxidative stress in intact cells, *J. Biol. Chem.*, 257, 12419, 1982.
26. **McCord, J. M.**, Oxygen-derived free radicals in postischemic tissue injury, *N. Engl. J. Med.*, 312, 159, 1985.
27. **Freeman, B. A. and Crapo, J. D.**, Free radicals and tissue injury, *Lab. Invest.*, 47, 412, 1982.
28. **Turrens, J. F., Freeman, B. A., and Crapo, J. D.**, Hyperoxia increases H$_2$O$_2$ release by lung mitochondria and microsomes, *Arch. Biochem. Biophys.*, 217, 411, 1982.
29. **Jaeschke, H., Smith, C. V., and Mitchell, J. R.**, Reactive oxygen species during ischemia-reflow injury in isolated perfused rat liver, *J. Clin. Invest.*, 81, 1240, 1988.
30. **Lesnefsky, E. J., Repine, J. E., and Horwitz, L. D.**, Oxidation and release of glutathione from myocardium during early reperfusion, *Free Radical Biol. Med.*, 7, 31, 1989.
31. **Lauterburg, B. H., Smith, C. V., Hughes, H., and Mitchel, J. R.**, Biliary excretion of glutathione and glutathione disulfide in the rat: regulation and response to oxidative stress, *J. Clin. Invest.*, 73, 124, 1984.
32. **Bucher, J. R., Tien, M., Morehouse, L. A., and Aust, S. D.**, Redox cycling and lipid peroxidation: the central role of iron chelates, *Fundam. Appl. Toxicol.*, 3, 222, 1983.
33. **Sies, H. and Akerboom, T. P. M.**, Glutathione disulfide (GSSG) efflux from cells and tissues, *Methods Enzymol.*, 105, 445, 1984.
34. **Fariss, M. W. and Reed, D. J.**, High-performance liquid chromatography of thiols and disulfides: dinitrophenol derivatives, *Methods Enzymol.*, 143, 101, 1987.
35. **Stein, A. F., Dills, R. L., and Klaassen, C. D.**, High-performance liquid chromatographic analysis of glutathione and its thiol and disulfide degradation products, *J. Chromatogr.*, 381, 259, 1986.
36. **Alpert, A. J. and Gilbert, H. F.**, Detection of oxidized and reduced glutathione with a recycling postcolumn reaction, *J. Chromatogr.*, 144, 553, 1985.
37. **Gilbert, H. F.**, Molecular and cellular aspects of thiol/disulfide exchange, *Adv. Enzymol.*, 63, 69, 1989.
38. **Sies, H.**, Biochemistry of oxidative stress, *Angew. Chem. Int. Ed. Engl.*, 25, 1058, 1986.
39. **Smith, C. V.**, Evidence for the participation of lipid peroxidation and iron in diquat-induced hepatic necrosis in vivo, *Mol. Pharmacol.*, 32, 417, 1987.
40. **Smith, C. V. and Jaeschke, H.**, Effect of acetaminophen on hepatic content and biliary efflux of glutathione disulfide in mice, *Chem. Biol. Interact.*, 70, 241, 1989.
41. **DiMonte, D., Bellomo, G., Thor, H., Nicotera, P., and Orrenius, S.**, Menadione-induced cytotoxicity is associated with protein thiol oxidation and alteration in intracellular Ca^{2+} homostasis, *Arch. Biochem. Biophys.*, 235, 343, 1984.
42. **Smith, C. V., Hughes, H., Lauterburg, B. H., and Mitchell, J. R.**, Oxidant stress and hepatic necrosis in rats treated with diquat, *J. Pharmacol. Exp. Ther.*, 235, 172, 1985.
43. **Imlay, J. A. and Linn, S.**, DNA damage and oxygen radical toxicity, *Science*, 240, 1302, 1988.

44. **Davies, K. J. A., Delsignore, M. E., and Lin, S. W.**, Protein damage and degradation by oxygen radicals. II. Modification of amino acids, *J. Biol. Chem.*, 262, 9902, 1987.
45. **Smith, L. L. and Johnson, B. H.**, Biological activities of oxysterols, *Free Radical Biol. Med.*, 7, 285, 1989.
46. **Pryor, W. A.**, Oxy-radicals and related species: their formation, lifetimes, and reactions, *Annu. Rev. Physiol.*, 48, 657, 1986.
47. **Porter, N.**, Mechanisms for the autoxidation of polyunsaturated lipids, *Accounts Chem. Res.*, 19, 262, 1986.
48. **Hughes, H., Smith, C. V., and Mitchell, J. R.**, Quantitation of lipid peroxidation products by gas chromatography-mass spectrometry, *Anal. Biochem.*, 152, 107, 1986.
49. **Hughes, H., Smith, C. V., Horning, E. C., and Mitchell, J. R.**, HPLC and GC-MS determination of specific lipid peroxidation products *in vivo*, *Anal. Biochem.*, 130, 431, 1983.
50. **Priest, R. E., Smuckler, E. A., Iseri, O. A., and Benditt, E. P.**, Liver lipid peroxide levels in carbon tetrachloride poisoning, *Proc. Soc. Exp. Biol. Med.*, 111, 50, 1962.
51. **Shimizu, T., Kondo, K., and Hayaishi, O.**, Role of prostaglandin endoperoxides in the serum thiobarbituric acid reaction, *Arch. Biochem. Biophys.*, 206, 271, 1981.
52. **Pryor, W. A., Stanley, J. P., and Blair, E.**, Autoxidation of polyunsaturated fatty acids. II. A suggested mechanism for the formation of TBA-reactive materials from prostaglandin-like endoperoxides, *Lipids*, 11, 370, 1976.
53. **Frankel, E. N. and Neff, W. E.**, Formation of malonaldehyde from lipid oxidation products, *Biochim. Biophys. Acta*, 754, 264, 1983.
54. **Ohkawa, H., Oshishi, N., and Yagi, K.**, Reaction of linoleic acid hydroperoxide with thiobarbituric acid, *J. Lipid Res.*, 19, 1053, 1978.
55. **Gardner, H. W.**, Oxygen radical chemistry of polyunsaturated fatty acids, *Free Radical Biol. Med.*, 7, 65, 1989.
56. **Frankel, E. N.**, Volatile lipid oxidation products, *Prog. Lipid Res.*, 22, 1, 1982.
57. **Frankel, E. N.**, Chemistry of free radical and singlet oxidation of lipids, *Prog. Lipid Res.*, 23, 197, 1985.
58. **Frankel, E. N.**, Lipid oxidation, *Prog. Lipid Res.*, 19, 1, 1980.
59. **Smith, C. V. and Anderson, R. E.**, Methods for determination of lipid peroxidation in biological samples, *Free Radical Biol. Med.*, 3, 341, 1987.
60. **Morrow, J. D., Harris, T. M., and Roberts, L. J., II**, Noncyclooxygenase oxidative formation of a series of novel prostaglandins: analytical ramifications for measurement of eicosanoids, *Anal. Biochem.*, 184, 1, 1990.
61. **Riely, C. A., Cohen, G. A., and Lieberman, M.**, Ethane evolution: a new index of lipid peroxidation, *Science*, 183, 208, 1974.
62. **Reif, D. W., Beales, I. L. P., Thomas, C. E., and Aust, S. D.**, Effect of diquat on the distribution of iron in rat liver, *Toxicol. Appl. Pharmacol.*, 93, 506, 1988.
63. **Smith, C. V. and Reilly, M. H.**, Formation of pentane versus 1-pentanol in the ferrous sulfate initiated decomposition of 15-hydroperoxyeicosatetraenoic acid in hypoxic and hyperoxic conditions, *Biochem. Pharmacol.*, 38, 1362, 1989.
64. **Daugherty, M. S., Ludden, T. M., and Burk, R. F.**, Metabolism of ethane and pentane to carbon dioxide by the rat, *Drug Metab. Disp.*, 16, 666, 1988.
65. **Frommer, U., Ullrich, V., and Staudinger, H.**, Hydroxylation of aliphatic compounds by liver microsomes. I. The distribution pattern of isomeric alcohols, *Hoppe-Seyler's Z. Physiol. Chem.*, 351, 903, 1970.
66. **Smith, C. V.**, Correlations and apparent contradictions in assessment of oxidant stress status *in vivo*, *Free Radical Biol. Med.*, 10, 217, 1991.

Chapter 2

HEMOGLOBIN-INDUCED OXIDANT DAMAGE TO THE CENTRAL NERVOUS SYSTEM

Sayed M. H. Sadrzadeh and John W. Eaton

TABLE OF CONTENTS

I. INTRODUCTION

The full consequences of trauma to the central nervous system (CNS) may not be evident until long after the primary insult. Hemorrhagic injury to the CNS often is associated with the delayed development of seizure states and paralysis.[1,2] The mechanisms leading to these sequelae of hemorrhagic trauma are not fully understood, although they may involve inflammatory and oxidative reactions.[3] The CNS may be particularly vulnerable to oxidant damage because neuronal membranes contain relatively high concentrations of polyunsaturated fatty acids and several oxidant defense mechanisms are poorly represented in the CNS.[4]

The general hypothesis to be explored in this chapter is that hemoglobin, released from extravasated red cells into the CNS interstitium, acts to promote secondary inflammation and oxidative damage to sensitive nervous tissues. More specifically, we believe that the proximate cause of this damage is iron released from hemoglobin and that iron-mediated oxidative reactions are important contributors to posthemorrhagic CNS damage. It should be emphasized that the following is an imbalanced summation of this area, no doubt over-emphasizing our own earlier work and ignoring a substantial body of excellent work by other investigators.

II. DOES HEMOGLOBIN PROMOTE CNS DAMAGE?

Our own interest in possible hemoglobin-mediated CNS damage was initially piqued by observations on patients with familial idiopathic epilepsy (i.e., seizure disorders occurring without evident precedent cause and affecting two or more members of the same family).[5] We found that, in several of the kindreds being studied, the affected individuals and some of their first-degree relatives had very low or absent levels of the serum protein, haptoglobin. We were struck by this finding because such low levels of haptoglobin are very rare in the general population. Haptoglobin is an α_2 glycoprotein which binds free hemoglobin with very high affinity and is thought to speed its removal by the reticuloendothelial system.[6] We hypothesized that the very low levels of serum haptoglobin in affected individuals might impair the normal process of clearance of free hemoglobin from the CNS. If so, residual free hemoglobin might then promote subsequent CNS dysfunction, eventuating in seizure disorders.

In partial confirmation of this hypothesis, the clearance of radiolabeled hemoglobin, previously injected into the brains of hypohaptoglobinemic mice, is substantially less than that from the brains of animals with normal haptoglobin levels.[5] Importantly, this defect in hemoglobin clearance is corrected by prior complexation of hemoglobin with purified haptoglobin (see Figure 1).

The possible consequences of the presence of free hemoglobin within the

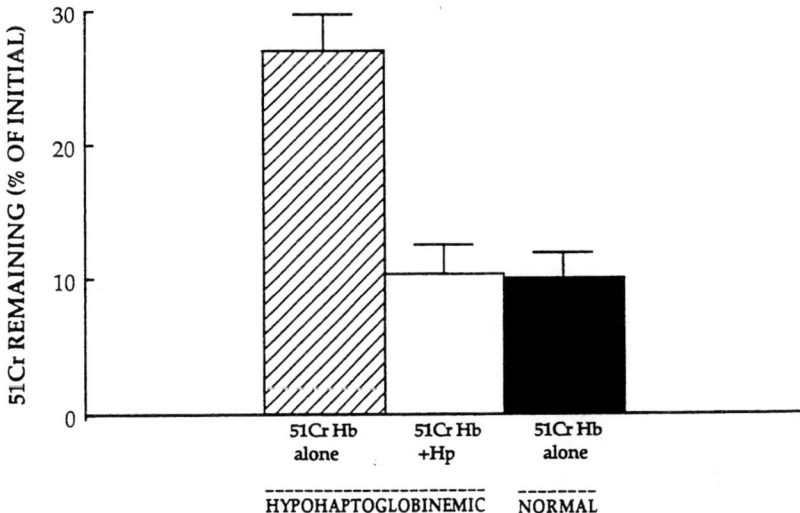

FIGURE 1. Clearance of radiolabeled hemoglobin from the brains of normal and hypohapto-globinemic mice. [^{51}Cr]-labeled hemoglobin, either alone or as complex with purified human haptoglobin (+Hp), was injected intracerebrally into normal and mice previously made hypo-haptoglobinemic via repeated intraperitoneal injections of purified hemoglobin. Twenty-four hours after injection, the animals were sacrificed and the amounts of radioactivity within the injected hemisphere as well as the liver (to control for variations in total amounts injected) were measured by gamma counting. Values for hypohaptoglobinemic animals injected with hemo-globin alone are significantly different from both normal and hypohaptoglobinemic animals injected with the complex of hemoglobin and haptoglobin ($p < 0.001$; $n = 10$ in all groups). (Redrawn from Panter, S. S. et al., *J. Exp. Med.*, 161, 748, 1985.)

CNS were suggested by earlier observations indicating that the intracortical instillation of either iron salts[7] or hemoglobin[8] would cause the appearance of epileptiform seizures and/or electroencephalographic abnormalities. Consequently, we launched investigations to determine whether the presence of hemoglobin or derived ferruginous compounds might promote inflammatory damage to the CNS.

III. DOES HEMOGLOBIN CONTRIBUTE TO THE OXIDATION OF CNS COMPONENTS?

A major hazard posed by iron-containing compounds is in facilitating the formation of more reactive forms of activated oxygen. In this regard, the hydroxyl radical (·OH) is most often mentioned as an important toxic species produced by reactions between iron and activated oxygen:[9,10]

$$O_2^- + Fe^{2+} \longrightarrow Fe^{3+} + O_2 \tag{1}$$

$$2O_2^- + 2H^+ \longrightarrow H_2O_2 + O_2 \tag{2}$$

$$H_2O_2 + Fe^{2+} \longrightarrow \cdot OH + OH^- + Fe^{3+} \tag{3}$$

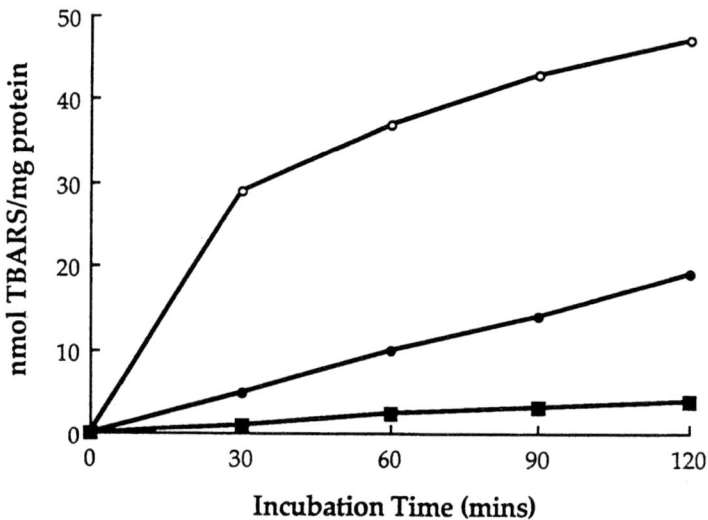

FIGURE 2. Hemoglobin- and iron-mediated peroxidation of polyunsaturated fatty acids in murine brain homogenates. Fresh murine brain homogenates (protein concentration: 2.7 mg/mL) suspended in 50 mM Tris buffer, pH 7.4, were incubated with 50 μM purified hemoglobin (as tetramer) (●), 50 μM FeCl$_2$ (○), or no addition (■). The concentration of thiobarbituric acid-reactive substances (TBARS) was determined at the indicated times. (Redrawn from Sadrzadeh, S. M. H. et al., *J. Clin. Invest*, 79, 662, 1987.)

As shown above, the final reaction between ferrous iron and H$_2$O$_2$, the so-called Fenton reaction,[11,12] leads to the formation of ·OH.

In an attempt to determine whether the iron within the heme group of intact hemoglobin might similarly generate ·OH, we incubated purified hemoglobin with a source of O$_2^-$ and H$_2$O$_2$ (hypoxanthine/xanthine oxidase). Under these conditions, hemoglobin catalyzes the formation of one or more reactive species resembling ·OH (Reference 13) (i.e., capable of generating formaldehyde and methane through reaction with dimethyl sulfoxide).[9,14,15] However, it appears likely that this reaction is not truly catalytic, with little reductive cycling of the oxidized methemoglobin back to oxyhemoglobin.[13] Furthermore, it is also likely that the activated oxygen species being detected by reaction with dimethyl sulfoxide is more likely a heme iron:activated oxygen complex (such as ferryl or perferryl radical) rather than free ·OH.

Perhaps more importantly, free hemoglobin, in the presence of xanthine/ xanthine oxidase, will also promote the peroxidation of arachidonic acid and of unsaturated fatty acids within normal cell membranes.[13] Furthermore, hemoglobin or red cell lysates also cause brisk peroxidation of crude murine brain homogenates. In this latter case, the peroxidation (as evidenced by accumulation of thiobarbituric acid-reactive aldehydic by-products of poly-unsaturated fatty acid oxidation, TBARS) proceeds in the absence of added xanthine/xanthine oxidase (Figure 2). In both cases, however, hemoglobin

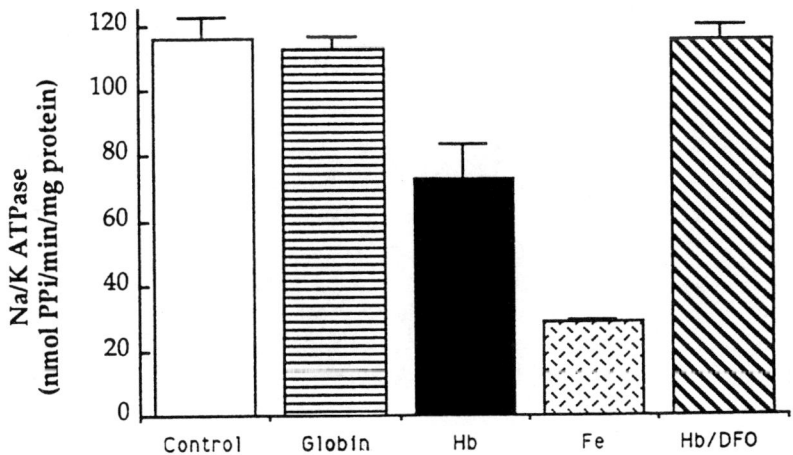

FIGURE 3.   Activity of Na^+/K^+-ATPase in areas of feline spinal cords 24 h following the stereotaxic injection of heme-free globin (110 nmol monomer in 20 μl total volume) ($n = 2$), purified hemoglobin (110 nmol monomer in 20 μl total volume ($n = 3$), $FeCl_2$ (500 nmol in 5 μl total volume) ($n = 10$) or a combination of hemoglobin and desferrioxamine B (110 nmol and 18 μmol, respectively, in a total volume of 20 μl) ($n = 2$). Error bars for globin and hemoglobin/desferrioxamine show range of both values. (Data from Sadrzadeh, S. M. H. et al., *J. Biol. Chem.* 259, 14354, 1984.)

per se is probably not responsible for the promotion of these oxidative reactions. Such hemoglobin-driven peroxidation is blocked by iron chelators such as desferrioxamine, indicating that iron, somehow released from heme, is responsible.[13]

Unfortunately, simple measurement of "malondialdehyde" is not adequate for detection of *in vivo* oxidative processes because aldehydic products of lipid peroxidation are small, water soluble, metabolizable, and readily lost from the locus of formation. However, hemoglobin promoted oxidation of CNS tissue leaves at least one other, and more stable, footprint. *In vitro*, free hemoglobin or iron will inhibit Na^+/K^+-ATPase in brain and spinal cord homogenates.[16] Desferrioxamine blocks this hemoglobin-induced enzyme inhibition, once again implicating free iron released from hemoglobin.

The facile hemoglobin-mediated inhibition of Na^+/K^+-ATPase in CNS homogenates suggested that inhibition of this enzyme might reflect the occurrence of similar processes *in vivo*. Therefore, to investigate the possible involvement of free hemoglobin in CNS injury *in vivo*, we measured the activity of Na^+/K^+-ATPase in feline spinal cords following the stereotactic injection of free hemoglobin and other agents. The results (Figure 3) indicate that both free iron and purified hemoglobin will cause substantial inhibition of neuronal Na^+/K^+-ATPase, specifically in the area of injection.[16] Perhaps most importantly, the addition of micromolar amounts of the powerful iron chelator, desferrioxamine, prevents inhibition of neuronal ATPase in feline spinal cords, again indicating the involvement of oxidative reactions driven by free iron.

IV. SPECIFIC MECHANISMS OF HEMOGLOBIN-MEDIATED CNS DAMAGE

The precise steps involved in damage to CNS tissues catalyzed by he-moglobin are not yet clear. It is likely that, within the CNS interstitium, heme will spontaneously dissociate from hemoglobin because oxyhemoglobin loses its heme quite readily.[17,18] In some experiments, free heme, following its loss from hemoglobin, has been trapped by albumin.[17] In the presence of high concentrations of phospholipid (such as those within the CNS), the heme is likely to insert into the lipid. Indeed, we and others have found that free heme is avidly taken up by the membranes of intact cells such as erythrocytes and endothelial cells.[19,20] This behavior of free heme is entirely predictable given the fact that it is an intensely hydrophobic molecule which readily enters organic solvents.[19]

Once admixed with unsaturated fatty acids or phospholipids containing unsaturated fatty acids, heme is evidently destroyed, probably in a reaction with partially oxidized (hydroperoxy?) fatty acids. Support for this is supplied by our visual observation that when micromolar amounts of free hemoglobin are added to crude murine brain homogenates, the pink color of the heme protein is lost within 30 to 60 min. In a possible experimental analog of this, when free hemoglobin is added to commercially obtained arachidonic acid, which has spontaneously formed peroxidative intermediates, heme is also rapidly destroyed as indicated by a loss of heme-specific absorption at 412 nm.

A predictable consequence of this (probably oxidative) destruction of the heme group is the release of free iron, perhaps within the lipid bilayer itself. The released iron may not be very reactive; it is likely to be Fe^{3+} and may, therefore, require reducing agents in order to participate in truly catalytic oxidation events (see, for example, the possibly analogous iron:activated oxygen reactions 1 through 3). Not surprisingly, water-soluble fractions of crude murine brain homogenates contain agents which facilitate the rapid reduction of Fe^{3+} to Fe^{2+}.[21] Furthermore, although the addition of either free iron or purified hemoglobin to brain homogenates stimulates brisk lipid per-oxidation, if the membranes from such homogenates are first washed to remove water-soluble substances, added hemoglobin (or iron) no longer has any effect (see Table 1).[21]

It appears that ascorbic acid, which is particularly abundant in the mam-malian CNS, is most important in the continued reduction of free iron and in fostering iron-dependent lipid peroxidation.[21] This is most directly proved by the observation that preincubation of crude brain homogenates with the enzyme ascorbate oxidase (which enzymatically depletes ascorbate-reducing equivalents) will prevent lipid peroxidation upon subsequent addition of either free iron or purified hemoglobin (Table 2). In further support of this conclu-sion, when physiologic concentrations of ascorbic acid are added to washed

TABLE 1
Cytoplasmic Factor Mediates Hemoglobin-Driven Peroxidation of Murine Brain Lipids

Sample	TBARS (nmol/mg protein/h)	n
Untreated homogenate	2.9 ± 1.6[b]	3
Homogenate + metHb (50 μ*M*)	72.0 ± 4.2[a]	3
Washed membrane + metHb (50 μ*M*)	2.4 ± 0.5[b]	4

Note: Incubations contained 20% (wt/v) brain homogenates (untreated or washed) in 50 m*M* Tris, pH 7.4, plus the additions indicated.

[a] Differs from [b], $p < 0.01$ (*t*-test, two-tailed).

Data from Sadrzadeh, S. H. M. and Eaton, J. W., *J. Clin. Invest.*, 82, 1510, 1988.

TABLE 2
Ascorbate Oxidase Blocks Methemoglobin and Iron-Mediated Peroxidation of Murine Brain Lipids

Sample	TBARS (nmol/mg protein/h)	n
Untreated homogenate (no addition)	2.93 ± 1.6[a]	1
Homogenate + metHb (50 μ*M*)	22.0 ± 4.2[b]	2
Homogenate + metHb (50 μ*M*) + ascorbate oxidase (1 U/ml)	3.3 ± 1.7[a]	3
Homogenate + $FeCl_3$ (10 μ*M*)	11.4 ± 3.1[c]	4
Homogenate + $FeCl_3$ (10 μ*M*) + ascorbate oxidase (1 U/ml)	2.4 ± 1.5[a]	4

Note: Brain homogenates (1:4 brain:50 m*M* Tris buffer, pH 7.4, (wt/v) were incubated with ascorbate oxidase (1 U/ml) for 20 min at 25°C where indicated, after which methemoglobin (50 μ*M*, tetramer) and $FeCl_3$ (10 μ*M*) were added. Samples were then incubated at 25°C for 1 h.

[a] Differs from [b], $p < 0.01$ (*t*-test, two-tailed).

membranes from murine brain homogenates, the susceptibility of these membranes to hemoglobin-iron-mediated peroxidation is restored.[21] In fact, this effect of ascorbic acid is entirely predictable in view of earlier reports of pro-oxidant activities of ascorbate toward mitochondrial lipids,[22] other tissues,[23] and the brain.[24-26]

Surprisingly, the precise mechanisms through which free iron catalyzes the peroxidation of unsaturated fatty acids have not yet been fully worked out (see, for example, References 27 to 29). The reaction appears to require both Fe^{2+} and Fe^{3+}, optimally in equimolar concentrations.[28-31] The ferric and ferrous iron may combine to form a ferrous:dioxygen:ferric complex, the

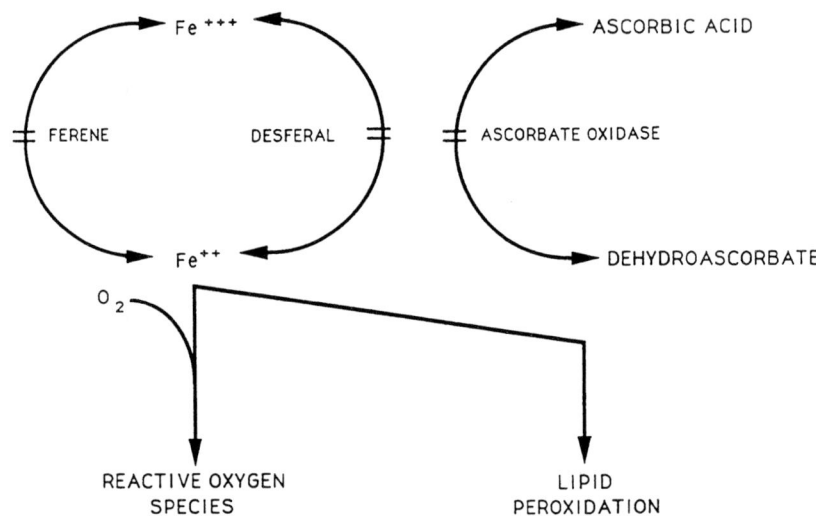

FIGURE 4. Simplified scheme showing the proposed reactions between hemoglobin-derived iron and various CNS components. As shown, once ferric iron is released from heme, it can be reduced by CNS ascorbate. Prevention of this reductive step (by treatment of brain homogenates with ascorbate oxidase) blocks subsequent lipid peroxidation. As also indicated, inclusion of either ferric (desferrioxamine) or ferrous ("ferene") iron chelators will interrupt iron catalysis of CNS oxidation. (From Sadrzadeh, S. H. M. and Eaton, J. W., *J. Clin. Invest.*, 82, 1510, 1988. With permission.)

structure of which is not yet defined. Furthermore, the Fe^{2+} may also be important in the homolytic decomposition of fatty acid hydroperoxides.[32] Regardless of the precise mechanism of reactions between iron, fatty acids, oxygen, and reducing agents, sufficient evidence does exist to support the scheme shown in Figure 4. The iron-catalyzed oxidation of CNS components likely requires the involvement of both Fe^{2+} and Fe^{3+} because chelators specific for Fe^{3+} (e.g., desferrioxamine) and Fe^{2+} (e.g., "ferene S") are equally effective in blocking the peroxidation of CNS polyunsaturated fatty acids. Furthermore, the continued reduction of Fe^{3+} to Fe^{2+} is also required as indicated by the observations that (1) washed membranes of CNS homogenates do not show iron-driven peroxidation, (2) the addition of physiologic concentrations of ascorbic acid restores peroxidation and, most directly, (3) pretreatment of CNS homogenates with ascorbate oxidase will block the hemoglobin-iron-mediated peroxidation of CNS polyunsaturated fatty acids.

V. CONCLUSIONS

A variety of evidence supports the general concept that an important fraction of posthemorrhagic damage to the CNS may involve oxidative reactions catalyzed by free hemoglobin released from extravasated erythrocytes.

More specifically, it appears that heme, which dissociates from extracellular hemoglobin, may insert into the hydrophobic milieu of CNS membranes, break down, and release free iron. Hemoglobin iron may then catalyze redox reactions which further damage the CNS (which is particularly at risk because these tissues have poor oxidant defense). This damage is reflected in the extensive peroxidation of CNS polyunsaturated fatty acids and inhibition of crucial membrane transport enzymes such as Na^+/K^+-ATPase and, perhaps, Ca^{2+}-ATPase. There is a certain irony in the fact than an important partner in these oxidative events is evidently CNS-derived ascorbic acid. CNS ascorbate normally functions as an antioxidant, preventing the oxidative decomposition of easily oxidized neurotransmitter substances. However, in the presence of free, hemoglobin-derived iron, this ascorbate may become part of the engine of destruction of these irreplaceable and very oxidant sensitive tissues. Our results suggest that interception of free iron with appropriate chelators may be a therapeutically useful approach to limiting the extent of posthemorrhagic injury to the CNS.

ACKNOWLEDGMENTS

We thank Diane Konzen for singlehandedly preparing this manuscript. Supported in part by grants from the National Institutes of Health (AI 25625) and the Minnesota Medical Foundation.

REFERENCES

1. **Gumnit, R. J.**, *The Epilepsy Handbook*, Raven Press, New York, 1983.
2. **Loiseau, P. and Jallon, P.**, Post-traumatic epilepsy, *Prog. Neurol. Surg.*, 190, 323, 1981.
3. **Means, E. E. and Anderson, D. K.**, Neurophagia by leukocytes in experimental spinal cord injury, *J. Neuropathol. Exp. Neurol.*, 42, 707, 1983.
4. **Cohen, G.**, Oxidative stress in the nervous system, in *Oxidative Stress*, Sies, H., Ed., Academic Press, San Diego, 1985, 383.
5. **Panter, S. S., Sadrzadeh, S. M. H., Hallaway, P. E., Haines, J., Anderson, V. E., and Eaton, J. W.**, Hypohaptoglobinemia: association with familial epilepsy, *J. Exp. Med.*, 161, 748, 1985.
6. **Javid, J.**, Human haptoglobins, *Curr. Topics Hematol.*, 1, 151, 1978.
7. **Willmore, L. J., Sypert, G. W., Munson, J. B., and Hurd, R. W.**, Chronic focal epileptiform discharges induced by injection of iron into rat and cat cortex, *Science*, 200, 1501, 1978.
8. **Rosen, A. D. and Frumin, N. V.**, Focal epileptogenesis after intracortical hemoglobin injection, *Exp. Neurol.*, 66, 277, 1979.
9. **Repine, J. E., Eaton, J. W., Anders, M. W., Hoidal, J. R., and Fox, R. B.**, Generation of hydroxyl radical by enzymes, chemicals and human phagocytes *in vitro*: detection using the anti-inflammatory agent — dimethyl sulfoxide (DMSO), *J. Clin. Invest.*, 64, 1642, 1979.

10. **Rosen, H. and Klebanoff, S. J.**, Role of iron and ethylenediaminetetraacetic acid in the bactericidal activity of a superoxide anion-generating system, *Arch. Biochem. Biophys.*, 208, 512, 1981.

11. **Walling, C.**, Fenton's reagent revisited, *Accounts Chem. Res.*, 8, 125, 1975.

12. **Koppenol, W. H., Butler, J., and Van Leeuwen, J. W.**, The Haber-Weiss cycle, *Photochem. Photobiol.*, 28, 655, 1978.

13. **Sadrzadeh, S. M. H., Graf, E., Panter, S. S., Hallaway, P. E., and Eaton, J. W.**, Hemoglobin: a biologic Fenton reagent, *J. Biol. Chem.*, 259, 14354, 1984.

14. **Klein, S. M., Cohen, G., and Cederbaum, A. I.**, Production of formaldehyde during metabolism of dimethylsulfoxide by hydroxyl radical generating systems, *Biochem. Pharmacol.*, 20, 6006, 1981.

15. **Klein, S. M., Cohen, G., and Cederbaum, A. I.**, The interaction of hydroxyl radicals with dimethylsulfoxide produces formaldehyde, *FASEB Letts.*, 117, 220, 1980.

16. **Sadrzadeh, S. M. H., Anderson, D. K., Panter, S. S., Hallaway, P. E., and Eaton, J. W.**, Hemoglobin potentiates central nervous system damage, *J. Clin. Invest.*, 79, 662, 1987.

17. **Bunn, H. F. and Jandl, J. H.**, Exchange of heme among hemoglobins and between hemoglobin and albumin, *J. Biol. Chem.*, 248, 465, 1968.

18. **Hebbel, R. P., Morgan, W. T., Eaton, J. W., and Hedlund, B. E.**, Accelerated autoxidation and heme loss due to instability of sickle hemoglobin, *Proc. Natl. Acad. Sci. U.S.A.*, 85, 237, 1988.

19. **Hebbel, R. P. and Eaton, J. W.**, Pathobiology of heme interactions with the erythrocyte membrane, *Semin. Hematol.*, 26, 136, 1989.

20. **Balla, G., Vercellotti, G., Eaton, J. W., and Jacob, H. S.**, Heme uptake by endothelium synergizes PMN-mediated damage, *Trans. Assoc. Am. Physicians*, CIII, 174, 1990.

21. **Sadrzadeh, S. H. M. and Eaton, J. W.**, Hemoglobin-mediated oxidant damage to the central nervous system requires endogenous ascorbate, *J. Clin. Invest.*, 82, 1510, 1515, 1988.

22. **Ottolenghi, A.**, Interaction of ascorbic acid and mitochondrial lipids, *Arch. Biochem. Biophys.*, 79, 353, 1959.

23. **Barber, A. A.**, Lipid peroxidation in rat tissue homogenates: interaction of iron and ascorbic acid as the normal catalytic mechanism, *Lipids*, 1, 146, 1966.

24. **Sharma, S. K. and Krishna Murti, C. R.**, Production of lipid peroxides by brain, *J. Neurochem.*, 15, 147, 1968.

25. **Sharma, O. P. and Krisha Murti, C. R.**, Ascorbic acid: a naturally occurring mediator of lipid peroxide formation in rat brain, *J. Neurochem.*, 27, 299, 1976.

26. **Zaleska, M. M. and Floyd, R. A.**, Regional lipid peroxidation in rat brain *in vitro:* Possible role of endogenous iron, *Neurochem. Res.*, 10, 397, 1985.

27. **Aust, S. D., Morehouse, L. A., and Thomas, C. E.**, Role of metals in oxygen radical reactions, *Free Radical Biol. Med.*, 1, 3, 1985.

28. **Braughler, J. M., Duncan, L. A., and Chase, R. L.**, The involvement of iron in lipid peroxidation: importance of ferric to ferrous ratios in initiation, *J. Biol. Chem.*, 261, 10282, 1986.

29. **Bucher, J. R., Tien, M., and Aust, S. D.**, The requirement for ferric in the initiation of lipid peroxidation by chelated ferrous iron, *Biochem. Biophys. Res. Commun.*, 111, 777, 1983.

30. **Minotti, G. and Aust, S. D.**, The requirement for ferric in the initiation of lipid peroxidation by ferrous and hydrogen peroxide, *J. Biol. Chem.*, 262, 1098, 1987.

31. **Bucher, J. R. and Tien, M.**, et al., Redox cycling and lipid peroxidation: the central role of iron chelates, *Fundam. Appl. Toxicol.*, 3, 222, 1983.

32. **Aust, S. D. and Svingen, B. A.**, The role of iron in enzymatic lipid peroxidation, in *Free Radicals in Biology*, Pryor, W. A., Ed., Academic Press, San Diego, 1982, 1.

Chapter 3

RADICALS AND OXIDANTS IN ETHANOL-INDUCED LIVER INJURY

Bernhard H. Lauterburg and Brigitte de Quay

TABLE OF CONTENTS

I. BACKGROUND

In urban areas of the U.S. cirrhosis of the liver secondary to excessive alcohol consumption is now the third most frequent cause of death among those 25 to 64 years of age. Alcohol kills many more people since alcohol intoxication is involved in a large portion of all traffic accidents. It is estimated that alcohol-related diseases cost the taxpayer $116 billion per year. Thus, alcohol represents a major health and economic problem, which, at first sight, appears simple to solve: one just prohibits alcohol. Unfortunately, as history shows, this approach does not work. Moreoever, prohibition is not an attractive solution since the majority of people consume alcohol with much enjoyment and without hazard to themselves or others.

Indeed, major health problems occur only with a daily consumption of approximately 30 g/d of pure ethanol for females and 80 g/d for males over prolonged periods of time (Figure 1). Why females are more susceptible to the toxic effects of ethanol is not totally clear. Females may be exposed to higher circulating concentrations of ethanol for a given ingested amount due to decreased first-pass metabolism by gastric alcohol dehydrogenase.[1] Even when these epidemiologically critical amounts of ethanol are ingested on a daily basis, only a fraction of heavy drinkers will develop alcoholic hepatitis and eventually cirrhosis of the liver with its life-threatening complications. Alcoholic beverages cannot be banned from our society. Therefore, the major aims of alcohol research are to identify and subsequently counsel the persons at risk of becoming alcoholics and to better understand the mechanisms by which alcohol leads to cell injury, hoping that these adverse effects might eventually be prevented by pharmacological interventions.

II. METABOLISM OF ETHANOL

The metabolism of ethanol might offer clues regarding the mechanism of toxicity of the substance. Ethanol is mainly metabolized in the liver by cytosolic alcohol dehydrogenase, which exists in multiple molecular forms and requires NAD^+ as cofactor (Figure 2). The resulting acetaldehyde is further oxidized to acetate by acetaldehyde dehydrogenases. The kinetic properties of the different acetaldehyde dehydrogenases indicate that acetaldehyde oxidation will mainly occur in mitochondria, where, consequently, early lesions of ethanol-related cell injury are seen. The intracellular redox shift resulting from the accumulation of NADH is, in part, responsible for some of the metabolic consequences of alcohol consumption, such as the accumulation of triglycerides (for review see Reference 2).

Ethanol is also oxidized by cytochrome P-450-dependent reactions to acetaldehyde. The contribution of this microsomal ethanol-oxidizing system (MEOS) to overall ethanol metabolism is not clear, but probably amounts to less than 10%. Nevertheless, the microsomal electron transfer may be an

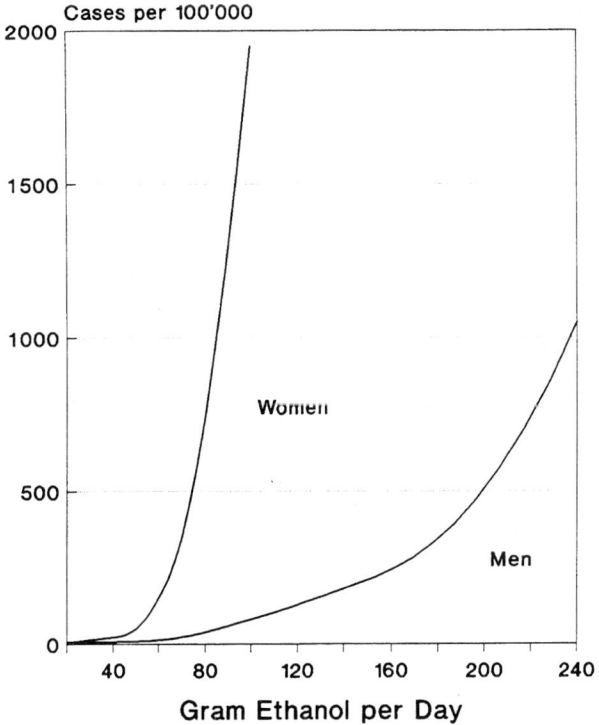

FIGURE 1.   Risk of cirrhosis as a function of the daily consumption
of pure ethanol. Adapted from Lelbach, in *Progress in Liver Disease,*
Popper, H. and Schaffner, F., Eds., Grune and Stratton, New York,
1976, 494.

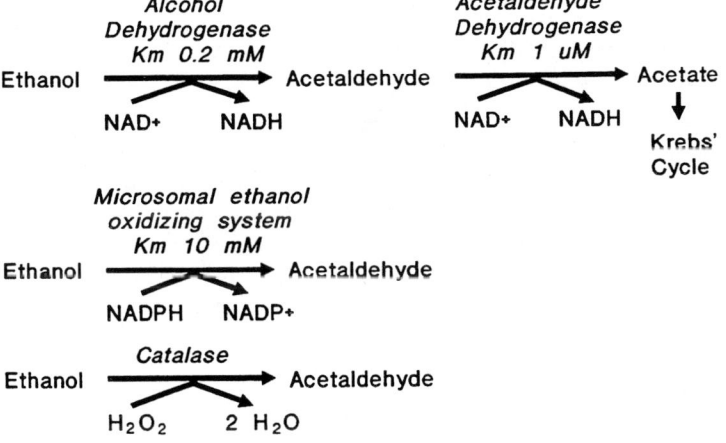

FIGURE 2.   Metabolism of ethanol.

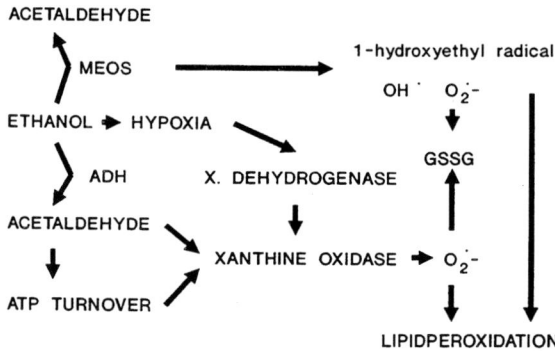

FIGURE 3.   Potential sources of reactive oxygen species and radicals during the metabolism of ethanol.

important source of reactive oxygen species and carbon-centered, ethanol-derived radicals.[3] Furthermore, induction of the microsomal ethanol-oxidizing system by chronic alcohol consumption will affect the metabolism of other xenobiotics that are metabolized by the same system. If, as in the case of many carcinogens and the analgesic paracetamol, metabolic activation is involved in the toxicity of the compound, chronic alcohol consumption may render such compounds much more toxic.[4]

There is general agreement that the metabolism of ethanol by catalase, the third ethanol-oxidizing system, plays a minimal role *in vivo*.

III. PATHOGENESIS OF ALCOHOLIC LIVER INJURY

The mechanisms by which ethanol injures the cell are not clearly understood. One major reason for the lack of reliable mechanistic information is the absence of a readily available animal model for the alcoholic liver disease seen in man. The typical lesions of heavy drinkers are difficult to reproduce in rats or mice. Only some baboons develop a clinical picture similar to alcoholic liver disease when fed ethanol over prolonged periods of time. Many hypotheses regarding the mechanism of injury have been put forward. Nutritional deficiencies may play an important role because the alcoholic receives much of his caloric requirements from ethanol rather than with a well-balanced diet. Genetic factors and immunological reactions are probably important and could explain why only a small percentage of heavy drinkers develop serious alcoholic liver disease.[5] Interactions of the reactive metabolite of ethanol, acetaldehyde, with critical cell components could also contribute to cell injury.[6] Finally, reactive oxygen metabolites and radicals might be involved (Table 1).

TABLE 1
Pathogenesis of Alcoholic Liver
Disease

Nutritional deficiency
Genetic predisposition
Immunologic reaction
Toxicity of acetaldehyde
Radicals and/or reactive oxygen species?

IV. ROLE OF RADICALS AND OXIDANTS IN ETHANOL-INDUCED CELL INJURY

Whether alcohol administration is associated with lipid peroxidation and subsequent injury *in vivo* has been a long-standing debate.[7,8] Indices of lipid peroxidation have been found to increase following acute and chronic administration of ethanol to experimental animals and in alcoholic patients by many, but not all, investigators. In some studies, evidence for the peroxidation of lipids was found to be associated with the metabolism of ethanol itself. In others, lipid peroxidation did not depend on the metabolism of ethanol but only occurred when acetaldehyde was metabolized by acetaldehyde dehydrogenases. Some investigators found evidence for a role of xanthine oxidase in the generation of reactive oxygen species (ROS) following the administration of ethanol, whereas the experiments of others refute a role for this enzyme. So far, a causal relationship between the occurrence of indices of lipid peroxidation and cell injury due to ethanol has not been established. Ethanol-associated alterations in metabolic processes may influence the decomposition of lipid hydroperoxides and the metabolism of the breakdown products, thereby increasing the steady-state concentrations of some indices of lipid peroxidation without actually increasing the generation of peroxidized fatty acids in membranes. Moreover, dead cells are prone to undergo autooxidation. The appearance of indices of lipid peroxidation together with cell injury may thus be a secondary event that is not related to the mechanism of injury. The following indices of lipid peroxidation suggesting the generation of free radicals from ethanol have been observed *in vivo*.

A. MALONDIALDEHYDE (MDA)

Increased concentrations of malondialdehyde have been described by many investigators in experimental models of alcoholic liver injury[8,9] and in human volunteers following the ingestion of ethanol.[10] In rats chronically fed alcohol somewhat lower, but still substantial, doses of ethanol are required to produce an increase in MDA than in rats given ethanol acutely.[9] The problems with the interpretation of increased MDA are well recognized. There is some evidence that increases in MDA following ethanol tend to overestimate the formation of lipid hydroperoxides measured enzymatically.[11] It is also likely that the metabolism of MDA is inhibited by ethanol, which could result in an increase of the steady-state concentration without an increased production of lipid hydroperoxides. Due to the poor specificity of the MDA assay *in vivo,* increased MDA concentrations following ethanol must be interpreted with caution.

B. CONJUGATED DIENES

In chronic alcoholics, increased concentrations of conjugated dienes have been found in liver biopsies.[12] Changes in the absorbance of lipids are rather difficult to interpret when appropriate control lipids are not available, as in the case when chronic alcoholics are compared with healthy controls eating a different diet. An increased absorption of a lipid fraction at 233 nm might well be due to chromophores other than conjugated dienes in unsaturated lipids. In chronic alcoholics, the circulating concentrations of a diene-conjugated nonperoxide isomer of linoleic acid (9,11-LA') and the ratio of 9,11-LA' to 9,12-LA' in phospholipids are increased.[13] Upon abstaining from alcohol, the abnormal values tend to normalize within days. No oxygen-free conjugated dienes from arachidonic acid or docosahexaenoic acid have been observed. There is some concern that 9,11-LA' may not be generated during free radical-induced injury. Possible other sources of the fatty acid are the diet and hydrogenation of 9,12-LA' by certain intestinal bacteria.[14]

In experimental models of alcohol-induced liver injury, the observations regarding conjugated dienes are controversial. In two models that are associated with cell injury resembling alcoholic liver disease in man, no increases in conjugated dienes have been found.[15,16] Others have found conjugated dienes in microsomes and mitochondria following acute and chronic administration of ethanol.[9,17,18]

C. ALKANES IN BREATH

In rats, large doses of ethanol, 5 g/kg, result in an increased exhalation of ethane and pentane.[7,19,20] However, large doses of ethanol inhibit the metabolism of ethane and pentane.[20] Thus, increased concentrations of these gases in breath may predominantly reflect decreased metabolism rather than increased production.

D. OTHER INDICES OF LIPID PEROXIDATION

In rats chronically fed ethanol, as well as in alcoholics, the hepatic concentration of tocopherol is decreased.[21] Chronic ethanol feeding of rats receiving a diet low in vitamin E results in increased concentrations of α-tocopheryl quinone, which is generated from tocopherol by radical reactions.[22] One possible interpretation of this observation could be that ethanol feeding is associated with the generation of free radicals.

E. DEPLETION OF GLUTATHIONE

Depletion of hepatic glutathione, following the administration of ethanol, has been construed as evidence for an ethanol-induced oxidant stress.[8] Large doses of ethanol given to rats lower the hepatic concentration of glutathione by up to 50%.[23] However, ethanol acutely inhibits the synthesis of glutathione in rats and increases the efflux from the liver.[24-26] There is no evidence that the increased efflux is due to the efflux of glutathione disulfide rather than reduced glutathione. The biliary excretion of glutathione disulfide, which is a sensitive index of oxidant stress *in vivo,* does not increase following ethanol. The concentration of glutathione disulfide in liver does not increase markedly either, supporting the conclusion that the decreased hepatic concentration of reduced glutathione does not reflect an oxidant stress. In rats fed ethanol chronically, glutathione concentrations have been reported to increase, decrease, or remain constant. The concentration of glutathione in mitochondria appears to be preferentially decreased compared to cytosol, which may be of relevance for the decreased mitochondrial oxidation of acetaldehyde in chronic alcoholics.[27] There is, however, no convincing evidence that the generation of glutathione disulfide, which is a sensitive marker of oxidant stress *in vivo,* is increased in mitochondria.

In man, chronic alcoholism leads to lower hepatic and circulating concentrations of glutathione.[28,29] As in animals, a single dose of ethanol may increase the plasma concentration of glutathione simultaneously with an increase in the circulating concentration of thiobarbituric acid-reactive substances (TBARS).[10] The decrease in hepatic glutathione may be critical for the capacity of alcoholics to detoxify electrophilic intermediates of drugs, such as the toxic intermediate of acetaminophen,[4,28] but it is not likely that the depletion is sufficient to promote lipid peroxidation.

V. POTENTIAL SOURCES OF RADICALS AND OXIDANTS

If we accept the present data as evidence for lipid peroxidation, we have to ask the question what steps in the metabolism of ethanol could generate radicals ultimately causing the peroxidation process.

A. XANTHINE OXIDASE

One candidate is the generation of superoxide anion radicals by xanthine oxidase, resulting in an oxidant stress and oxidative cell injury. Under physiological conditions, most of the xanthine oxidase is a NAD^+-dependent dehydrogenase and not an oxidase. The conversion of the dehydrogenase form into the oxidase form occurs during hypoxia, possibly as a consequence of increased cytosolic calcium and the activation of proteases. Moreover, the increase in NADH content during the metabolism of ethanol might favor the oxygen-dependent oxidase. Whether the conversion takes place in alcoholic liver disease is not known. Following chronic administration of ethanol to rats, total oxidase activity does not increase.[30] However, with long-standing hypoxia and episodes of ischemia, the conversion could well take place in human liver. Alcohol consumption is associated with an increased consumption of oxygen, which is only partially compensated for by an increase in blood flow. Thus, the more central portions of the liver lobule which, by the way, are the prime targets for alcoholic liver injury, will be deprived of oxygen.

Two sources of substrates for xanthine oxidase may be relevant to alcoholic cell injury. The metabolism of ethanol results in a marked increase in the turnover of ATP. Most of the breakdown products will be reutilized for the resynthesis of ATP, but part will escape and enter the purine-uric acid pathway. Consequently, the availability of purine degradation products, which are substrates for xanthine oxidase, increases. This is the major reason for the increase in uric acid and the podagra seen in many alcoholics.[31] This mechanism has been postulated as a major factor in the formation of malondialdehyde in rat liver following ethanol.[32] Moreover, acetaldehyde itself is a substrate for xanthine oxidase. Acetaldehyde is relatively innocuous to cells *in vitro*. In the presence of xanthine oxidase, however, acetaldehyde is toxic and promptly damages lymphocytes *in vitro*.

B. OXIDANT STRESS FOLLOWING ETHANOL?

Superoxide anion radicals generated intracellularly are efficiently detoxified by superoxide dismutase and glutathione peroxidase. The resulting glutathione disulfide is, in part, reduced back to glutathione by glutathione reductase but, in part, is excreted into bile and also appears in plasma. Thus, the biliary excretion of glutathione disulfide is a sensitive index of oxidant stress *in vivo*.[33,34] If the generation of superoxide anion radicals by xanthine oxidase was relevant to the pathogenesis of alcoholic cell injury, substantial amounts of glutathione disulfide (GSSG) should be formed and should be excreted into bile. Several studies, including our own, however, were not able to demonstrate an increased biliary excretion of GSSG under these conditions.[24,35] Even when glutathione reductase was inhibited by bis(2-chloroethyl)-1-nitrosourea) (BCNU), no increase in GSSG was found in rat bile following 5 g/kg of ethanol.[35a]

C. ROLE OF IRON

Even small amounts of superoxide, however, might be relevant in alcoholic liver injury because the concentration of iron in the liver is increased in many alcoholics. This may be related to an increased dietary intake. Some wines contain high concentrations of iron. In many alcoholics, an abnormal transferrin lacking terminal sugars circulates in plasma, and there is some evidence that this transferrin transfers iron more readily to the liver than does the normal protein.[36] Iron overload is a well-characterized model of liver injury due to radicals. In experimental animals, acutely administered ethanol leads to a redistribution of iron such that more nonheme iron may be available for reaction with reactive oxygen species.[37]

D. ROLE OF THE MICROSOMAL ETHANOL-OXIDIZING SYSTEM

The microsomal ethanol-oxidizing system might be a source of radicals.[38,39] Whether the generation of radicals observed *in vitro* is relevant to the situation *in vivo* is difficult to assess. By electron paramagnetic resonance spectrometry, carbon-centered radicals have been observed *in vivo* following ethanol.[3] No correlation between the generation of radicals and biological effects, however, has been established. Moreover, ethanol, as well as acetaldehyde, could act as a radical scavenger under certain circumstances and thus play a protective role.

E. ROLE OF NEUTROPHILS

The accumulation of neutrophils is a characteristic feature of alcoholic hepatitis. Oxidants generated by activated neutrophils might well contribute to liver injury in alcoholics. Rat hepatocytes produce a chemoattractant substance upon exposure to ethanol. The biological relevance of this finding, however, is questionable since neutrophils are conspicuously absent in the livers of rats treated with ethanol *in vivo*. Acetaldehyde at high concentrations can bind to proteins by reacting with lysine residues, thus forming a Schiff base.[6] Plasma membranes of hepatocytes altered by acetaldehyde have been shown to stimulate neutrophils to produce superoxide anion radicals.[40] Clinically, neutrophils are often seen at sites of cell necrosis. Thus, it is tempting to speculate that the metabolism of ethanol alters the plasma membrane of the cell and thereby triggers its self-destruction by stimulating neutrophils to release reactive oxygen species and proteolytic enzymes.

The issue of whether ethanol consumption is associated with lipid peroxidation is far from settled. Even less clear is the question of the pathogenetic significance of lipid peroxidation in alcoholic liver disease. In most models where indices of lipid peroxidation have been observed, there is no evidence for alcohol-related cell injury. Conversely, in models demonstrating cell injury associated with the administration of ethanol, there is no evidence for lipid peroxidation.[15,16] Particularly in patients with alcoholic liver disease, where the pathologic process has evolved over a number of years, it is difficult to resolve the issue of whether the appearance of indices of lipid peroxidation, such as conjugated dienes, precedes cell injury or rather reflects a secondary event subsequent to the autooxidation of fatty acids in dead cells. Clearly, there is an urgent need for more sophisticated and more specific methodology to demonstrate the presence or absence of lipid peroxidation following ethanol.

ACKNOWLEDGMENT

Supported by grant Nr 3.812-0.87 from the Swiss National Foundation for Scientific Research.

REFERENCES

1. **Frezza, M., Dipadova, C., Pozzato, G., Terpin, M., Baraona, E., and Lieber, C. S.**, High blood alcohol levels in women — the role of decreased gastric alcohol dehydrogenase activity and 1st-pass metabolism, *N. Engl. J. Med.*, 322, 95, 1990.
2. **Lieber, C. S.**, Mechanism of ethanol induced hepatic injury, *Pharm. Ther.*, 46, 1, 1990.
3. **Reinke, L. A., Lai, E. K., DuBose, C. M., and McCay, P. B.**, Reactive free radical generation *in vivo* in heart and liver of ethanol-fed rats: correlation with radical formation *in vitro*, *Proc. Natl. Acad. Sci. U.S.A.*, 84, 9223, 1987.
4. **Seeff, L. B., Cuccherini, B. A., Zimmerman, H. J., Alder, E., and Benjamin, S. B.**, Acetaminophen hepatotoxicity in alcoholics, *Ann. Intern. Med.*, 104, 399, 1986.
5. **Devor, E. J., Reich, T., and Cloninger, C. R.**, Genetics of alcoholism and related end-organ damage, *Semin. Liver Dis.*, 8, 1, 1988.
6. **Lauterburg, B. H. and Bilzer, M.**, Mechanisms of acetaldehyde hepatotoxicity, *J. Hepatol.*, 7, 384, 1988.
7. **Muller, A. and Sies, H.**, Alcohol, aldehydes and lipid peroxidation. Current notions, *Alcohol Alcohol Suppl.*, 22, 67, 1987.
8. **Videla, L. A. and Valenzuela, A.**, Alcohol ingestion, liver glutathione and lipoperoxidation: metabolic interrelations and pathological implications, *Life Sci.*, 31, 2395, 1982.
9. **Shaw, W., Jayatilleke, E., Ross, W. A., Gordon, E. R., and Lieber, C. S.**, Ethanol-induced lipid peroxidation: potentiation by long-term alcohol feeding and attenuation by methionine, *J. Lab. Clin. Med.*, 98, 417, 1981.
10. **Vendemiale, G., Altomare, E., Grattagliano, I., and Albano, O.**, Increased plasma levels of glutathione and malondialdehyde after acute ethanol ingestion in humans, *J. Hepatol.*, 9, 359, 1989.
11. **Papatheofanis, F. J. and Lands, W. E. M.**, Overestimation of lipid peroxidation during injury to liver by ethanol, *Hepatology*, 4, 1025, 1984.
12. **Shaw, S., Rubin, K. P., and Lieber, C. S.**, Depressed hepatic glutathione and increased diene conjugates in alcoholic liver disease, *Dig. Dis. Sci.*, 28, 585, 1983.
13. **Fink, R., Marjot, D. H., Cawood, P., Iversen, S. A., Clemens, M. R., Patsalos, P., Norden, A. G., and Dormandy, T. L.**, Increased free-radical activity in alcoholics, *Lancet*, ii, 291, 1985.
14. **Smith, M. T. and Thompson, S.**, Free radicals and alcoholics, *Lancet*, ii, 774, 1985.
15. **Inomata, T., Rao, G. A., and Tsukamoto, H.**, Lack of evidence for increased lipid peroxidation in ethanol-induced centrilobular necrosis of rat liver, *Liver*, 7, 233, 1987.
16. **Speisky, H., Bunout, D., Orrego, H., Giles, H. G., Gunasekara, A., and Israel, Y.**, Lack of changes in diene conjugate levels following ethanol induced glutathione depletion or hepatic necrosis, *Res. Commun. Chem. Pathol. Pharmacol.*, 48, 77, 1985.
17. **Sippel, H. W.**, Effect of an acute dose of ethanol on lipid peroxidation and on the activity of microsomal glutathione S-transferase in rat liver, *Acta Pharmacol. Toxicol.*, 53, 135, 1983.
18. **MacDonald, C. M.**, The effects of ethanol on hepatic lipid peroxidation and on the activities of glutathione reductase and peroxidase, *FEBS Lett.*, 35, 227, 1973.
19. **Koster, U., Albrecht, D., and Kappus, H.**, Evidence for carbon tetrachloride- and ethanol-induced lipid peroxidation *in vivo* demonstrated by ethane production in mice and rats, *Toxicol. Appl. Pharmacol.*, 41, 639, 1977.
20. **Frank, H., Hintze, T., Bimboes, D., and Remmer, H.**, Monitoring lipid peroxidation by breath analysis: endogenous hydrocarbons and their metabolic elimination *Toxicol. Appl. Pharmacol.*, 56, 337, 1980.
21. **Bjorneboe, G. E. A., Bjorneboe, A., Hagen, B. F., Morland, J., and Drevon, C. A.**, Reduced hepatic alpha tocopherol content after long-term administration of ethanol to rats, *Biochim. Biophys. Acta*, 918, 236, 1987.
22. **Kawase, T., Kato, S., and Lieber, C. S.**, Lipid peroxidation and antioxidant defense systems in the rat liver after chronic ethanol feeding, *Hepatology*, 10, 815, 1989.

23. **Videla, L. A., Ugarte, F. G., and Valenzuela, A.**, Effect of acute ethanol intoxication on the content of reduced glutathione of the liver in relation to its lipoperoxidative capacity in the rat, *FEBS Lett.*, 111, 6, 1980.

24. **Lauterburg, B. H., Davies, S., and Mitchell, J. R.**, Ethanol suppresses hepatic glutathione synthesis in rats *in vivo*, *J. Pharmacol. Exp. Ther.*, 230, 7, 1984.

25. **Speisky, H., MacDonald, A., Giles, G., Orrego, H., and Israel, Y.**, Increased loss and decreased synthesis of hepatic glutathione after acute ethanol administration, *Biochem. J.*, 225, 565, 1985.

26. **Speisky, H., Kera, Y., Penttilä, K. E., Israel, Y., and Lindros, K. O.**, Depletion of hepatic glutathione by ethanol occurs independently of ethanol metabolism, *Alcoholism Clin. Exp. Res.*, 12, 224, 1988.

27. **Fernandez-Checa, J. C., Ookhtens, M., and Kaplowitz, N.**, Effect of chronic ethanol feeding on rat hepatocytic glutathione, *J. Clin. Invest.*, 80, 57, 1987.

28. **Lauterburg, B. H. and Velez, M. E.**, Glutathione deficiency in alcoholics: risk factor for paracetamol hepatotoxicity, *Gut*, 29, 1153, 1988.

29. **Jewell, S. A., Di Monte, D., Gentile, A., Guglielmi, A., Altomare, E., and Albano, O.**, Decreased hepatic glutathione in chronic alcoholic patients, *J. Hepatol.*, 3, 1, 1986.

30. **Abbondanza, A., Battelli, M. G., Soffritti, M., and Cessi, C.**, Xanthine oxidase status in ethanol-intoxicated rat liver, *Alcoholism Clin. Exp. Res.*, 13, 841, 1989.

31. **Puig, J. G. and Fox, I. H.**, Ethanol-induced activation of adenine nucleotide turnover. Evidence for a role of acetate, *J. Clin. Invest.*, 74, 936, 1984.

32. **Kato, S., Kawase, T., Aldermann, J., Inatomi, N., and Lieber, C. S.**, Role of xanthine oxidase in ethanol-induced lipid peroxidation in rats, *Gastroenterology*, 98, 203, 1990.

33. **Lauterburg, B. H., Smith, C. V., Hughes, H., and Mitchell, J. R.**, Biliary excretion of glutathione and glutathione disulfide in the rat, *J. Clin. Invest.*, 73, 124, 1984.

34. **Adams, J. D., Lauterburg, B. H., and Mitchell, J. R.**, Plasma glutathione and glutathione disulfide in the rat: regulation and response to oxidative stress, *J. Pharmacol. Exp. Ther.*, 227, 749, 1983.

35. **Vendemiale, G., Jayatilleke, E., Shaw, S., and Lieber, C. S.**, Depression of biliary glutathione excretion by chronic ethanol feeding in the rat, *Life Sci.*, 34, 1065, 1984.

35a. **Lauterburg, B. H.**, unpublished observation.

36. **Regoeczi, E., Chindemi, P. A., and Debanne, M. T.**, Transferrin glycans: a possible link between alcoholism and hepatic siderosis, *Alcoholism Clin. Exp. Res.*, 8, 287, 1984.

37. **Rouach, H., Houze, P., Orfanelli, M.-T., Gentil, M., Bourdon, R., and Nordmann, R.**, Effect of acute ethanol administration on the subcellular distribution of iron in rat liver and cerebellum, *Biochem. Pharmacol.*, 39, 1095, 1990.

38. **Winston, G. W. and Cederbaum, A. I.**, A correlation between hydroxyl radical generation and ethanol oxidation by liver, lung and kidney microsomes, *Biochem. Pharmacol.*, 31, 2031, 1982.

39. **Klein, S. M., Cohen, G., Lieber, C. S., and Cederbaum, A. I.**, Increased microsomal oxidation of hydroxyl radical scavenging agents and ethanol after chronic consumption of ethanol, *Arch. Biochem. Biophys.*, 223, 425, 1983.

40. **Williams, A. J. K. and Barry, R. E.**, Superoxide anion production and degranulation of rat neutrophils in response to acetaldehyde-altered liver cell membranes, *Clin. Sci.*, 71, 313, 1986.

Chapter 4

LIPID PEROXIDE-DEPENDENT MODIFICATIONS OF LIPOPROTEINS IN ATHEROSCLEROSIS

Margaret E. Haberland and Charles V. Smith

TABLE OF CONTENTS

I. INTRODUCTION

Coronary heart disease is the leading cause of death and disability in the U.S. and accounts for approximately 500,000 deaths each year. The annual economic costs associated with this disease have been estimated at $60 billion in direct health care costs, lost wages, and loss in productivity.[1] Epidemiologic studies[2] have established that the higher the total plasma or low-density lipoprotein (LDL) cholesterol, the greater the risk for development of coronary heart disease. The Lipid Research Clinical Coronary Primary Prevention Trial[3] has shown that the incidence of coronary heart disease can be reduced by lowering plasma levels of LDL cholesterol. These and other findings from clinical and experimental studies have led to the concept that LDL plays a causal role in atherosclerosis, the basic pathophysiological process leading to coronary heart disease.

The pathogenesis of atherosclerosis is marked by the focal accumulation of macrophages, smooth muscle cells, and extracellular matrix, and by the intracellular and extracellular deposition of lipoprotein-derived cholesterol in the tunica intima of arteries.[4-6] The sequence of cellular events in atherogenesis has been investigated by detailed morphologic studies in experimental animals with dietary-induced hypercholesterolemia,[7,8] and in humans.[9] It is now generally accepted that the early cellular events of atherosclerosis are marked by monocyte adherence to arterial endothelium at sites of predilection, transendothelial migration, and rapid localization of the leukocytes to the subendothelial matrix, and appearance of foam cells filled with cytoplasmic droplets of cholesteryl ester.[1,4-9] Morphologic and immunocytochemical studies have shown that subendothelial macrophages, originally derived from the blood monocytes recruited to sites of predilection, are progenitors of many of these arterial foam cells.[4,5,7] Lesion progression is characterized by continued monocytic invasion together with proliferation of vascular smooth muscle cells recruited from the tunica media. The advanced fibrotic plaque is marked by a high proportion of smooth muscle cell-derived foam cells, calcification, and a necrotic core comprised of cholesteryl ester, crystalline cholesterol, and cellular debris.[1,4-6,9] A surprising degree of luminal narrowing due to the atheromatous reaction can occur without significant clinical effects and, unfortunately, without significant clinical warning signals. Eventual rupture or fissure of an atherosclerotic plaque can cause acute ischemia in the tissue served by the artery, particularly in the presence of thrombus formation initiated by the fissure. Blockage of a coronary artery in this way is the most common mechanism of acute myocardial infarction.[1]

A major risk factor for the development of atherosclerosis is an elevated plasma concentration of LDL.[2,10] Human LDL is a spherical particle of molecular weight 2.5 to 3 million; the lipoprotein is a complex of lipids in dynamic association with a single polypeptide chain known as the apoB-100 protein, and contains approximately 78% lipids and 22% protein by weight.[11]

The basic organization of LDL includes a hydrophobic core comprised of the neutral lipids, cholesteryl esters and triglycerides, which is surrounded and stabilized by a monolayer of amphipathic molecules comprised of the apoB-100 protein, free cholesterol, and phosphatidylcholine-predominating phospholipids. In addition to imparting stability to the particle, the apoB-100 protein (molecular weight 515,000) contains recognition site(s) that mediate binding by the LDL receptor.[10] While the fatty acyl chain composition of LDL lipids is subject to interindividual variations and dietary intake, the major polyunsaturated fatty acids present in LDL are linoleate and arachidonate.[12] Linoleate accounts for 80 to 90% of the polyunsaturates and occurs chiefly in cholesteryl esters. Conversely, most of the arachidonate is found in the phospholipids. Smaller amounts of other polyunsaturates, such as docosahexaenoate, are usually present.

Since plasma lipoproteins serve as the major source of cholesteryl ester deposited in atheroma, attention has focused upon both the pathways of lipoprotein metabolism and changes in function which might explain the biochemical and cellular events of atherosclerosis. The classic studies of Brown and Goldstein[10] have established that the receptor-dependent uptake of LDL provides the major route for cellular acquisition of cholesterol from the milieu. LDL serves as the primary carrier of plasma cholesterol to cells for direct incorporation into membranes as well as for synthesis of steroid hormones in the adrenals and synthesis of bile acids in the liver. The activity of the LDL receptor is regulated by intracellular levels of cholesterol and this modulation serves as a sensitive control for the maintenance of cellular cholesterol homeostasis. Mutations which produce loss or deficiency of function of LDL receptors result in the genetic disease familial hypercholesterolemia (Type II hyperlipoproteinemia).[10] Early, spontaneous atherosclerosis and dramatically elevated levels of plasma LDL are cardinal features of this autosomal dominant disorder in homozygotic patients and the Watanabe heritable hyperlipidemic (WHHL) rabbit, an animal model of this disease. This evidence has revealed an apparent paradox in the pathogenesis of atherosclerosis in familial hypercholesterolemia, namely, that the deposition of lipoprotein-derived cholesterol within arterial cells is accelerated by the virtual absence of LDL receptor-mediated uptake of lipoprotein lipid. Thus, attention has turned to mechanisms independent of the LDL receptor that might account for the development of atherosclerosis.

Recent evidence favors modification of LDL by lipid peroxides as a prerequisite to the initiation of atherosclerosis. The demonstration that probucol, a hypocholesterolemic drug with antioxidant properties, retards lesion formation in WHHL rabbits[13,14] has reinforced the view that lipid peroxidation contributes to the pathogenesis of this disease. It has been proposed that different products of lipid peroxidation are responsible for key events including ability to trigger critical, early cellular events of atherogenesis,[15-20] to chemically modify the apoB-100 protein of LDL as a prerequisite to formation of macrophage-derived foam cells,[21-27] and to produce cytotoxicity.[28-31] This chapter will highlight experimental evidence that has led to development of these concepts, and point the interested reader to original publications and review articles that discuss these topics in greater detail. So that the reader can gain a sense of the evolution and refinement of concepts that have been more than a decade in development, we have taken a historical approach in discussing the role of lipid peroxide-modified LDL in the initiation and the progression of atherosclerosis.

II. ROLE OF LIPID PEROXIDATION IN FORMATION OF MACROPHAGE-DERIVED FOAM CELLS

A critical lead in elucidating the mechanistic link between elevated plasma concentrations of LDL and formation of macrophage-derived foam cells was provided by Goldstein et al.[32] in 1979. The key observation was that certain chemical modifications of normal LDL led to recognition by a separate receptor, later termed the scavenger receptor, which was not subject to feedback inhibition by intracellular cholesterol. The cloning of the scavenger receptor by Kodama et al.[33] in 1990 has validated cell biology studies that this binding site is genetically and physically distinct from the LDL receptor. Goldstein et al.[32] demonstrated that, in the presence of modest concentrations of modified LDL, macrophages *in vitro* accumulated large amounts of LDL-derived cholesterol in the form of cytoplasmic droplets of cholesteryl ester and developed histologic features similar to the foam cells that are observed in atherosclerotic lesions *in vivo*. This report initiated a flurry of research activity into the nature of the chemical modifications necessary and sufficient for uncontrolled uptake of LDL and foam cell formation. The chemical modification employed in the initial report was acetylation, which was accomplished *in vitro* by incubation of LDL with acetic anhydride.[32] Since acetylation of the majority of lysines of apoB-100 protein is required to produce recognition by the scavenger receptor,[34] it is unlikely that acetylation of LDL is a physiologically significant event *in vivo* in atherogenesis. Nevertheless, the changes in functional properties of the LDL particle produced by acetylation may reflect the essential changes in LDL that contribute to foam cell formation *in vivo*.

In 1980, Fogelman et al.[21] demonstrated that derivatization of LDL by malondialdehyde (MDA) produced the accumulation of LDL-derived cholesteryl esters in human monocyte-macrophages *in vitro,* and suggested that MDA released by macrophages or platelets or as a secondary product of lipid peroxidation could be a physiologically significant source of LDL modification. Fogelman et al.[21] further hypothesized that modification of native LDL by products of lipid peroxidation may be a prerequisite to the accumulation of cholesteryl esters within the cells of the atherosclerotic reaction. Haberland et al.[23] subsequently demonstrated that MDA derivatization of LDL led to progressive loss of recognition by the LDL receptor, and produced threshold recognition by the scavenger receptor upon modification of 16% of the lysine residues of the apoB-100 protein. If LDL particles were first adsorbed to heparin-Sepharose, MDA derivatization produced conversion of LDL to a form recognized by the scavenger receptor upon modification of 10% of the lysyl groups and concomitantly produced release of the modified lipoprotein from the matrix. These findings suggest that solid-phase adsorption of LDL particles facilitates modification of lysyl residues critical to formation of the scavenger receptor binding determinants.

Ligands recognized by the scavenger receptor characteristically are anionic. The primary change caused by derivatization with lysine-specific reagents, such as acetic anhydride, succinic anhydride, or MDA, is an increase in the net negative charge of the lipoprotein particle.[21,35] Acetylation serves to convert α- and ϵ-amino groups, which are protonated at physiological pH and possess a net positive charge, to neutral amide groups. Mono- and difunctional aldehydes are thought to interact with LDL primarily through Schiff base formation with the ϵ-amino group on lysine. Reaction of MDA with lysine in a 1:2 molar ratio yields an N,N'-disubstituted 1-amino-3-iminopropene crosslink and a decrease in positive charge of the apoprotein.[23] Fogelman et al.[21] showed that incubation of LDL with MDA *in vitro* did not cause a significant change in the molecular weight of the modified lipoprotein, indicating the majority of the lysyl crosslinks formed by MDA were intramolecular. Conformational changes induced in the apoB-100 protein by crosslinking pairs of lysyl amino groups might prove to be a more significant factor in macrophage uptake by the scavenger receptor, but the relative contributions of charge and conformation in the derivatization of LDL remain unknown. For a comprehensive review of ligand recognition by the scavenger receptor, please refer to the article by Brown and Goldstein.[35]

In 1981, Henriksen et al.[22] demonstrated that incubation of LDL with cultured endothelial cells for 24 h produced a more negatively charged lipoprotein. Like acetyl-LDL, incubation of macrophages *in vitro* with endothelial cell-modified LDL produced intracellular cholesteryl ester deposition.[22,24] Modification of LDL also can be accomplished by incubation with vascular smooth muscle cells or macrophages.[25,29,36] The cell-mediated alteration of the LDL was found to be inhibited uniformly by antioxidants or by metal chelators, implicating peroxidation of the lipids in LDL as the mechanism of this cellular alteration.[29,37] In 1987, Steinbrecher et al.[24] demonstrated that an apparent requirement for redox-active transition metals, particularly copper or iron, was a common feature of these studies. Subsequent investigations revealed that addition of relatively small amounts of transition metals, particularly copper, markedly stimulated LDL oxidation and its conversion into a form recognized by the scavenger receptor[12,27,38] and possibly a second binding site for oxidized LDL.[39,40] Studies in a number of laboratories have shown that similar changes in the physicochemical properties of LDL can be produced by simple incubation in the presence of oxygen.[12] This process is inhibited by chain-terminating antioxidants, such as α-tocopherol, probucol, or butylated hydroxyanisole (BHA), and is also inhibited by sufficient concentrations of transition metal chelators.[12]

Esterbauer and his colleagues[12,41] have identified a number of aldehydes produced as a result of transition metal-induced oxidation of LDL *in vitro*. However, aldehydes measured in these studies only account for about 2% of the polyunsaturated fatty acids consumed by the oxidation. Lenz et al.[42] have recently demonstrated that most of the fatty acids consumed in the oxidation of LDL in the presence of 5 μ*M* copper sulfate *in vitro* were present as lipid hydroperoxides or lipid hydroxy acids. Lipid hydroperoxides are the initial products of radical chain oxidation of unsaturated fatty acids, formed through abstraction of bis-allylic H atoms, coupling of the alkyl radical with O_2, and abstraction of another H atom by the peroxyl radical to generate the product hydroperoxide and propagate the radical chain. However, the accumulation of lipid hydroperoxides in this system was somewhat unexpected inasmuch as these compounds are unstable in the presence of redox-active transition metals, such as iron and copper.[43-45] Apparently, some form of compartmentalization of the copper and the lipid hydroperoxides keeps the two separated from each other in this system. Whether a similar pattern of products is observed in cell-mediated oxidation of LDL *in vitro* is not known at this time, but there is evidence that even cell-mediated modification of LDL has a critical requirement for small quantities of iron or copper ions.[12] The participation of trace amounts of reactive chelates of iron or copper ions in the autoxidation of LDL and in the cell-mediated oxidation of LDL is probably responsible for many of the apparently contradictory results that have been obtained. For an excellent discussion of this topic, please see the recent review by Esterbauer et al.[12]

In the copper-catalyzed autoxidation of LDL in aqueous buffer, a lag phase is observed between the initiation of the exposure to the oxidizing conditions and the onset of the rapid phase of consumption of polyunsaturated fatty acids.[12] The duration of this lag phase is determined, in part, by the rate of production of radicals capable of initiating lipid peroxidation and, in part, by the consumption of endogenous chain-breaking antioxidants, such as α-tocopherol. The subsequent oxidation of polyunsaturated fatty acids is accompanied by other changes in LDL composition.[37] These include marked fragmentation of the apoB-100 protein attributed to protein scission by free radicals released during oxidation, production of lysophosphatidylcholine due to a platelet-activating factor acetylhydrolase associated with LDL,[46] increased particle density due to the loss of lipid from the lipoprotein, and loss of about 50% of the lysyl ε-amino groups attributed to derivatization by chemically reactive lipid peroxides. In contrast to LDL modified by lysine-specific reagents, such as acetic anhydride or MDA, copper-oxidized LDL also induces chemotaxis of human monocytes and arrests migration of murine peritoneal macrophages. This chemotactic activity has been attributed to lysophosphatidylcholine,[47] generated by platelet-activating factor acetylhydrolase after selective cleavage of oxidized fatty acid residues at the *sn*-2 position of phosphatidylcholine.[46,48] The functional properties of oxidized LDL and their proposed roles in the initiation and progression of atherosclerosis have recently been reviewed by Steinberg et al.[49] and Steinbrecher et al.[50]

The measurement of thiobarbituric acid-reactive substances (TBARS) was used to estimate lipid peroxidation in the majority of the studies of LDL lipid peroxidation. In this method of analysis, the experimental sample or lipid fraction is heated with 2-thiobarbituric acid (TBA) in an acidic medium. Measurement of absorbance at 532 nm is employed to determine the formation of a 2:1 adduct between TBA and MDA. MDA is thought to arise during conduct of the assay from scission reactions of endoperoxides formed during radical chain peroxidation of polyunsaturated fatty acids.[51] The method of analysis is sensitive and simple to perform, and has therefore dominated studies of the peroxidation of biological lipids for several decades. However, MDA itself is quantitatively a minor product of lipid peroxidation. A growing appreciation of this fact and that a number of biological substances other than peroxidized lipids will form the red chromophore with TBA have dampened enthusiasm for the use of this assay. While Esterbauer et al.[12] have made a strong case for measurement of conjugated dienes by absorption at 234 nm as an index of lipid peroxide production, many publications acknowledge the lack of specificity of the TBA test and then proceed to present data and interpretations that rely solely upon the measurement of TBARS as a determination of lipid peroxidation.

Despite the lack of chemical specificity of the TBA test, an increase in TBARS does demonstrate a change in the chemical composition of the material being studied. As such, this change may offer a clue to the mechanisms that are responsible for whatever biological effects are being studied. Kosugi and Kikugawa[52] have examined some of the chemical complexities involved in producing the red pigment from non-MDA precursors and have shown that TBARS can be formed from a variety of lipid-derived intermediates under the normal conditions of the TBA test through transformations that involve the presence of O_2, water, and trace amounts of redox-active transition metals. These and related studies may enable us to derive additional information from TBARS determinations. It is still important, nonetheless, to include other indices of lipid peroxidation, such as the routine measurement of conjugated dienes.[12]

Because transition metal-catalyzed transformations are important in the conversion to MDA and/or TBARS of lipid hydroperoxides and other substances, many of which may not be lipid hydroperoxides,[52] the role of iron and copper in lipid hydroperoxide disposition needs to be considered. The virtual absence of consumption of fatty acids in the attempted autoxidation of LDL in buffer in the absence of iron or copper indicates that chelates of these metal ions are critical in the peroxidation of the lipids.[12] Iron and copper may contribute to cell-mediated oxidation of LDL *in vitro* through similar mechanisms, but this point has not been investigated adequately at this time. The contributions of metal-catalyzed processes to LDL oxidation *in vivo* are virtually unexplored. In publications referring to addition of transition metal to LDL *in vitro*, the distinction between dose and free concentration is an important one. Free concentrations of metal ions are almost never measured, and the quantity quoted refers to the moles of ion added per liter of incubation solution. The step-response effect of amount of metal ion added on the oxidation of LDL that has been observed by a number of investigators suggests occupation by the metal ion of distinct binding sites on the lipoprotein.[12] Below 1 μM added copper or iron, little oxidation is observed, while between 3 and 10 μM added transition metal, the conjugated dienes and TBARS are formed at similar rates. Recently, Kalyanaraman et al.[53] have reported the existence of two copper binding sites on LDL: one of these has been identified as a complex of copper with residual EDTA trapped in LDL, while the other is a high molecular weight cupric complex bound to apoB-100 protein. Whether any of these sites bind iron or copper to a significant extent *in vivo*, either normally or under pathophysiological stresses, and whether any such chelates exert biologically significant effects are important questions. In addition, the manner in which metal chelation on the apoprotein affects peroxidation of the LDL lipid may be very different from the effects on oxidation, and fragmentation, of the protein.

It has been shown that incubation of LDL with purified soybean lipoxygenase plus phospholipase A_2 effects most of the alterations induced by cellular incubations.[54] In addition, a number of lipoxygenase inhibitors have been shown to inhibit cell-mediated oxidation of LDL.[55] However, most lipoxygenase inhibitors presently available also exhibit substantial capacity to function as simple chemical antioxidants (see Taylor and Shappell, Chapter 5, this book), and similar pharmacological studies with nonantioxidant lipoxygenase inhibitors are needed. Investigations by Chisolm and colleagues[30,56] illustrate the complexities of distinguishing lipoxygenase-dependent from superoxide anion-mediated oxidation of LDL in cell culture. Studies are needed in which the specific products formed during oxidation by cells are characterized, in order to distinguish product patterns characteristic of specific enzyme-mediated oxidations from the pattern of products formed by radical chain autoxidation. If the oxidation of the lipids in LDL is truly mediated by cellular lipoxygenases, it is not clear how the reactions should be inhibited by metal chelators, unless the lipoxygenase products simply serve as initiators of radical chain oxidation. In this circumstance, the reaction of product hydroperoxides with free or loosely bound iron or copper ions would function to provide the organic radical species to initiate the chain reaction in the LDL particle. The product profile would reflect this mode of reaction, despite the critical role played by cellular lipoxygenases.

A substantial body of evidence has now been presented that LDL found in atherosclerotic lesions both in rabbits and in humans has been modified by products of lipid peroxidation. In 1988, Haberland et al.[57] presented immunohistochemical evidence for modification of LDL apoprotein by MDA in atheroma from WHHL rabbits. The material reactive with the monoclonal antibody to the MDA-lysine adduct was found to be extracellular and to colocalize with monoclonal antibody to apoB-100 protein. Despite the presence of MDA-modified LDL in atherosclerotic lesions, no modification of LDL in the circulating plasma or in healthy arterial tissue was observed. A series of investigations by Steinberg and collaborators[58,59] with antibodies to MDA-lysine and 4-hydroxynonenal-LDL adducts have confirmed and extended these observations. Studies by Chait et al.[60] have shown that a monoclonal antibody specific for an epitope common to copper-oxidized LDL, acetyl-LDL, and MDA-LDL demonstrates immunoreactivity in atheroma. Ylä-Hertualla et al.[59] have, furthermore, demonstrated that arterial LDL extracted from lesions demonstrates many features in common with oxidized LDL, including enhanced electronegative charge of the particles, fragmentation of apoB-100 protein, increased particle density, ability to stimulate cholesteryl ester synthesis in macrophages, and ability to induce monocyte chemotaxis. In a recent study, Ylä-Hertualla et al.[61] have shown that oxidation-specific lipid-protein adducts colocalize with messenger RNA for 15-lipoxygenase and for the scavenger receptor in macrophage-rich human atherosclerotic lesions. Taken together, these findings indicate that apoB-100 protein of LDL is modified by chemically reactive products of lipid peroxidation *in vivo* and that the arterial wall is the dominant site for oxidative alteration.

Palinsky et al.[58] have also demonstrated that autoantibodies specific for MDA-lysine adducts and 4-hydroxynonenal-protein adducts are present in the circulation of rabbits and humans. The similarity in antibody titers from normal New Zealand white and WHHL rabbits with spontaneous atherosclerosis suggests that these immune responses may be raised by mechanisms generally operating *in vivo,* for example, by release of MDA and other lipid peroxide products during thrombotic events[21] and subsequent modification of thrombus-entrapped proteins. Nonetheless, the presence of autoantibodies directed against lipid peroxide adducts offers yet another potential route for macrophage uptake of lipid peroxide-modified LDL as immune complexes via the F_c receptor.[35] Whether this mechanism operates in atheroma *in vivo* has yet to be established.

III. CYTOTOXICITY OF OXIDIZED LOW-DENSITY LIPOPROTEIN

The oxidative alteration of LDL has been studied most extensively for the mechanisms leading to foam cell formation. It has long been appreciated from studies by Chisolm and colleagues.[29-31,62-64] that oxidized LDL also can be cytotoxic and, ultimately, the formation of an acellular core region of cholesterol-rich lipid and cellular debris is observed in more mature atherosclerotic lesions. Cell death observed in atherosclerotic plaques may arise in part from regional hypoxia secondary to macrophage infiltration, smooth muscle cell proliferation, and foam cell formation in the intimal lesion. However, a significant contribution to cell necrosis from products of lipid peroxidation is also to be expected.

While oxidized LDL is capable of killing cells *in vitro*, little is presently known about the mechanisms through which this oxidized LDL exerts its cytotoxicity. Kosugi et al.[64] have demonstrated that toxicity of oxidized LDL (iron-catalyzed) to cultured fibroblasts is selective for the S phase of the cell cycle. Recently, Cathcart et al.[56] have shown that treatment of LDL with soybean lipoxygenase, a 15-lipoxygenase, can convert the lipoprotein to a cytotoxin. Kuzuya et al.[65] have shown that the cell death caused by LDL that had been oxidized *in vitro* (iron-catalyzed) was preceded by depletion of intracellular glutathione (GSH) and was enhanced by the treatment of the cells with buthionine sulfoximine, which inhibits the first step of GSH synthesis. The cytotoxicity of oxidized LDL was attenuated by addition of L-2-oxothiazolidine-4-carboxylate, which is capable of supplying cysteine residues to support cellular GSH resynthesis.[45] The oxidation of LDL has been shown to cause the formation of 4-hydroxynonenal and related species that are cytotoxic and form GSH-derived conjugates in a reaction catalyzed by glutathione *S*-transferases. By analogy with other reactive intermediates that are substrates for glutathione *S*-transferases and show a GSH-dependent toxic dose threshold,[45] it would be reasonable to propose that cells exposed to 4-hydroxynonenal would conjugate the α,β-unsaturated aldehyde with GSH until its availability was depleted, at which time protein thiols would be alkylated. Alteration of a sufficient number of protein thiols critical to cell function might then comprise cell viability.

With regard to the chemically reactive intermediates in oxidized LDL (copper-catalyzed), lipid hydroperoxides are present in much greater abundance than are enals.[42] If the hydroperoxides are predominantly responsible for the cytotoxicity of oxidized LDL, it is possible that they are first converted to α,β-unsaturated aldehydes or ketones or other alkylating species. Nevertheless, it is reasonable to first investigate the more straightforward possibility that the lipid hydroperoxides interact with target cells through oxidative processes. The characteristic product of this oxidative interaction is glutathione disulfide (GSSG).[45] However, the fundamental data regarding the nature of the interaction of oxidized LDL with cells in culture are simply not available at this time. The studies of Kuzuya et al.[65] strongly implicate a glutathione-dependent cellular defense by endothelial cells, but offer no insight into the mechanisms by which the glutathione is consumed. Studies are needed on the mechanisms through which these interactions occur. Similar investigations with vascular smooth muscle cells and with monocyte-macrophages are also needed, because there could be significant qualitative or quantitative difference in the responses of these different cell types.

Products derived from the oxidation of cholesterol or the cholesterol moiety in cholesteryl esters, often termed oxysterols, are to be expected in processes involving nonenzymatic lipid peroxidation in LDL. However, the formation of oxysterols during LDL lipid peroxidation has received far less attention than the oxidation of the polyunsaturated fatty acids in LDL.[66] This proportionation of research effort probably has resulted more from the ease of measuring TBARS or absorbance at 234 nm than from the demonstration of the relative biological relevance of the two processes. In a recent report, Guyton et al.[31] have described studies of the effects of focal crystalline deposits of cholestane-3b,5a,6b-triol, 25-hydroxycholesterol, or cholesterol, overlaid with collagen gel, on vascular smooth muscle cells in culture. The oxysterols did not appear to be chemoattractants in this system, but did exhibit a persistent cytotoxicity that resulted in an acellular gap in the culture, not unlike the necrotic core characteristic of many advanced atherosclerotic lesions *in vivo*.

IV. ROLE OF LIPID PEROXIDATION IN TRIGGERING EARLY CELLULAR EVENTS

Investigations led by Fogelman and Berliner[15-20] in 1990 have developed the novel concept that mild peroxidation of polyunsaturated fatty acyl chains in LDL generates a lipoprotein particle with potent bioregulatory properties. These lipoproteins, denoted minimally modified LDL, retain recognition by the LDL receptor. Minimally modified LDL generated by mild iron oxidation, as well as the polar lipids extracted from minimally modified LDL,[15] display an array of modulatory effects after short-term incubation with endothelial cells or smooth muscle cells. The observed regulatory effects of minimally modified LDL (1 to 5 μg protein/ml) include cellular events characteristic of initiation of atherosclerosis: expression of a monocyte-specific adhesion site by endothelial cells,[15] induction of monocyte chemotactic protein (MCP-1) by endothelial cells and smooth muscle cells,[17] and increased mRNA levels for growth factors such as granulocyte-monocyte colony-stimulating factor and colony-stimulating factor.[16,18,20] Cellular delivery of the bioregulatory lipids may be accomplished through LDL receptor-mediated uptake of minimally modified LDL or through lipid transfer from the lipoprotein particles to the cellular membranes.[15] It is likely that each cellular effect is modulated by a different bioregulatory lipid.

Berliner et al.[15] have also examined cytotoxic effects of minimally modified LDL. At concentrations of minimally modified LDL tenfold higher than were required for enhancement of monocyte adherence, one strain of rabbit aortic endothelial cells was susceptible to cytotoxicity, whereas the other of the two strains tested in their study was not. The resistant strain could be made sensitive by pretreatment with cycloheximide and the susceptible strain could be rendered resistant by prior incubation with a subtoxic dose of minimally modified LDL. These observations implicate a cytoprotective mechanism that shows interanimal variation, requires protein synthesis, is inducible, and apparently is depletable. Intracellular GSH is a recognizable candidate, but direct measurements are needed of cellular GSH and products of its utilization in response to these treatments and pharmacological manipulation of the model by more selective inhibitors of GSH synthesis (buthionine sulfoximine). Other candidates include intracellular enzymes that participate in antioxidant mechanisms of defense such as catalase, superoxide dismutase, and glutathione peroxidase.[67]

The importance of attempting to reintegrate the individual elements of the arterial wall has been demonstrated by Navab et al.[68] through invention of an ingenious device[69] for the coculture of human aortic endothelial and smooth muscle cells separated by a layer of collagen. Studies performed with this model of the human artery wall have established the endothelial cell monolayer as the principal barrier to transport of micro- and macromolecules and have demonstrated maintenance of this barrier after transmigration of monocytes into the subendothelial space.[68,69] Recent investigations have convincingly shown that minimally modified LDL in the human artery wall model promotes monocyte chemotaxis and that MCP-1 elaborated by the endothelial and smooth muscle cells in response to minimally modified LDL accounts for the majority of the enhanced transmigration of monocytes into the subendothelial space.[17] The latest studies[18] demonstrate that cocultures of human endothelial cells and smooth muscle cells exhibit marked increases in production of growth factors and extracellular matrix proteins over and above the levels produced by the two cell types in separate cultures. The production of extracellular matrix is further enhanced upon transmigration of monocytes to the subendothelial space of the human artery wall model. These synergistic effects are likely due to enhanced formation of gap junctions and other factors important for communication among cells.[18]

The concept of minimally modified LDL is particularly attractive in providing an explanation of the biochemical and cellular events of the initiation of atherogenesis. The products of lipid peroxidation characterized in the more extensively modified LDL may play significant roles in the progression of the lesion. It is risky and probably incorrect to assume that the product profile of minimally oxidized LDL is similar to the product profile of more extensively modified LDL, but at lower conversion. Obviously, there are a number of unanswered questions regarding the correlations of the type and extent of modification of lipoprotein-derived lipids with the diverse cellular responses that have been observed, and future developments promise to be most interesting.

V. LIPOPROTEINS IN THE ATHEROSCLEROTIC LESION: INTERFACIAL RATHER THAN SOLUTION CHEMISTRY

In 1974, Smith[70] determined that the ratio of apoB-containing lipoproteins to other plasma proteins in human arterial plaques increases three- to tenfold, and, in 1982, Hoff and Gaubatz[71] provided detailed chemical and physico-chemical characterizations of arterial LDL isolated from human atheroma. In 1989, Schwenke and Carew[72] convincingly demonstrated that this accumulation and retention of LDL at sites of predilection precede the migration of monocytes into the artery wall in rabbits with diet-induced hypercholesterolemia. Frank and Fogelman[73] have examined the ultrastructure of intima from aortas of cholesterol-fed New Zealand white and WHHL rabbits prepared by ultrarapid freezing and freeze-etching. The subendothelial region of intima from cholesterol-fed New Zealand white rabbits was filled with clusters of lipid vesicles enmeshed in the complex extracellular matrix. The deposition of lipid particles occurred as early as 10 d after initiating dietary hypercholesterolemia. More than 75% of the lipid vesicles in the intima of cholesterol-fed rabbits had diameters between 20 and 68 nm. These diameters are within the range of β-VLDL or LDL particles. Similarly, the subendothelial region of the intima of WHHL rabbits was filled with collagen fibrils surrounding and entwined between clusters of discrete lipid vesicles. In contrast to the particle size distribution in the intima of cholesterol-fed rabbits, the diameters of 80% of the vesicle population in WHHL rabbit intima measured between 70 and 169 nm in diameter, and many of the individual lipid particles appeared to be in the process of fusing into larger vesicles.

These studies show that LDL particles, or perhaps lipid vesicles derived from LDL, are trapped in the acellular subendothelial matrix of the arterial intima in both New Zealand White rabbits within 10 d after initiating a cholesterol-containing diet as well as in WHHL rabbits.[73] This would represent the earliest morphologic alteration noted thus far at sites for predilection. The aggregation of the LDL particles in the intima may be associated with fundamental changes in the biochemical composition of the matrix elements.[74,75] Recent histochemical studies by Völker et al.[74] illustrate the typical changes in proteoglycan composition in human atherosclerotic tissues; by comparison to the normal arterial wall, these include a significant increase in chondroitin sulfate/dermatan sulfate and a loss of heparan sulfate.

Matrix entrapment of apoB-100-containing lipoproteins and the apparent vesicle coalescence might contribute to atherogenesis through one or more mechanisms. The sulfated glycosaminoglycans present in the fibrous network of the subendothelial space in the arterial wall demonstrate high-affinity binding of LDL both *in vivo* and *in vitro*,[75,76] and proteoglycan-LDL complexes formed *in vitro* with certain isolated arterial proteoglycans stimulate uptake and cholesteryl ester accumulation in macrophages.[75-77] The uptake of this proteoglycan-modified LDL appears to be mediated both by the classic LDL receptor pathway and by a separate pathway that is not inhibited competitively by native LDL.[75,77] Hurt-Camejo et al.[78] have recently reported that human arterial chondroitin sulfate proteoglycans are able to separate LDL particles into at least four subclasses based on differential affinities for complexation. The subfractions of LDL with the greatest affinity for the proteoglycans were found to have higher isoelectric points and were bound, internalized, and degraded more effectively by human monocyte-macrophages.

Vesicle fusion described in the ultrastructure studies by Frank and Fogelman[73] is not unlike the self-aggregation of LDL particles produced by Khoo et al.[79] after vortexing a suspension of LDL less than 1 min. The self-aggregated LDL particles stimulated LDL receptor-dependent phagocytosis and cholesteryl ester deposition in macrophages. In a recent ultrastructural study, lipoproteins aggregated by vortexing underwent fusion with the creation of two distinct lipid forms — lipid droplets rich in cholesteryl esters and vesicles comprised mostly of phospholipids and unesterified cholesterol.[80] These two lipid forms are very similar to lipid particles isolated from atheroma.[81,82] Furthermore, Mora et al.[82] have noted an increase in TBARS in the cholesteryl ester-rich, apoB-100-containing droplets. Whether products of lipid peroxidation contribute to arterial vesicle fusion, however, remains unknown.

Biochemical studies *in vitro* have implicated electrostatic interaction of lipoproteins with glycosaminoglycans and other matrix proteins as the primary factor accounting for lipoprotein accumulation in the artery wall. The extensive connections of matrix filaments with the lipid/lipoprotein particles that have been observed in ultrastructural analyses[73] suggest that physical entrapment plays a major role in retention of arterial lipoproteins and facilitates particle coalescence. Investigations of lipoprotein interaction with matrix molecules typically have been conducted in fluid phase. It may be more relevant to turn to interfacial chemistry in consideration of lipoprotein interactions with elements of the artery wall, and subsequent physicochemical and chemical modifications of the lipoprotein particles. As discussed earlier, solid-phase adsorption of LDL particles to heparin-Sepharose facilitates modification by MDA of lysyl residues critical to formation of the scavenger receptor binding determinants; furthermore, this derivatization produces release of the anionic, modified LDL particles from the anionic resin.[23] This observation from experiments *in vitro* has prompted Haberland et al.[23,83] to propose that a similar mechanism may operate *in vivo* to release entrapped, modified apoB-100 particles from the extracellular matrix in forms that can be cleared by scavenger receptors of subendothelial macrophages in atheroma.

VI. CONCLUSIONS

The balance of evidence supports the current concept that oxidation-mediated alterations in LDL contribute to both the initiation and the progression of atherosclerosis. Attempts to chemically define the radicals and reactive oxygen species that arise during lipoprotein oxidation and, therefore, might be contributing to the observed changes in atherosclerosis are confounded by the low chemical selectivity of these reactions, by the chemical complexity of the lipid constituents, and by the contributions of metabolic alterations of both the substrates and the products of these transformations. Purely chemical systems can be studied with greater precision, but lack the structural and cellular organization of atheroma, for example, the intimate association of protein with lipid during oxidant exposure and the association, chelation, and redox activity of transition metal ions potentially endogenous to the biological matrix of the developing lesion.

Entrapment of LDL particles by the subendothelial matrix may not only facilitate subsequent events of lipid peroxidation, but also alter transcellular eicosanoid metabolism.[83a] Arterial lipid droplets likely disrupt the transcellular transfer of critical molecular signals between endothelial cells and vascular smooth cells through sequestration of prostaglandins, leukotrienes, and eicosanoid precursors. Some of these lipid mediators are chemically reactive oxidants, particularly those retaining a hydroperoxide or endoperoxide functional group. These metabolites may function as autocoids, initiate radical chain peroxidation of the LDL lipid, particularly if reactive chelates of iron or copper are present; or react directly with the LDL apoprotein to modify specificity of receptor recognition.

There are a number of critical issues to be addressed in the role of lipid peroxide-modified LDL in the initiation and the progression of atherosclerosis. The time now is rich with opportunity for joint explorations by cell biologists and chemists of these complex and intriguing issues.

ACKNOWLEDGMENTS

Studies from Dr. Haberland's laboratory cited in this review have been supported in part by funding from USPHS HL 30568 and the Laubisch Fund. Dr. Smith gratefully acknowledges the support of HL45619.

REFERENCES

1. **Comai, K., Feldman, D. L., Goldstein, A. L., and Hamilton, J. G.,** Atherosclerosis: an overview, *Drug Dev. Res.,* 6, 113, 1985.
2. **Gordon, T., Castelli, W. P., Hjortland, M. C., and Kannel, W. B.,** The prediction of coronary heart disease by high-density and other lipoproteins: an historical perspective, in *Hyperlipidemia—Diagnosis and Therapy,* Rifkind, B. M. and Levy, R. I., Eds., Grune & Stratton, New York, 1977, 71.
3. Lipid Research Clinics Program, The Lipid Research Clinics Coronary Primary Prevention Trial Results. II. The relationship of reduction in incidence of coronary heart disease to cholesterol lowering, *J. Am. Med. Assoc.,* 251, 365, 1984.
4. **Davies, M. J.,** Atherosclerosis — a pathologists view, in *Atherosclerosis: Developments, Complications and Treatment,* Shepherd, J., Morgan, H. G., Packard, C. J., and Brownlie, S. M., Eds., Excerpta Medica, Amsterdam, 1987, 21.
5. **Wissler, R. W., Vesselinovitch, D., and Davis, H. R.,** Cellular components of the progressive atherosclerotic process, in *Atherosclerosis: Biology and Clinical Science,* Olsson, A. G., Ed., Churchill Livingstone, Edinburgh, 1987, 57.
6. **Adams, C. W. M.,** Disordered structure and function in the atherosclerotic artery, in *Atherosclerosis: Biology and Clinical Science,* Olsson, A. G., Ed., Churchill Livingstone, Edinburgh, 1987, 75.

7. **Gerrity, R. G., Naito, H. K., Richardson, M., and Schwartz, C. J.**, Dietary induced atherogenesis in swine, *Am. J. Pathol.*, 95, 775, 1979.

8. **Faggiotto, A., Ross, R., and Harker, L.**, Studies of hypercholesterolemia in the non-human primate. I. Changes that lead to fatty streak formation, *Arteriosclerosis*, 4, 323, 1984.

9. **Stary, H. C.**, The sequence of cell and matrix changes in atherosclerotic lesions of coronary arteries in the first forty years of life, *Eur. Heart J.*, 11 (Suppl. E), 3, 1990.

10. **Brown, M. S. and Goldstein, J. L.**, A receptor-mediated pathway for cholesterol homeostasis, *Science*, 232, 32, 1986.

11. **Havel, R. J. and Kane, J. P.**, Introduction: structure and metabolism of plasma lipoproteins, in *The Metabolic Basis of Inherited Disease*, Scriver, C. R., Beudet, A. L., Sly, W. S., and Valle, D., Eds., McGraw-Hill, New York, 1989, 1129.

12. **Esterbauer, H., Dieber-Rotheneder, M., Waeg, G., Streigl, G., and Jurgens, G.**, Biochemical, structural, and functional properties of oxidized low-density lipoprotein, *Chem. Res. Toxicol.*, 3, 77, 1990.

13. **Kita, T., Nagano, Y., Yokode, M., Ishi, K., Kume, N., Ooshima, A., Yoshida, H., and Kawai, C.**, Probucol prevents the progression of atherosclerosis in Watanabe heritable hyperlipidemic rabbit, an animal model for familial hypercholesterolemia, *Proc. Natl. Acad. Sci. U.S.A.*, 84, 5928, 1987.

14. **Carew, T. E., Schwenke, D. C., and Steinberg, D.**, Antiatherogenic effect of probucol unrelated to its hypocholesterolemic effect: evidence that antioxidants *in vivo* can selectively inhibit low density lipoprotein degradation in macrophage-rich fatty streaks and slow the progression of atherosclerosis in the Watanabe heritable hyperlipidemic rabbit, *Proc. Natl. Acad. Sci. U.S.A.*, 84, 7725, 1987.

15. **Berliner, J. A., Territo, M. C., Sevanian, A., Ramin, S., Kim, J. A., Bamshad, B., Esterson, M., and Fogelman, A. M.**, Minimally modified low density lipoprotein stimulates monocyte endothelial interactions, *J. Clin. Invest.*, 85, 1260, 1990.

16. **Rajavashisth, T. B., Andalibi, A., Territo, M. C., Berliner, J. A., Navab, M., Fogelman, A. M., and Lusis, J.**, Induction of endothelial cell expression of granulocyte and macrophage colony-stimulating factors by modified low density lipoproteins, *Nature (London)*, 344, 254, 1990.

17. **Cushing, S., Berliner, J. A., Valente, A. J., Territo, M. C., Navab, M., Parhami, F., Gerrity, R., Schwartz, C. J., and Fogelman, A. M.**, Minimally modified low density lipoprotein induces monocyte chemotactic protein 1 in human endothelial cells and smooth muscle cells, *Proc. Natl. Acad. Sci. U.S.A.*, 87, 5134, 1990.

18. **Navab, M., Liao, F., Hough, G. P., Ross, L. A., Van Lenten, B. J., Rajavashisth, T. B., Lusis, A. J., Laks, H., Drinkwater, D., and Fogelman, A. M.**, Interaction of monocytes with coculture of human aortic wall cells involves interleukins 1 and 6 with marked increase in connexin43 message, *J. Clin. Invest.*, 87, 1763, 1991.

19. **Fogelman, A. M., Berliner, J. A., Rajavashisth, T. B., Navab, M., Andalibi, A., Imes, S., Cushing, S. D., and Lusis, A. J.**, Molecular and cellular interactions in the early development of the atherosclerotic lesion, *J. Drug Dev.*, 3 (Suppl. 1), 143, 1990.

20. **Andalibi, A., Imes, S., Berliner, J. A., Fogelman, A. M., and Lusis, A. J.**, The effect of minimally modified low density lipoprotein on the expression of cytokines and adhesion molecules in endothelial cells, in *Molecular Biology of Atherosclerosis*, Attie, A. D., Ed., Elsevier, New York, 1990, 181.

21. **Fogelman, A. M., Shechter, I., Seager, J., Hokom, M., Child, J. S., and Edwards, P. A.**, Malondialdehyde alteration of low density lipoproteins leads to cholesteryl ester accumulation in human monocyte-macrophages, *Proc. Natl. Acad. Sci. U.S.A.*, 77, 2214, 1980.

22. **Henriksen, T., Mahoney, E. M., and Steinberg, D.**, Enhanced macrophage degradation of low density lipoprotein previously incubated with cultured endothelial cells: recognition by receptor for acetylated low density lipoproteins, *Proc. Natl. Acad. Sci. U.S.A.*, 78, 6499, 1981.

23. **Haberland, M. E., Fogelman, A. M., and Edwards, P. A.,** Specificity of receptor-mediated recognition of malondialdehyde-modified low density lipoproteins, *Proc. Natl. Acad. Sci. U.S.A.,* 79, 1712, 1982.

24. **Steinbrecher, U. P., Witzum, J. L., Parthasarathy, S., and Steinberg, D.,** Decrease in reactive amino groups during oxidation or endothelial cell modification of LDL: correlation with changes in receptor-mediated catabolism, *Arteriosclerosis,* 7, 135, 1987.

25. **Parthasarathy, S., Printz, D. J., Boyd, D., Joy, L., and Steinberg, D.,** Macrophage oxidation of low density lipoprotein generates a modified form recognized by the scavenger receptor, *Arteriosclerosis,* 6, 505, 1986.

26. **Steinbrecher, U.,** Oxidation of human low density lipoprotein results in derivatization of lysine residues of apolipoprotein B by lipid peroxide decomposition products, *J. Biol. Chem.,* 262, 3603, 1987.

27. **Steinbrecher, U., Lougheed, M., Kwan, W.-C., and Dirks, M.,** Recognition of oxidized low density lipoprotein by the scavenger receptor of macrophages results from derivatization of apolipoprotein B by products of fatty acid peroxidation, *J. Biol. Chem.,* 264, 15216, 1989.

28. **Morel, D. W., Hessler, J. R., and Chisolm, G. M., III,** Low density lipoprotein cytotoxicity induced by free radical peroxidation of lipid, *J. Lipid Res.,* 24, 1070, 1983.

29. **Morel, D. W., DiCorleto, P. E., and Chisolm, G. M., III,** Endothelial and smooth muscle cells alter low density lipoprotein *in vitro* by free radical oxidation, *Arteriosclerosis,* 4, 357, 1984.

30. **Cathcart, M. K., McNally, A. K., Morel, D. W., and Chisolm, G. M., III,** Superoxide anion participation in human monocyte-mediated oxidation of low-density lipoprotein and conversion of low-density lipoprotein to a cytotoxin, *J. Immunol.,* 142, 1963, 1989.

31. **Guyton, J. R., Black, B. L., and Seidel, C. L.,** Focal toxicity of oxysterols in vascular smooth muscle cell culture: a model of the atherosclerotic core region, *Am. J. Pathol.,* 137, 425, 1990.

32. **Goldstein, J. L., Ho, Y. K., Basu, S. K., and Brown, M. S.,** Binding site on macrophages that mediates uptake and degradation of acetylated low density lipoprotein, producing massive cholesterol deposition, *Proc. Natl. Acad. Sci. U.S.A.,* 76, 333, 1979.

33. **Kodama, T., Freeman, M., Rohrer, L., Zabrecky, J., Matsudaira, P., and Kreiger, M.,** Type I macrophage scavenger receptor contains α-helical and collagen-like coiled coils, *Nature (London),* 343, 531, 1990.

34. **Haberland, M. E., Olch, C. L., and Fogelman, A. M.,** Role of lysines in mediating interaction of modified low density lipoproteins with the scavenger receptor of human monocyte-macrophages, *J. Biol. Chem.,* 259, 11305, 1984.

35. **Brown, M. S. and Goldstein, J. L.,** Lipoprotein metabolism in the macrophage: implications for cholesterol deposition in atherosclerosis, *Annu. Rev. Biochem.,* 52, 223, 1983.

36. **Heinecke, J. W., Baker, L., Rosen, L., and Chait, A.,** Superoxide-mediated modification of low density lipoprotein by arterial smooth muscle cells, *J. Clin. Invest.,* 77, 757, 1986.

37. **Steinbrecher, U. P., Parthasarathy, S., Leake, D. S., Witztum, J. L., and Steinberg, D.,** Modification of low density lipoprotein by endothelial cells involves lipid peroxidation and degradation of low density lipoprotein phospholipids, *Proc. Natl. Acad. Sci. U.S.A.,* 81, 3883, 1984.

38. **Parthasarathy, S., Fong, L. G., Quinn, M. T., and Steinberg, D.,** Oxidative modification of LDL: comparison between cell-mediated and copper-mediated modification, *Eur. Heart J.,* 11 (Suppl. E), 83, 1990.

39. **Arai, H., Kita, T., Yokode, M., Narumiya, S., and Kawai, C.,** Multiple receptors for modified low density lipoproteins in mouse peritoneal macrophages: different uptake mechanisms for acetylated and oxidized low density lipoproteins, *Biochem. Biophys. Res. Commun.,* 159, 1375, 1989.

40. **Sparrow, C. P., Parthasarathy, S., and Steinberg, D.**, A macrophage receptor that recognizes oxidized low density lipoprotein but not acetylated low density lipoprotein, *J. Biol. Chem.*, 264, 2599, 1989.

41. **Esterbauer, H., Jurgens, G., Quenhenberger, O., and Koller, E.**, Autoxidation of human low density lipoprotein: loss of polyunsaturated fatty acids and vitamin E and generation of aldehydes, *J. Lipid Res.*, 28, 495, 1987.

42. **Lenz, M. L., Hughes, H., Mitchell, J. R., Via, D. P., Guyton, J. R., Taylor, A. A., Gotto, A. M., Jr., and Smith, C. V.**, Lipid hydroperoxy and hydroxy derivatives in copper-catalyzed oxidation of low density lipoprotein, *J. Lipid Res.*, 31, 1043, 1990.

43. **Frankel, E. N.**, Volatile lipid oxidation products, *Prog. Lipid Res.*, 22, 1, 1982.

44. **Frankel, E. N.**, Chemistry of free radical and singlet oxidation of lipids, *Prog. Lipid Res.*, 23, 197, 1985.

45. **Smith, C. V. and Mitchell, J. R.**, Pharmacological aspects of glutathione in drug metabolism, in *Coenzymes and Cofactors*, Dolphin, D., Poulson, R., and Avramovic, O., Eds., John Wiley & Sons, New York, 1989, 1.

46. **Steinbrecher, U. P. and Pritchard, P. H.**, Hydrolysis of phosphatidylcholine during LDL oxidation is mediated by platelet-activating factor acetylhydrolase, *J. Lipid Res.*, 30, 305, 1989.

47. **Quinn, M. T., Parthasarathy, S., and Steinberg, D.**, Lysophosphatidylcholine: a chemotactic factor for human monocytes and its potential role in atherogenesis, *Proc. Natl. Acad. Sci. U.S.A.*, 85, 2805, 1988.

48. **Stremler, K. E., Stafforini, D. M., Prescott, S. M., Zimmerman, S. M., and McIntyre, T. M.**, An oxidized derivative of phosphatidylcholine is a substrate for the platelet-activating factor acetylhydrolase from human plasma, *J. Biol. Chem.*, 264, 5331, 1989.

49. **Steinberg, D., Parthasarathy, S., Carew, T. E., Khoo, J. C., and Witztum, J. L.**, Beyond cholesterol: modifications of low-density lipoprotein that increase its atherogenicity, *N. Engl. J. Med.*, 320, 915, 1989.

50. **Steinbrecher, U. P., Zhang, H. F., and Lougheed, M.**, Role of oxidatively modified LDL in atherosclerosis, *Free Radical Biol. Med.*, 9, 155, 1990.

51. **Pryor, W. A., Stanley, J. P., and Blair, E.**, Autoxidation of polyunsaturated fatty acids. II. A suggested mechanism for the formation of TBA-reactive materials from prostaglandin-like endoperoxides, *Lipids*, 11, 370, 1976.

52. **Kosugi, H. and Kikugawa, K.**, Potential thiobarbituric acid-reactive substances in peroxidized lipids, *Free Radical Biol. Med.*, 12, 205, 1989.

53. **Kalyanaraman, B., Antholine, W. E., and Parthasarathy, S.**, Oxidation of low-density lipoprotein by Cu^{2+} and lipoxygenase: an electron spin resonance study, *Biochim. Biophys. Acta*, 1035, 286, 1990.

54. **Sparrow, C. P., Parthasarthy, S., and Steinberg, D.**, Enzymatic modification of low density lipoprotein by purified lipoxygenase plus phospholipase A_2 mimics cell-mediated oxidative modification, *J. Lipid Res.*, 29, 745, 1988.

55. **Parthasarathy, S., Wieland, E., and Steinberg, D.**, A role for endothelial cell lipoxygenase in the oxidative modification of low density lipoprotein, *Proc. Natl. Acad. Sci. U.S.A.*, 86, 1046, 1989.

56. **Cathcart, M., McNally, A. K., and Chisolm, G. M., III**, Lipoxygenase-mediated transformation of human low density lipoprotein to an oxidized and cytotoxic complex, *J. Lipid Res.*, 32, 63, 1991.

57. **Haberland, M. E., Fong, D., and Cheng, L.**, Malondialdehyde-altered protein occurs in atheroma of Watanabe heritable hyperlipidemic rabbits, *Science*, 241, 215, 1988.

58. **Palinski, W., Rosenfeld, M. E., Ylä-Herttuala, S., Gurtner, G. C., Socher, S. S., Butler, S. W., Parthasarathy, S., Carew, T. E., Steinberg, D., and Witzum, J. L.**, Low density lipoprotein undergoes oxidative modification in vivo, *Proc. Natl. Acad. Sci. U.S.A.*, 86, 1372, 1989.

59. **Ylä-Herttualla, S., Palinski, W., Rosenfeld, M. E., Parthasarathy, S., Carew, T. E., Butler, S., Witzum, J. L., and Steinberg, D.**, Evidence for the presence of oxidatively modified low density lipoprotein in atherosclerotic lesions of rabbit and man, *J. Clin. Invest.*, 84, 1086, 1989.

60. **Boyd, H. C., Gown, A. M., Wolfbauer, G., and Chait, A.**, Direct evidence for a protein recognized by a monoclonal antibody against oxidatively modified LDL in atherosclerotic lesions from a Watanabe heritable hyperlipidemic rabbit, *Am. J. Pathol.*, 135, 815, 1989.

61. **Ylä-Herttualla, S., Rosenfeld, M. E., Parthasarathy, S., Sigal, E., Särkioja, T., Witztum, J. L., and Steinberg, D.**, Gene expression in macrophage-rich human atherosclerotic lesions, *J. Clin. Invest.*, in press, 1991.

62. **Hessler, J. R., Robertson, A. L., Jr., and Chisolm, G. M., III**, LDL-induced cytotoxicity and its inhibition by HDL in human vascular smooth muscle and endothelial cells in culture, *Atherosclerosis*, 32, 213, 1979.

63. **Hessler, J. R., Morel, D. W., Lewis, L. J., and Chisolm, G. M., III**, Lipoprotein oxidation and lipoprotein-induced cytotoxicity, *Arteriosclerosis*, 3, 215, 1983.

64. **Kosugi, K., Morel, D. W., DiCorleto, P. E., and Chisolm, G. M., III**, Toxicity of oxidized low density lipoprotein to cultured fibroblasts is selective for S phase of the cell cycle, *J. Cell. Physiol.*, 130, 311, 1987.

65. **Kuzuya, M., Naito, M., Funaki, C., Hayashi, T., Asai, K., and Kuzuya, F.**, Protective role of intracellular glutathione against oxidized low density lipoprotein in cultured endothelial cells, *Biochem. Biophys. Res. Commun.*, 163, 1466, 1989.

66. **Smith, L. L. and Johnson, B. H.**, Biological activities of oxysterols, *Free Radical Biol. Med.*, 7, 285, 1989.

67. **Cross, C. E., Halliwell, B., Borish, E., Pryor, W. A., Ames, B. N., Saul, R. L., McCord, J. M., and Harman, D.**, Oxygen radicals and human disease, *Ann. Intern. Med.*, 107, 526, 1987.

68. **Navab, M., Hough, G. P., Stevenson, L. W., Drinkwater, D. C., Laks, H., and Fogelman, A. M.**, Monocyte migration into the subendothelial space of a coculture of adult human aortic endothelial and smooth muscle cells, *J. Clin. Invest.*, 82, 1853, 1988.

69. **Navab, M., Hough, G. P., Berliner, J. A., Frank, J. A., Fogelman, A. M., Haberland, M. E., and Edwards, P. A.**, Rabbit beta-migrating very low density lipoprotein increases endothelial macromolecular transport without altering electrical resistance, *J. Clin. Invest.*, 78, 389, 1986.

70. **Smith, E. B.**, The relationship between plasma and tissue lipids in human atherosclerosis, *Adv. Lipid Res.*, 12, 1, 1974.

71. **Hoff, H. F. and Gaubatz, J. W.**, Isolation, purification, and characterization of a lipoprotein containing apoB from the human aorta, *Atherosclerosis*, 42, 273, 1982.

72. **Schwenke, D. C. and Carew, T. E.**, Initiation of atherosclerotic lesions in cholesterol-fed rabbits. I. Focal increases in arterial LDL concentration precede development of fatty streak lesions, *Arteriosclerosis*, 9, 895, 1989.

73. **Frank, J. S. and Fogelman, A. M.**, Ultrastructure of the intima in WHHL and cholesterol-fed rabbit aortas prepared by ultra-rapid freezing and freeze-etching, *J. Lipid Res.*, 30, 967, 1989.

74. **Völker, W., Schmidt, A., Oortmann, W., Broszey, T., Faber, V., and Buddecke, E.**, Mapping of proteoglycans in atherosclerotic lesions, *Eur. Heart J.*, 11 (Suppl. E), 29, 1990.

75. **Radhakrishnamurthy, B., Srinivasan, S. R., Vijayagopal, P., and Berenson, G. S.**, Arterial wall proteoglycans—Biological properties related to pathogenesis of atherosclerosis, *Eur. Heart J.*, 11 (Suppl. E), 148, 1990.

76. **Srinivasan, S. R., Dolan, P., Radhakrishnamurthy, B., Pargaonkar, P. S., and Berenson, G. S.**, Isolation of lipoprotein-acid mucopolysaccharide complexes from human atherosclerotic lesions, *Biochim. Biophys. Acta*, 388, 58, 1975.

77. **Hurt, E., Bondjers, G., and Camejo, G.,** Interaction of LDL with human arterial proteoglycans stimulates its uptake by human monocyte-derived macrophages, *J. Lipid Res.*, 31, 443, 1990.

78. **Hurt-Camejo, E., Camejo, G., Rosengren, B., Lopez, F., Wiklund, O., and Bondjers, G.,** Differential uptake of proteoglycan-selected subfractions of low density lipoprotein by human macrophages, *J. Lipid Res.*, 31, 1387, 1990.

79. **Khoo, J. C., Miller, E., McLoughlin, P., and Steinberg, D.,** Enhanced macrophage uptake of low density lipoprotein after self-aggregation, *Arteriosclerosis*, 8, 348, 1988.

80. **Guyton, J. R., Klemp, K. F., and Mims, M. P.,** Altered ultrastructural morphology of self-aggregated low density lipoproteins: coalescence of lipid domains forming droplets and vesicles, *J. Lipid Res.*, in press, 1991.

81. **Chao, F-F., Amenda, L. M., Blanchette-Mackie, E. J., Scarlatos, S. I., Gamble, W., Resau, J. H., Mergner, W. T., and Kruth, H. S.,** Unesterified cholesterol-rich lipid particles in atherosclerotic lesions of human and rabbit aortas, *Am. J. Pathol.*, 131, 73, 1988.

82. **Mora, R., Simionescu, M., and Simionescu, N.,** Purification and partial characterization of extracellular liposomes isolated from the hyperlipidemic rabbit aorta, *J. Lipid Res.*, 31, 1793, 1990.

83. **Haberland, M. E., Fong, D., and Cheng, L.,** Malondialdehyde, modified lipoproteins, and atherosclerosis, *Eur. Heart J.*, 11 (Suppl. E), 100, 1990.

83a. **Haberland, M. E.,** unpublished.

Chapter 5

REACTIVE OXYGEN SPECIES, NEUTROPHIL AND ENDOTHELIAL ADHERENCE MOLECULES, AND LIPID-DERIVED INFLAMMATORY MEDIATORS IN MYOCARDIAL ISCHEMIA-REFLOW INJURY

Addison A. Taylor and Scott B. Shappell

TABLE OF CONTENTS

ABSTRACT

Reactive oxygen species have been implicated in the tissue damage that occurs during reperfusion of previously ischemic myocardium. This conclusion is based largely on observations of myocardial protection by agents with antioxidant properties. However, in animal models of ischemia-reflow injury, measurements of glutathione disulfide and of products of lipid peroxidation during early reperfusion have provided data that are not consistent with intracellular oxidant stresses of sufficient magnitude to compromise myocardial viability directly. Antioxidant therapy may attenuate tissue injury by modulating the formation of lipid-derived inflammatory mediators and/or expression of cell surface molecules necessary for adhesion of activated neutrophils to vascular endothelial cells and cardiac myocytes. Lipoxygenase products, which are potent activators of neutrophil adhesion, are increased in the myocardium following ischemia-reflow and may contribute to neutrophil adhesion and subsequent release of cytotoxic products, including the elaboration of hydrogen peroxide. Hydrogen peroxide is also a potent stimulus for endothelium-dependent neutrophil adhesion, probably by its capacity to stimulate production of the lipid-derived inflammatory mediator, platelet-activating factor (PAF). Details of the molecular nature of neutrophil-endothelial and neutrophil-myocyte interactions are gradually emerging and may allow for the development of specific therapeutic agents to reduce myocardial infarct size.

I. PATHOPHYSIOLOGY OF MYOCARDIAL ISCHEMIA AND REPERFUSION

The left and right coronary arteries, arising from the proximal aorta, supply the human heart with oxygenated blood. As they descend from the base to the apex of the heart on its epicardial or outer surface, progressively smaller perforating arteries exit the large coronary arteries at right angles and traverse the cardiac muscle from its epicardial to endocardial (inner) surface. Each segment of the myocardium in the normal human heart, and in certain other species, such as the pig, usually receives its blood supply from one of the terminal branches of a single large coronary artery. In species like the dog, or in human, whose coronary arteries are partially occluded by atherosclerotic plaques, branches (collaterals) from more than one coronary artery may supply blood to the same segment of the myocardium. Interruption of blood supply to a portion of the heart, as may occur following complete occlusion of a coronary artery by a thrombus (blood clot) that forms on injured vascular endothelium or an ulcerated atherosclerotic plaque, results in injury to the cardiac myocytes and other types of cells, such as the microvascular endothelium, vascular smooth muscle, and resident macrophages, that comprise that portion of the heart. Biochemical derangements, including a shift to anaerobic metabolism,[1,2] decreased ATP formation,[2,3] and a net influx of calcium secondary to decreased activity of the sarcolemmal calcium ATPase pump[4,6] occur within minutes when cardiac myocytes are subjected to ischemia. If not promptly reversed, these cells undergo structural changes characterized by nuclear chromatin clumping, swelling of mitochondria, especially those in the subsarcolemmal region,[7,8] edema with widening of myofibrillar bands,[9,10] and other disruptions of normal cellular architecture that accompany or contribute to cell death.[1,11-13]

The extent of the injury to that portion of the myocardium placed at risk by coronary artery occlusion is dependent upon both the duration and the severity of the ischemia.[13,14] Ischemic severity, in turn, is affected by several factors including the degree of microvascular plugging by leukocytes[15,16] or swollen endothelial cells[9,10] and by the extent of collateral circulation to the myocardium.[17] Thus, during a comparable period of coronary artery occlusion, the size of the myocardial infarction produced by occlusion of a major coronary artery in the guinea pig or the dog, species that have extensive collateral coronary circulation, is much smaller than that in pigs, rabbits, or humans, which have limited collateral coronary circulation.[17] Although there are a few collaterals in humans that have normal coronary arteries,[17] the pathophysiologic adaptations that accompany the relatively slow progression of atherosclerosis include dilation of existing small collaterals and the formation of new collateral vessels, thus providing some protection for the diseased heart against sudden coronary occlusion.

Whether occlusion is reversed by clot lysis (spontaneously or by thrombolytic agents) or by coronary artery bypass grafting, restoration of blood flow to a previously ischemic area of the myocardium prior to the occurrence of irreversible myocellular injury often results in myocardial stunning[18-20] (prolonged contractile dysfunction of the affected myocardium) that may persist for hours to weeks before recovery is complete.[21] If the ischemia is sufficiently prolonged, however, some of the myocardium, particularly in the subendocardial regions most distant from their subepicardial coronary blood supply, becomes irreversibly damaged and undergoes necrosis or infarction.[14] Myocardial necrosis may occur not only as a direct consequence of the initial ischemic insult but may also develop in focal areas due to narrowing of the capillary lumen by endothelial cells and plugging of the microvasculature by activated neutrophils, even though flow through the larger coronary vessels is restored. The amount of postischemic myocardium affected by this "no-reflow" phenomenon, which always occurs in areas of necrosis, appears to increase with the duration of reperfusion and can be reduced by treatments which reduce the degree of postischemic necrosis.[22,23] Observations in both experimental animals[14,24] and humans[25,26] that early restoration of blood flow to the ischemic myocardium improves cardiac muscle function, diminishes infarct size, and improves survival have prompted early intervention with thrombolytic agents to lyse clots and with coronary angioplasty or coronary artery bypass grafting to restore blood flow to the myocardium of patients with impending myocardial infarction.

Paradoxically, reperfusion of previously ischemic myocardium or reoxygenation of previously hypoxic myocardium may promote further myocardial injury.[27,28] Much attention in recent years has been directed toward identifying the biochemical derangements that contribute to this "reperfusion injury" with the aim of designing rational therapeutic strategies to preserve the myocardium that is vulnerable to further damage during reperfusion.

II. ROLE OF REACTIVE OXYGEN SPECIES (ROS) IN MYOCARDIAL ISCHEMIA-REPERFUSION INJURY

Reactive oxygen species (ROS) have been implicated in both the myocardial dysfunction that is observed during reperfusion following short periods of ischemia (the stunned myocardium) and the irreversible injury to cardiac myocytes that occurs during reperfusion after longer periods of ischemia. As recently summarized by Mullane,[29] four complementary lines of evidence indicate that ROS contribute to cardiac injury following ischemia and reperfusion: (1) ROS are generated when oxygen is restored to ischemic tissues; (2) ROS are toxic to cells and tissues; (3) *in vitro*, oxygen radical-generating systems cause biochemical, ultrastructural, and functional changes in cardiac preparations that resemble changes observed during ischemia and reperfusion; and (4) removal or suppression of ROS attenuates myocellular damage and improves cardiac function during early reperfusion *in vivo*. The remainder of

TABLE 1
Simplified Chemistry of Reactive Oxygen
Species Formation

1. $O_2 + e \rightarrow O_2^{\cdot -}$

2. $O_2^{\cdot -} + O_2^{\cdot -} + 2H^+ \;(SOD) \rightarrow H_2O_2 + O_2$

3. $H_2O_2 + e \rightarrow HO\cdot + OH^-$

4. $HO\cdot + e + H^+ \rightarrow H_2O$

this chapter will focus on the biochemical reactions that may produce ROS during ischemia and reperfusion, on the cell types that may be responsible for their production, and on the mechanisms by which ROS may alter the function of target cells such as the cardiac myocyte or vascular endothelium.

A. CHEMICAL REACTIONS INVOLVED IN THE METABOLISM OF MOLECULAR OXYGEN

In animal systems, the tetravalent reduction of molecular oxygen to water occurs efficiently during mitochondrial respiration. Alternatively, the metabolism of molecular oxygen may proceed via a series of four univalent reduction reactions. A simplified scheme depicting the formation of ROS at each of these four sequential steps in the conversion of molecular oxygen to water is presented in Table 1. The superoxide anion ($O_2^- \cdot$), hydrogen peroxide (H_2O_2), and hydroxyl radical ($HO\cdot$) formed have been variably referred to in the literature as oxygen-derived free radicals, oxygen radicals, and reactive oxygen species (ROS). Since several biologically important oxygen metabolites, including hydrogen peroxide, singlet oxygen, and lipid hydroperoxides, are technically not free radicals since they do not have unpaired electrons, the aforementioned metabolites of molecular oxygen will be referred to throughout this chapter as ROS.

Superoxide anions may be formed in the postischemic myocardium by a variety of cell types through biochemical reactions that are discussed in detail below. Two molecules of superoxide anion can spontaneously dismutate to hydrogen peroxide and molecular oxygen in the presence of two protons. This reaction occurs at a greatly accelerated rate in the presence of superoxide dismutase (SOD), a ubiquitous enzyme that exists both as a copper-zinc form and as a tetrameric manganese form in the cytosol and in mitochondria, respectively.

$$O_2^- \cdot + O_2^- \cdot + 2H^+ \xrightarrow[\text{SOD}]{} H_2O_2 + O_2$$

Hydrogen peroxide is converted to molecular oxygen and water by catalase, a heme-containing enzyme, or it reacts with glutathione (GSH) in the presence of glutathione peroxidase to form water and glutathione disulfide (GSSG).

$$H_2O_2 + H_2O_2 \xrightarrow{\text{catalase}} 2H_2O + O_2$$

$$H_2O_2 + 2GSH \xrightarrow{\text{GSH peroxidase}} 2H_2O + GSSG$$

Transition metals, particularly Fe^{2+}/Fe^{3+}, can catalyze the formation of hydroxyl radicals from superoxide anion and hydrogen peroxide by a two-step reaction, termed a transition metal-catalyzed Haber-Weiss reaction, as illustrated below, or may produce ferryl or perferryl Fe-O· complexes; both radicals are potent oxidants.

$$O_2^- \cdot + Fe^{3+}\text{-complex} \longrightarrow Fe^{2+}\text{-complex} + O_2$$

$$H_2O_2 + Fe^{2+}\text{-complex} \longrightarrow HO\cdot + HO^- + Fe^{3+}\text{-complex}$$

Hydrogen peroxide can also react with halides in the presence of the granulocyte-specific enzyme, myeloperoxidase, to form hypohalous acids that also are very strong oxidants.

$$H_2O_2 + Cl^- \xrightarrow{\text{myeloperoxidase}} HOCl + Ho^-$$

The tissue injury produced by HO· or HOCl is substantially greater than that ascribed to either $O_2^- \cdot$ or H_2O_2, since the former are not only more reactive with most biological molecules but there also are no specific antioxidant defenses against their biological effects.

Chemical species such as HO· that are sufficiently reactive to abstract an allylic hydrogen atom from polyunsaturated fatty acids can produce carbon centered radicals that, in turn, react with O_2 to form peroxyl radicals. The peroxyl radicals produced by this reaction can then abstract another hydrogen atom from the fatty acid side chain of an adjacent lipid molecule. This series of chain reactions, which constitute the propagation phase of lipid peroxidation, will continue until chain termination occurs by formation of nonradical products. Lipid peroxidation may affect the permeability and structural integrity of lipid bilayers that comprise cellular membranes and also may alter the proteins associated with these lipid bilayers. The reader is referred to several excellent reviews for a more thorough discussion of the complex chemical reactions that occur during lipid peroxidation.[30-32]

B. SOURCES OF REACTIVE OXYGEN SPECIES IN MYOCARDIAL ISCHEMIA-REFLOW

Several biochemical changes that accompany tissue ischemia predispose to the formation of ROS when O_2 is reintroduced at the time of reperfusion. Included among the numerous potential sources for the generation of reduced oxygen metabolites in the postischemic myocardium are: (1) the mitochondrial electron transport chain,[33] as well as a cyanide-insensitive NADH oxidase located in this same cellular organelle;[34-36] (2) xanthine oxidase,[37-42] which in human hearts is primarily of endothelial cell origin; (3) the membrane-bound NADPH oxidase complex of infiltrating neutrophils;[43-47] (4) autoxidation of catecholamines[48,49] and oxidation of heme proteins such as myoglobin;[50,51] and (5) metabolism of arachidonic acid in both resident and nonresident cells found in the inflamed myocardium.[52-56] The formation of ROS by invading neutrophils or from xanthine oxidase-catalyzed reactions is considered to be quantitatively the most important and has received the most attention experimentally,[29,41,57,58] particularly with regard to pharmacologic interventions.

1. Uncoupling of Mitochondrial Electron Transport Chain

Electron transport by the mitochondria is normally a tightly coupled process resulting in efficient four electron reduction of molecular oxygen to two molecules of water. Under normoxic conditions, only about 2% of mitochondrial oxygen consumption results in the generation of superoxide anion by "leakage" of electrons.[33] However, as the adenine nucleotide pool declines under conditions of ischemia, mitochondrial electron transport centers become increasingly reduced. Decreased transport through the terminal cytochrome oxidase due to depleted ADP levels and compromised membrane structural integrity during ischemia[8] may contribute to a burst of electron leakage and superoxide generation upon the return of normal oxygen tensions with reperfusion.[41] In addition, a cyanide-insensitive mitochondrial NADH oxidase activity, located at the outer surface of the inner mitochondrial membrane and unrelated to the respiratory chain,[59,60] may represent a novel source of reactive oxygen during reflow following myocardial ischemia.[34] Using a cytochemical technique to detect hydrogen peroxide production in a canine model of regional myocardial ischemic injury, the activity of this NADH oxidase was shown to be highest in the mitochondria from the periinfarct region.[34] Even if mitochondrial generation of ROS during either ischemia or reperfusion is insufficient to cause overt necrosis by overwhelming all cellular antioxidant defense mechanisms, this process could contribute to mitochondrial injury, a well-documented histologic consequence of myocardial ischemia,[9,10,61] because of the marked decrease in specific mitochondrial antioxidant defense systems during myocardial ischemia.[41,62] Furthermore, any ROS generated within myocytes may contribute to myocyte metabolic and contractile dysfunction, through inhibition of substrate catabolism and oxidative phosphorylation, as well as through oxidation of ion transport proteins. Further consideration will be given to these latter processes, as well as peroxide-dependent myocyte modulation of neutrophil-myocyte adhesion.

2. Superoxide Production from Xanthine Oxidase

Xanthine dehydrogenase, an enzyme present in numerous cells including the capillary endothelial cells,[37,38] is involved in the transfer of electrons to

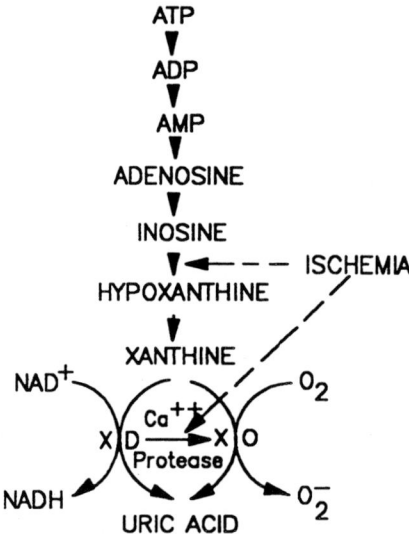

FIGURE 1.   Under normoxic conditions, high-energy phosphates are normally metabolized to uric acid by a xanthine dehydrogenase (XD)-catalyzed conversion of xanthine to uric acid. NAD^+ serves as an electron acceptor. Ischemia induces the conversion of XD to xanthine oxidase (XO), a calcium-dependent process that involves sulfhydryl oxidation or limited proteolysis. The nucleotide metabolites that have accumulated during ischemia are converted to uric acid by XO with the consumption of O_2 and the formation of superoxide anion.

NAD^+ during the oxidation of xanthine or hypoxanthine to uric acid under normoxic conditions (Figure 1). However, the enzyme can be converted to xanthine oxidase (XO), which is capable of using molecular oxygen as an electron acceptor, by the oxidation of thiol groups (reversible) or limited proteolysis (irreversible), perhaps by a Ca^{2+}-dependent protease such as calpain.[63-65] Upon restoration of molecular oxygen to the ischemic cells, xanthine oxidase metabolizes purine substrates, such as hypoxanthine, that have accumulated during ischemia due to the depletion of high-energy phosphates (ATP, ADP, and AMP) with the formation of superoxide anions from molecular oxygen[39] (Figure 1). The superoxide anion can then be spontaneously or enzymatically dismutated to hydrogen peroxide which, in the presence of additional $O_2^- \cdot$ and an available source of Fe^{3+}, can form the potent oxidant $HO\cdot$ via a metal-catalyzed Haber-Weiss reaction.[66,67] Xanthine oxidase activity, in turn, may be inhibited by high concentrations of O_2 or its metabolites, either self- or neutrophil-generated.[68]

A role for xanthine oxidase-generated $O_2^- \cdot$ in myocardial ischemia-reperfusion injury has been suggested by observations that inhibitors of xanthine oxidase, allopurinol and oxypurinol, reduce myocardial infarct size or blunt postischemic contractile dysfunction in the dog[69-73] and rat[74-78] and reduce myocardial free radical concentrations as determined by electron spin resonance measurements of the spin-trap adduct formed with 5,5-dimethyl-1-pyrroline-*N*-oxide (DMPO).[78] These species have high concentrations of xanthine oxidase/xanthine dehydrogenase in the myocardium.[79] The failure of allopurinol to reduce infarct size in a rabbit model of myocardial ischemia and reperfusion[80] coupled with the demonstration of little or no xanthine oxidase in the hearts of rabbits, pigs, or humans,[79,81-85] have been cited as evidence against xanthine oxidase-derived superoxide anion as a significant determinant of myocardial injury during reperfusion in these species. Furthermore, observations *in vitro* suggest that allopurinol and oxypurinol may preserve myocardial function and/or reduce infarct size by inhibiting purine metabolism, thus making more substrate available for regeneration of ATP upon reintroduction of molecular oxygen during reperfusion,[78,86] or by scavenging the hydroxyl radical.[87,88] However, a recent study demonstrated that concentrations of allopurinol in extracellular fluid that inhibited xanthine oxidase activity did not inhibit myoglobin-catalyzed linolenic acid peroxidation or enhance hypochlorous acid scavenging.[89] This study raises doubt about whether the concentrations of allopurinol achieved following its administration *in vivo* are sufficiently high to effectively scavenge hydroxyl radicals. It is of interest that allopurinol hastens the return of contractile function during reperfusion of the globally ischemic rabbit heart[90] even though it fails to reduce cardiac postischemic reperfusion-induced injury *in vivo* in this species.[80] The mechanism by which this beneficial effect occurs has not been defined.

3. Metabolism of Arachidonic Acid

An analysis of the involvement of arachidonic acid metabolism in pathological conditions in which reactive oxygen metabolites are implicated must involve a consideration not only of fatty acid oxygenases (cyclooxygenase and multiple lipoxygenases) as sources of radical species, but also of the modulation of the activity of these oxygenase enzymes by peroxides, since many of their products have well-defined receptor-mediated functions in inflammation. This latter point is frequently overlooked when interpreting the mechanisms of anti-inflammatory action of antioxidants.[91] Both cyclooxygenase and lipoxygenase enzymes incorporate molecular oxygen into polyunsaturated fatty acids, the most important physiologic substrate being arachidonic acid (5,8,11,14-eicosatetraenoic acid) (Figure 2). In the case of cyclooxygenase (prostaglandin H synthase), following hydrogen atom abstraction, O_2 attaches initially at C-11 and then to C-9 forming the dioxygen bridge in a transformation accompanied by carbon–carbon bond formation between C-8 and C-12, to give bicyclic endoperoxide.[92] A second molecule of O_2 is incorporated at C-15 and the peroxyl radical is converted to the 15-OOH derivative, prostaglandin G_2 (PGG_2). PGG_2 is converted to the corre-

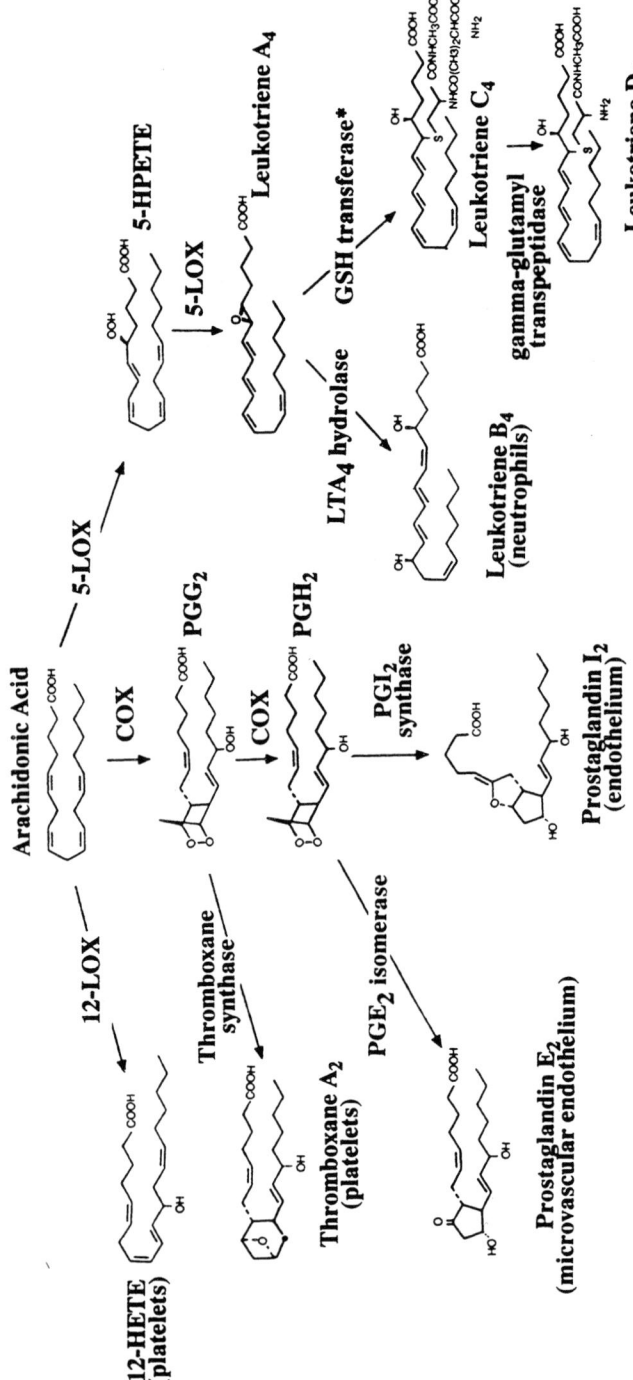

FIGURE 2.   Arachidonic acid is metabolized via the cyclooxygenase (COX), 5-lipoxygenase (5-LOX), and 12-lipoxygenase (12-LOX) enzymes to form several products that are of importance in postischemic myocardial inflammation. The enzymes involved in the formation of these products and the principal cellular source of their production are illustrated in the diagram. Asterisk indicates mononuclear phagocytes, mast cells, neutrophil-endothelial, and neutrophil-platelet interactions.

sponding 15-OH derivative, PGH_2, the immediate precursor for the biologically important prostaglandins and thromboxanes, by a peroxidase activity that copurifies with cyclooxygenase and thus appears to be the same protein[92] (Figure 2). The hydroperoxidase activity of this enzyme is very nonspecific, and this enzyme is capable of reducing a variety of hydroperoxides and co-oxidizing a variety of substrates;[55,93-96] GSH apparently is the major reducing cofactor in cells.[97] It has been suggested that oxidizing equivalents can be released by the hydroperoxidase action of this enzyme,[98,99] since release of a potent oxidizing species during PGG_2 reduction has been observed.[99]

Lipoxygenase (LOX, e.g., 5-, 12-, and 15-lipoxygenases) activity, as illustrated in Figure 2, also involves hydrogen atom abstraction and addition of molecular oxygen to the resulting pentadienyl radicals, leading to formation of 5-, 12-, or 15-hydroperoxyeicosatetraenoic acids (HPETEs).[92,100-102] These species are reduced to the corresponding hydroxy derivatives (HETEs), apparently by glutathione peroxidase.[97,100,103] In addition, the leukocyte 5-lipoxygenase enzyme that forms 5-HPETE also catalyzes its conversion to 5,6-oxido-7,9,11,14-eicosatetraenoic acid, or leukotriene A_4 (LTA_4),[104-106] the precursor for the chemoattractant leukotriene B_4, LTB_4, and the smooth muscle constricting cysteinyl-leukotrienes, LTC_4 and LTD_4 (Figure 2). The relative formation of 5-HETE vs the much more biologically potent inflammatory leukotrienes depends on stimulus conditions *in vitro*[100] and on complex, cell-specific regulatory mechanisms under physiologic or pathologic conditions *in vivo*.[107]

Both purified ram seminal vesicle microsomal cyclooxygenase and soybean lipoxygenase were found to produce superoxide anion in the presence of NADH or NADPH.[54] The formation of superoxide was dependent on the peroxidase activity. In the case of cyclooxygenase, superoxide generation occurred when arachidonic acid, linoleic acid, or PGG_2, but not PGH_2, was used as a substrate. The generation of superoxide by cyclooxygenase under these conditions occurs as a result of a side-chain reaction initiated by the hydroperoxidase, and is dependent upon the presence of a suitable reducing substrate.[54] Such cosubstrates must be capable of being oxidized by the hydroperoxidase in a manner that yields a free radical that subsequently reacts with molecular oxygen at a rate faster than the radical spontaneously dismutates. Although NADH and NADPH are suitable cofactors for this reaction, GSH is not.[54] Hence, under normal conditions where GSH is abundant, such a mechanism may not represent a significant source of reactive oxygen generation. It is not clear if such a mechanism occurs more readily under ischemic conditions. Myocytes themselves do not appear to significantly metabolize arachidonic acid[52,108] by either cyclooxygenase[109-111] or lipoxygenase[112,113] pathways.

Arachidonic acid metabolism in other cell types that are potentially relevant sources of reactive oxygen in myocardial ischemia-reperfusion injury, such as endothelial cells and infiltrating leukocytes, is probably much more important for the receptor-mediated actions of the products formed. Thus, arachidonic acid metabolism by fatty acid oxygenases is not generally considered to be an important source of ROS during postischemic myocardial reperfusion. Instead, the potential effects of peroxide generation or removal on the activity of these enzymes and thus on the amounts of pro- or anti-inflammatory arachidonic metabolites formed is likely much more relevant, although of still uncertain significance *in vivo*.

Several biologically active arachidonic acid metabolites may be involved as mediators in the inflammatory response that is initiated during myocardial ischemia and augmented by reflow. The metabolites that are likely candidates due to their cellular sources and known biologic actions are thromboxane A_2 derived from platelets, prostacyclin (PGI_2) and prostaglandin E_2 (PGE_2), formed by endothelial cells, and the leukotrienes, LTB_4, LTC_4, LTD_4, and LTE_4, formed from the epoxide intermediate LTA_4 by leukocytes (Figure 2). Since a detailed description of the enzymes involved in the generation of these substances, their full range of biological functions, and the evidence for their participation in certain aspects of myocardial injury is beyond the scope of this chapter, interested readers are referred to several recent reviews.[52,92,100,102,114-116] Brief descriptions of some of the biologic properties of these mediators are included in order to illustrate the complex mechanisms by which their formation might be modulated by oxidative processes that can occur upon reperfusion and by antioxidant interventions that are detailed below.

Platelets metabolize arachidonic acid via 12-lipoxygenase to produce 12-HETE, of uncertain biologic function, via cyclooxygenase to produce PGG_2, PGH_2, and hydroxyheptadecenoate (HHT), and via thromboxane synthases to produce TXA_2.[92,97] PGG_2, PGH_2, and TXA_2 are potent inducers of platelet aggregation and stimulate vasoconstriction. Although platelets are a component of the thrombus that may produce initial coronary occlusion or reocclusion following thrombolytic therapy,[117] there is little evidence to suggest they contribute to reperfusion injury in the more distal microvasculature and myocardium.[118] Endothelial cells metabolize arachidonic acid via cyclooxygenase and prostacyclin synthase to produce PGI_2 primarily in the large epicardial arteries; PGE_2 is produced mainly in the microvasculature.[119] These metabolites produce vasodilatation, inhibit platelet adhesion, aggregation, and secretion, and inhibit neutrophil adhesion, superoxide production, and arachidonate metabolism (by activation of platelet and neutrophil adenylate cyclase).[120-126] Hence, these substances constitute an important part of the endothelial defense mechanism and the barrier to inflammatory cell emigration into the subendothelial space and adjacent tissue.

Infiltrating leukocytes are the most likely source for LTB_4 (neutrophils)[127,128] and for LTC_4 and LTD_4 (mononuclear cells).[52,53,129,130] In addition, the latter vasoactive agents, which probably contribute to myocardial contractile dysfunction,[131-133] may be formed by resident leukocytes (especially mast cells),[133-135] by coronary arteries,[136] and by a novel transcellular metabolic process that involves neutrophil-generated LTA_4 and an endothelial glutathione S-transferase.[137] This latter process, which has also been described for neutrophil-platelet interactions,[138-140] may be particularly favored during substantial flux through the neutrophil 5-LOX pathway (as the hydrolase converting LTA_4 to LTB_4 is apparently rate-limiting) and during cell-cell adhesive interactions.[141]

Formation of prostaglandins and lipoxygenase products is regulated not only by the availability of substrate arachidonic acid, but also by the requirement for peroxide activators of the fatty acid oxygenases[56,142-144] and by autoinactivation of the oxygenases as described below.[56,145,146] Low levels (e.g., 10^{-8} M) of peroxides are required for initiation of the cyclooxygenase reaction, with lipid peroxides being more potent than hydrogen peroxide itself.[142] These concentrations are substantially lower than the K_m for the peroxidase activity of the PGH synthase (e.g., 10^{-5} M).[143] A trace amount of peroxide may be required to react with iron(III) heme at the enzyme's active site, resulting in the formation of a peroxyl radical that is able to stereospecifically abstract an allylic hydrogen from arachidonic acid to initiate PGG_2 formation. Hence, addition of glutathione to cell-free preparations of cyclooxygenase prevents the immediate oxidation of substrate arachidonate.[145,147] Addition of lipid peroxides, including PGG_2 or lipoxygenase products or higher concentrations of H_2O_2, reduces the time required for the enzyme to reach maximum catalytic activity.[56,142] Similar behavior is exhibited by cell-free preparations of 5-lipoxygenase, with hydroperoxides containing long-chain fatty acid moieties being much more effective than H_2O_2 or *t*-butyl-hydroperoxide.[148-150] In addition, a requirement for continued presence of hydroperoxide is demonstrated by the fact that addition of glutathione peroxidase causes a lag in the time for cyclooxygenase to reach its maximal catalytic activity.[56]

In contrast to its stimulation by low levels of peroxides, including those from stimulated leukocytes,[151] cyclooxygenase activity is inhibited by higher levels of peroxides.[144,152-154] If the peroxidase activity of PGH synthase does not have available an oxidizable substrate during reduction of PGG_2 to PGH_2, the enzyme is irreversibly inactivated due to its oxidation by an apparent oxygen-centered radical species.[99] Under certain conditions, oxygen radical scavengers can enhance prostaglandin synthesis by preventing this enzyme inactivation.[99] As indicated above, the peroxidase activity of PGH synthase is very nonselective and can reduce a variety of other peroxides. Hence, the oxidative inactivation of cyclooxygenase may be accelerated by the presence of large quantities of peroxides. This dual regulation has lead to the concept of modulation of cyclooxygenase and lipoxygenase activity in intact cells by the prevailing "peroxide tone,"[155] a balance established by the capacity of the cell to form and remove peroxides.

Modulation of cyclooxygenase and lipoxygenase product formation by intact cells exposed to exogenous peroxides also has been examined.[151,156,157] The generation of hydrogen peroxide by a glucose-glucose oxidase system or by phorbol myristate acetate (PMA)-stimulated neutrophils induced a time- and dose-dependent release of PGI_2 (measured as the stable metabolite, 6-

keto-PGF$_{1\alpha}$) from human and bovine endothelial cell monolayers[156] that was inhibited by catalase but not by SOD. In contrast, a subsequent study found that H$_2$O$_2$ produced a dose-dependent inhibition of PGI$_2$ synthesis from exogenous arachidonic acid[157] by endothelial monolayers. This inhibitory effect appeared to be mediated by modulation of cyclooxygenase, was maximum within 1 min after H$_2$O$_2$ exposure, and occurred at micromolar levels. In contrast, endothelial cytotoxicity (as assessed by ^{51}Cr release) required much higher levels of H$_2$O$_2$ (e.g., 1 to 10 mM) and prolonged incubation times (e.g., >1 h). Detection of stimulation or inhibition of endothelial cyclooxygenase product formation may depend on subtle differences in experimental conditions and variations in endogenous peroxide and glutathione levels leading to differences in the "peroxide tone."

In addition, the situation in intact cells is complicated by the fact that exogenous peroxides can affect other enzymes involved in arachidonic acid metabolism either directly or indirectly. For example, lipid peroxides such as 15-HPETE inactivate prostacyclin synthase.[158-161] Furthermore, exogenous H$_2$O$_2$ and possibly also lipid peroxides cause Ca^{2+} influx into endothelial cells under noncytotoxic conditions.[162-164] Increased cytosolic Ca^{2+} can activate phospholipases (e.g., phospholipase A$_2$) resulting in the liberation of arachidonate. This view is consistent with the observation in one study that increased endothelial PGI$_2$ production in response to H$_2$O$_2$ was attenuated by the phospholipase inhibitor mepacrine,[156] and that exogenous H$_2$O$_2$ induced simultaneous endothelial production of PGI$_2$ and platelet-activating factor (PAF) in a manner dependent upon extracellular Ca^{2+} (i.e., activation of phospholipase A$_2$ with hydrolysis of arachidonic acid from the *sn*-2 position of 1-*O*-alkyl-2-acyl-*sn*-glycerol-3-phosphocholine, simultaneously liberating the immediate precursor of PAF). The 5-lipoxygenase enzyme also requires Ca^{2+} for activity[165] and undergoes translocation from the cytosol to membranes in a Ca^{2+}-dependent manner as part of its activation in intact cells,[166,167] an important factor to keep in mind regarding potential mechanisms whereby exogenous peroxides can stimulate formation of lipoxygenase metabolites in intact cells. In addition, 5-lipoxygenase enzyme preparations require the presence of ATP for maximal activity.[165] The complexity of the potential regulatory interaction of peroxides with the 5-lipoxygenase cascade in intact cells is demonstrated by a recent report showing that H$_2$O$_2$ inhibits 5-LOX metabolism in stimulated rat alveolar macrophages by ATP depletion, despite stimulation of arachidonate release and thromboxane A$_2$ formation over the same range of peroxide concentrations.[168]

Finally, ROS not only may modulate the formation of lipoxygenase products, but also, under certain circumstances, leukocyte- and mononuclear phagocyte-derived ROS may participate in the oxidative degradation of biologically active leukotrienes. At high concentrations, H$_2$O$_2$ alone can inactivate LTC$_4$ and LTD$_4$; at much lower concentrations it also can rapidly inactivate LTB$_4$, LTC$_4$, LTD$_4$, and LTE$_4$ in the presence of added granulocyte myeloperoxidase or eosinophil peroxidase and halides.[169] PMA-stimulated eosinophils[170] and neutrophils[171] degrade LTC_4 in an apparently peroxidase-dependent manner, with formation of 5-*(S)*,12-*(R)*-6-*trans*- and 5-*(S)*,12-*(S)*-6-*trans* isomers of LTB_4, as well as sulfoxide derivatives.[170,171] The inactivation of LTB_4 and LTC_4 by the hydroxyl radical has also been reported,[172] and this mechanism has been demonstrated for the inactivation of LTC_4 by stimulated neutrophils from patients deficient in myeloperoxidase.[173] Greater amounts of LTB_4 and LTC_4 have been recovered from the supernatants of stimulated neutrophils of patients with chronic granulomatous disease,[173] whose neutrophils are unable to undergo a respiratory burst due to defects in the neutrophil NADPH oxidase (see below). This observation indicates not only that stimulation of the neutrophil respiratory burst does not appear to be required (to provide peroxide activators) for 5-lipoxygenase activity, but also suggests that this system may contribute to leukotriene removal in inflammation homeostasis.

4. Autoxidation of Catecholamines and Oxidation of Heme Proteins

Infusions of high concentrations of the catecholamines epinephrine or norepinephrine into experimental animals are known to produce myocellular mitochondrial swelling, myofibrillar disruption, plasma membrane blebbing, and myocardial necrosis.[48] It has been suggested that these cardiotoxic effects result not from the catecholamines themselves but from the production of $O_2^- \cdot$ and H_2O_2 formed by a complicated series of reactions during the autoxidation of catecholamines, a process that is greatly accelerated by the presence of transition metal ions.[29] Support for this hypothesis is provided by the observation that rats rendered vitamin E deficient were more sensitive to the cardiotoxic effects of isoproterenol whereas myocardial damage induced by this synthetic catecholamine was reduced when the diet was supplemented with vitamin E.[48] Although increased circulating concentrations of norepinephrine and, to a lesser degree, epinephrine may occur in animals and man during myocardial ischemia, their contribution to myocellular injury via the formation of ROS remains unclear. Furthermore, the results of studies demonstrating protection by antioxidants against catecholamine-induced myocardial necrosis *in vivo* must be interpreted with caution since accumulation of neutrophils, a major source of oxygen radicals, has been observed in this model.[48]

5. Formation of Reactive Oxygen Species by Neutrophils

The production of ROS by neutrophils not only is essential for the destruction of most microorganisms during phagocytic ingestion, but also appears to contribute to injury of host tissues during a variety of inflammatory conditions.[174,175] Neutrophils produce superoxide anion upon stimulation with a variety of soluble and particulate stimuli by virtue of a multicomponent, transmembrane electron transport system — NADPH oxidase.[174] This enzyme complex is inactive in resting cells, but undergoes assembly and/or activation under appropriate conditions and utilizes cytoplasmic NADPH as the electron donor for one-electron reduction of molecular oxygen in the extracellular fluid

or in the interior of invaginated phagosomes during phagocytosis. Reducing
equivalents (NADPH) are provided by simultaneous activation of the hexose-
monophosphate shunt.[174] Electron transfer to molecular oxygen occurs through
a series of reactions, the nature of which are still incompletely characterized.
Likely constituents include a flavoprotein that accepts electrons from
NADPH,[176,177] a quinone intermediate,[174] and a *b*-cytochrome, which is probably the terminal oxidase capable of directly reducing molecular oxygen.[174]
In unstimulated neutrophils, both the *b*-cytochrome and flavoprotein are present to some degree in the plasma membrane, with the remainder located in
specific granules.[178,179] Following stimulation of the neutrophil under conditions that lead to activation of the respiratory burst, there is translocation
of the cytochrome and the flavoprotein from the specific granules to the plasma
membrane (i.e., granule membrane and plasma membrane fusion during ex-
ocytosis).[179-184] In addition, studies on NADPH oxidase activation in cell-free
systems[185-187] have revealed the requirement for multiple cytosolic components.[188] Two of these factors are proteins, p47 and p67, which appear to
undergo translocation to the plasma membrane upon stimulation of the neutrophil respiratory burst.[184,189] These proteins are absent in distinct forms of
autosomal recessively inherited chronic granulomatous disease (CGD),[188,190-192]
a condition in which patients' neutrophils fail to undergo a respiratory burst
upon stimulation with soluble or particulate stimuli due to a deficiency in one
or more components of the NADPH oxidase complex. Although a strict linear
correlation between extent of cytochrome "upregulation" and quantity of
superoxide production may not exist, translocation of oxidase components
from intracellular granule pools to the membrane is consistent with the assembly of the electron transport apparatus upon neutrophil stimulation. Furthermore, a continuous supply of components may be particularly necessary
under conditions of prolonged respiratory burst activation.

A GTP-binding protein appears to be involved in the regulation of NADPH
oxidase activity in neutrophils.[193,194] This GTP-binding protein may be located
in the cytosol of unstimulated neutrophils.[195] The p47 and p67 proteins, which
are components of the NADPH oxidase complex, may be GTP-binding proteins since p47 has an apparent nucleotide binding domain in its deduced
amino acid sequence,[190,196] and both p47 and p67 bind to GTP-agarose.[191] A
Ras-related protein that has recently been demonstrated to be associated with
the *b*-cytochrome in stimulated human neutrophils also may function as a
GTP-binding oxidase regulatory protein.[197] Partial purification of cytochrome
b from stimulated neutrophils resulted in coisolation of a 22-kDa protein with
an amino terminal sequence similar to the human *Ras* proteins and which was
recognized by a monoclonal antibody directed against the GTP-binding domain of *Ras* proteins.[197] Screening a cDNA library with oligonucleotides based
on the partial amino acid sequence resulted in the isolation of a cDNA clone
identical to *Ras*-related rap-1.[197] Furthermore, immunoaffinity purification
experiments revealed copurification of both cytochrome subunits and the 22-
kDa *Ras*-related protein with either anticytochrome antibody or anti-*Ras* an-

tibody-conjugated matrices.[197] Hence, this close association of a *Ras*-like protein and the cytochrome of the neutrophil NADPH oxidase makes the former a potential candidate for the GTP-binding factor implicated in oxidase activation. Whether or not this *Ras*-related protein is also located in the cytosol of unstimulated neutrophils is not yet established.

In addition to the possible regulatory function of guanine nucleotides, substantial evidence exists for a role of protein kinase C (PKC) (Ca^{2+}, phospholipid-dependent protein kinase) in the activation of the respiratory burst in neutrophils. Phorbol esters, such as PMA, are potent but nonphysiologic promoters of neutrophil reactive oxygen secretion.[198] PMA stimulates cells by activating PKC, the apparent receptor for the tumor-promoting phorbol esters.[199] Activation of PKC may result in stimulation of the neutrophil respiratory burst by the phosphorylation of certain components of the NADPH oxidase.[198] The predicted amino acid sequence of the 47-kDa cytosolic protein that functions in regulation of the oxidase contains several potential PKC phosphorylation sites.[190,196] Stimulation of human neutrophils with PMA, opsonized latex beads, the calcium ionophore A23187, or high concentrations of the chemotactic bacterial peptide formylmethionyl-leucyl-phenylalanine (fMLP) results in the phosphorylation of a 47-kDa protein.[200] The kinetics of phosphorylation of the 47-kDa protein correspond to the kinetics of respiratory burst activation, and both are inhibited by trifluoperazine and chlorpromazine, agents capable of inhibiting both PKC and calmodulin.[200] It is uncertain whether phosphorylation of p47 is essential for translocation to the membrane or activation of this factor once it is in the membrane, but stimulation of intact neutrophils with PMA results in a subsequent deficiency of the cytosolic component (vs. that of resting cells) in the cell-free oxidase system.[184]

Physiologic regulation of PKC is presumed to be mediated by the formation of diacylglycerols (DAG), which compete with phorbol esters for binding to PKC and which, like phorbol esters, increase the affinity of the enzyme for calcium, so that in the presence of phospholipids, such as phosphatidylserine, the kinase is active at physiological levels of calcium. The biochemistry of PKC is beyond the scope of this review, but the reader is referred to several recent reviews for more information.[199,201-205] The neutrophil respiratory burst is also activated by synthetic, membrane-permeable diacylglycerols such as 1-oleyl-2-acetylglycerol (OAG),[198] and the superoxide generation achieved in response to PMA or OAG is blocked by inhibitors of PKC, such as the isoquinoline sulfonamide derivatives H7 and C1.[206,207] However, in contrast to agents like PMA and OAG, chemotactic factors, such as fMLP, LTB_4, interleukin-8 (IL-8), and complement fragment C5a (for which neutrophils have specific membrane receptors), elicit little reactive oxygen from neutrophils stimulated in suspension, even at concentrations two to three orders of magnitude higher than those required for the stimulation of maximal chemotaxis and adhesion.[208,209] Furthermore, the small respiratory burst elicited by these receptor-mediated stimuli has been largely found not to be inhibited by PKC inhibitors that block superoxide in response to PMA.[207]

Such results have been taken to indicate the existence of multiple pathways for activating the neutrophil respiratory burst, although it should be borne in mind that most investigators find phorbol esters to be much more potent activators of neutrophil superoxide and hydrogen peroxide generation than are very high concentrations of chemotactic agents such as fMLP.

As illustrated in Figure 3, the neutrophil plasma membrane receptors for chemotactic factors such as fMLP, C5a, and LTB_4, although as yet incompletely characterized at a molecular level, appear to be coupled by a pertussis toxin-sensitive G (guanine nucleotide-binding heterotrimeric) protein to a phospholipase C enzyme.[210] Stimulation of these receptors results in the hydrolysis of polyphosphoinositides (particularly phosphatidylinositol 4,5-bisphosphate or PIP_2) with formation of *sn*-1,2-diacylglycerol and inositol 1,4,5-triphosphate (IP_3), which mobilizes intracellular calcium (see Reference 201 for review). However, the formation of DAG under such circumstances is extremely transient, due, at least in part, to its activation and conversion to phosphatidic acid by diacylglycerol kinase,[211,212] and may be insufficient for sustained stimulation of the NADPH oxidase complex and persistent production of superoxide or hydrogen peroxide. Consistent with this notion is the fact that pretreatment of neutrophils with the fungal metabolite, cytochalasin B, markedly augments production of superoxide and concomitant DAG accumulation upon subsequent exposure to chemotactic factors.[212] These pathways for the activation of neutrophil reactive oxygen production, particularly with regard to potential physiologic mechanisms of activation, will be considered again below in the context of neutrophil-endothelial interactions.

III. EVIDENCE FOR FORMATION OF REACTIVE OXYGEN SPECIES DURING MYOCARDIAL ISCHEMIA-REPERFUSION

A. CARDIOPROTECTIVE EFFECTS OF ANTIOXIDANT ENZYMES AND CHEMICALS

Studies demonstrating a reduction in infarct size or improvement in post-ischemic myocardial contractile function after treatment with antioxidant enzymes (superoxide dismutase or catalase[213-218]) or with chemicals that scavenge ROS (α-tocopherol,[219] mannitol,[220] N-2-mercaptopropionylglycine,[221,222] or dimethylthiourea[223,224]) or that reduce ROS formation (allopurinol, oxypurinol,[69-72,75,88,225] or deferoxamine[226-228]) have provided indirect support for the participation of ROS in myocardial injury or functional derangements resulting from regional ischemia and reperfusion *in vivo*. The similar cardioprotective benefit achieved by agents that scavenge hydroxyl radicals (or reduce their formation) compared to those that reduce superoxide anion production argues against $O_2^-\cdot$ as the only cytotoxic ROS and implicates metabolites of superoxide in the myocellular injury mediated by this oxygen radical. Precise identification of a single specific cytotoxic ROS, its cellular source, or its exact intra- or extracellular site of action from studies *in vivo* is difficult.

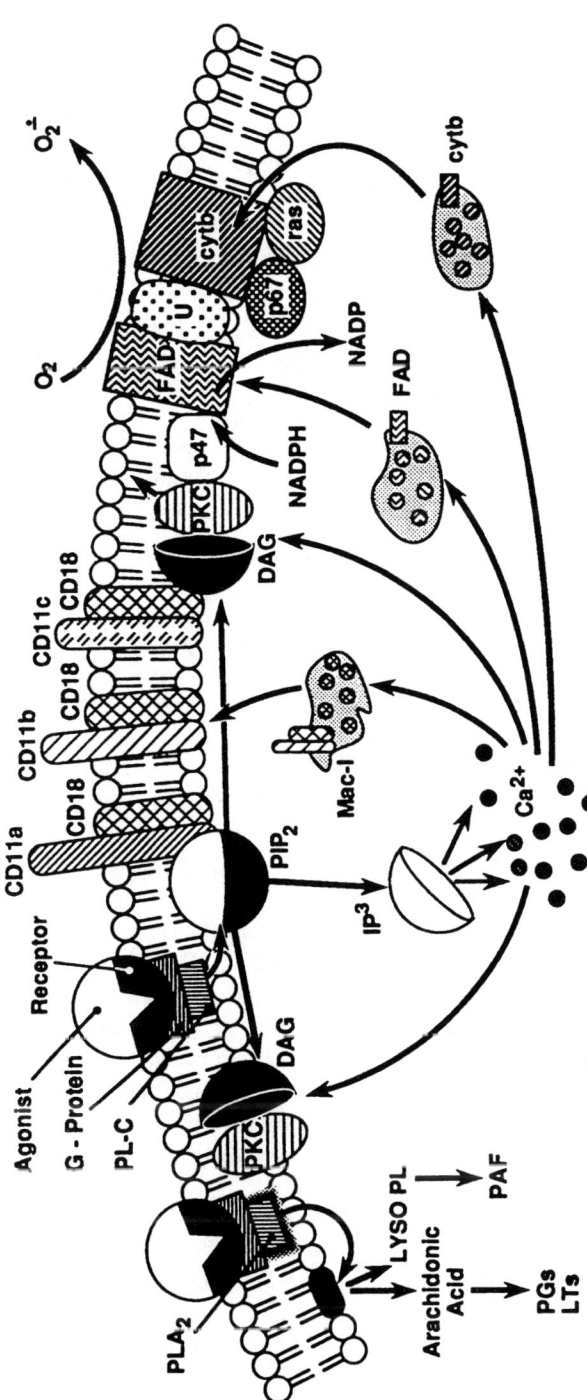

FIGURE 3.   Binding of agonists such as fMLP, C5a, and LTB₄ to plasma membrane surface receptors causes the activation of phospholipase C (PL-C) by a process coupled to a guanine nucleotide binding protein (G-protein). PL-C catalyzes the hydrolysis of phosphoinositides, such as phosphatidylinositol 4,5-bisphosphate (PIP₂), with the release of diacylglycerol (DAG) and inositol 1,4,5-triphosphate (IP₃). DAG activates protein kinase C (PKC) whereas IP₃ promotes an increase in cytosolic calcium concentration. A flavoprotein (FAD), a possible ubiquinone intermediate (U), and a b-cytochrome (cyt b) contained in the plasma membrane and in translocatable-specific cytoplasmic granules combine with other components such as the p47 and p67 proteins to form the activated NADPH oxidase system of the neutrophil that is responsible for the formation of ROS. A PKC-dependent process also appears to be responsible for the activation of phospholipase A₂ (PLA₂) that catalyzes the release of arachidonic acid, the precursor to prostaglandins (PGs) and leukotrienes (LTs), and the formation of platelet-activating factor (PAF) from lysophospholipids. Increases in cytosolic calcium induce the translocation of CD11b/CD18 adherence glycoproteins (Mac-1) stored in cytosolic (secondary and/or tertiary) storage granules to the plasma membrane. CD11a/CD18 (LFA-1) is not found in storage granules but is expressed and likely regulated in the plasma membrane.

For example, freely diffusible antioxidants like dimethylthiourea[224] or N-2-mercaptopropionylglycine[221,222] may effectively scavenge several ROS including H_2O_2, HOCl and HO· under *in vitro* conditions.[229,230] In addition, the production of ROS by resident leukocytes and macrophages[134] or by activated neutrophils that have accumulated at sites of postischemic inflammation[15,231,232] may contribute to the cytotoxic effects of ROS produced by the vascular endothelium[233] and cardiac myocytes.[33-36,59,234] Thus, antioxidants could ameliorate the effects of ROS generated by either resident cardiac cells (especially endothelial cells) or accumulated neutrophils. Finally, direct effects of antioxidants on neutrophil activation and subsequent release of ROS have not been examined rigorously. Such a mechanism probably explains the cardioprotective actions of perfluorochemicals[235,236] in reperfusion injury and may even contribute to the salvage of postischemic myocardium by 5-lipoxygenase inhibitors like BW 755C,[237] nafazatrom,[238] and nordihydroguaiaretic acid,[239-241] which are also potent antioxidants.[242]

Studies examining protection from myocardial necrosis following infusion of SOD are of interest because this rather large molecule of approximately 32 kDa should not readily enter normal cells and thus should have a predominantly extracellular site of action. This enzyme, with or without catalase,[22,214-218,243,244] when infused during early reperfusion following relatively short periods (≤ 90 min) of ischemia[215,217] has been reported to reduce infarct size measured up to 24 h later. Catalase alone appears to exert little cardioprotective effect.[214] If ischemia is sufficiently prolonged to permit extensive irreversible myocellular injury, then SOD + catalase administered during early reperfusion appears to provide no measurable salvage of myocardium or improvement in recovery of ventricular function.[245,246] This cardioprotective action of SOD *in vivo* implies either that superoxide anion or a metabolite in the extracellular fluid contributes significantly to the observed postischemic myocellular injury, that increased plasma membrane permeability as a consequence of ischemic injury may facilitate entry of the enzyme into the cell, that SOD is actively transported into the cell, or that SOD is acting by mechanisms that do not involve dismutation of superoxide anion. In the absence of significant accumulation of exogenous SOD within the cell, the most likely site of antioxidant action extracellularly is the interface between neutrophil and endothelium since the stimulated neutrophil adherent to endothelium can produce micromolar concentrations of ROS.[209,247] However, recent demonstrations of intracellular localization of bovine liver SOD in postischemic reperfused cardiac myocytes of the isolated working rat heart after addition of bovine SOD to the perfusate[248] and the uptake of SOD by cultured pulmonary artery endothelial cells[249] suggest the possibility of an intracellular site of action of exogenously administered SOD. The possible contribution of intracellular mechanisms in the cardioprotective actions of SOD against the myocellular injury produced by O_2^-· *in vivo* merits further investigation.

Beneficial effects of antioxidant enzymes and chemicals on biochemical and morphological indices of myocellular injury and on ventricular function also have been observed in isolated perfused heart preparations subjected to global ischemia or hypoxia followed by reperfusion or reoxygenation.[90,213,219,250-252] The results of such studies provide further evidence for the formation and cytotoxic action of ROS by resident cardiac cells since the tissues in these experiments were perfused with buffers free of blood or its cellular components.

B. DIRECT MEASUREMENTS OF OXYGEN RADICAL FORMATION

Recently, physicochemical techniques such as electron paramagnetic resonance (EPR) spectroscopy of free radicals[253-257] or of their spin-trap adducts,[258-261] and chemiluminescence[262] have provided more direct evidence for free radical formation in hearts subjected to ischemia and reperfusion. In studies from one laboratory, EPR spectra of rabbit hearts freeze-clamped after perfusion *ex vivo* with oxygenated buffer or after 10 min of global ischemia followed by reperfusion with oxygenated buffer for up to 60 s demonstrated three different signals; one signal was similar to those generated by alkyl peroxyl or superoxide oxygen-centered free radicals, another was identical to the spectrum of a carbon-centered ubisemiquinone associated with the mitochondrial electron transport chain, and a third was consistent with a nitrogen-centered free radical of the peroxyl amine type.[255-257] Signal intensity of the oxygen- and nitrogen-centered radicals increased twofold during ischemia, whereas the intensity of the carbon-centered radical declined. Reperfusion caused a further marked increase in the oxygen-centered free radical signal intensity that peaked 10 s after the onset of reperfusion. Administration of SOD during reperfusion almost completely inhibited this burst of oxygen-derived free radical formation[256,257] and was associated with improved recovery of cardiac contractile function.[256] Other investigators have questioned whether the superoxide-derived alkyl peroxyl radical and the nitrogen-centered radical tentatively identified in these studies might have been generated artifactually by the grinding of frozen myocardial tissue.[253,263,264] Additional signals consistent with coenzyme Q_{10} of mitochondrial origin[264] or a reduced iron-sulfur center associated with mitochondrial dehydrogenases[263] were tentatively identified in other experiments that avoided grinding of the tissues. The ESR signal for this latter radical increased during progressively longer periods of ischemia up to 60 min.[263] Although the exact chemical nature and cellular origin of the detected free radicals remain to be resolved, these studies provide direct evidence for the formation of micromolar concentrations of free radicals in the heart during both ischemia and the early stages of postischemic reperfusion. Additional studies will be required to determine the specific chemical nature and cellular and intracellular source(s) of these free radicals, to define the specific biochemical reactions responsible for their generation, and to establish any causative role for these radicals on the deterioration of cardiac contractile function with which their formation is temporally correlated.

C. EPR SPIN TRAPPING OF OXYGEN RADICALS

Studies utilizing electron spin resonance spectroscopy to detect nitroso or nitrone spin-trap adducts of free radicals have implicated ROS in both the prolonged contractile dysfunction without cellular necrosis that is the hallmark of the "stunned" myocardium and in the irreversible cellular injury that occurs during reperfusion following more prolonged periods of ischemia. DMPO-OH and a carbon-centered radical adduct of α-phenyl-N-butylnitrone (PBN) have been detected in the effluent of the isolated perfused working rat heart model during the first 3 to 4 min of reperfusion, regardless of the duration of the preceding global ischemia.[254,258] Free-radical spin-adduct signals could be inhibited by adding SOD to the perfusate[258] or delayed by reperfusing the heart initially with anoxic buffer.[254] The spin-trap reagent was added directly to the perfusate in these studies and no attempt was made to correlate the appearance of free-radical spin adducts with functional or histological evidence of myocardial injury.

Relevant to the question of whether these free radicals are responsible for reperfusion-induced myocardial injury, both DMPO[261] and PBN[260] have been reported to hasten the recovery of ventricular function and DMPO to reduce the incidence of reperfusion-induced ventricular arrhythmias[261] when added to the perfusate during reperfusion of the working rat heart perfused *ex vivo*[261] or of the canine "stunned myocardium" model *in vivo*.[260] These observations indicate that spin-trap reagents may themselves have the capacity to modify cardiac functional derangements induced by reperfusion. To avoid exposing the heart directly to the spin-trap reagent, Kramer and colleagues[265] rapidly mixed PBN with effluent from the isolated working rat heart during postischemic reperfusion. They observed multiple bursts of radical formation during the first 12 min of reperfusion using this technique and also noted that the severity of contractile dysfunction during reperfusion in this model was correlated with the intensity of the ESR signal of a radical spin adduct, thought to be derived from an alkoxyl radical. Although these findings implicate free radicals in the contractile dysfunction observed in these models, it should be remembered that the detection of free radicals during early reperfusion does not establish a causal relationship between their formation and the functional myocardial injury observed in this model.

D. CHEMILUMINESCENCE MEASUREMENTS OF OXYGEN RADICALS

Although considerably less specific than either direct EPR or ESR spin-trap adduct determinations, chemiluminescence has also been measured as an index of oxygen radical formation. When phorbol myristate-activated neutrophils were added to the oxygenated buffer perfusing rat hearts *ex vivo*, the increase in chemiluminescence, presumed secondary to the formation of ROS by these neutrophils, correlated with a decline in left ventricular contractile function and a reduction in coronary flow.[262] Both the augmented chemiluminescence and the deterioration in ventricular function could be abolished by addition of thiourea (HO· radical scavenger), SOD, and catalase to the suspension of activated neutrophils with which the hearts were perfused.

IV. ROLE OF NEUTROPHILS IN MYOCARDIAL ISCHEMIA-REPERFUSION INJURY

A. ROLE OF NEUTROPHILS IN POSTISCHEMIC MYOCARDIAL "STUNNING"

Reactive oxygen species make an important contribution to the prolonged contractile dysfunction, myocardial stunning,[20] that follows short periods of myocardial ischemia insufficient to produce necrosis. This conclusion is based on numerous reports of accelerated recovery of myocardial contractility after treatments with scavengers of ROS [SOD and catalase,[266-268] dimethylthiourea,[223] captopril,[269] and N-2-mercaptopropionylglycine[270]] or with chemicals that reduce ROS formation (deferoxamine,[228,271] allopurinol, or oxypurinol[70,71]). There is some controversy, however, as to whether neutrophils are a source of the ROS that contribute to this dysfunction. For example, the use of extracorporeal filters to remove more than 90% of the neutrophils (and platelets) in the blood perfusing the left anterior descending coronary artery hastened the recovery of contractile function after a 15- to 20-min coronary occlusion in the dog;[272,273] recovery was even more dramatic when the filter was used only during reperfusion.[273] In contrast, another investigation failed to find an improvement in contractile function following extracorporeal removal of neutrophils.[274] Furthermore, neutrophil depletion in dogs by use of an antineutrophil antiserum also failed to decrease myocardial contractile dysfunction.[275] In addition, if neutrophil-derived oxidants or other cytotoxic mechanisms are causally involved in the stunned myocardium, one would perhaps expect an increase in neutrophil accumulation during reperfusion after even brief ischemic episodes. However, by use of either [111]indium-labeled neutrophils[276] or assay for myocardial myeloperoxidase content,[277] techniques that have been successfully employed to demonstrate neutrophil accumulation in ischemia-reperfusion protocols sufficient to result in infarction, no neutrophil accumulation has been detected in myocardial regions subjected to brief (e.g., 12 or 15 min) ischemia.

Antineutrophil pharmacologic interventions, which have been successful in reducing infarct size following prolonged ischemia and reperfusion, have failed to prevent myocardial stunning. Nafazatrom, an antioxidant lipoxygenase inhibitor, that may also be capable of directly inhibiting neutrophil motility and secretory functions,[238] has been shown to reduce infarct size in both rabbit and canine models of ischemia-reperfusion injury.[238,278,279] In contrast, this agent failed to reduce postischemic myocardial contractile dysfunction.[280] Neutrophil monoclonal antibody 904, directed against neutrophil CD11b (Mac-1 or Mo1) and is discussed below in terms of its ability to inhibit human and canine neutrophil adherence-dependent production of H_2O_2, reduced infarct size when administered as an IgG in a canine model of left circumflex artery occlusion for 90 min and reperfusion for 6 h[281] or when administered as a F(ab')$_2$ fragment preparation in a canine model with reperfusion for 72 h.[282] In contrast, treatment with 904 F(ab')$_2$ did not prevent myocardial stunning in a canine protocol monitoring systolic wall thickening during 15 min of left anterior descending coronary artery occlusion and 3 h of reperfusion.[277] Hence, the bulk of the evidence would appear to indicate that neutrophils do not constitute a significant source of the ROS in myocardial stunning. Differences in the results from studies involving reperfusion of neutrophil-depleted blood may be due to differences in experimental parameters, such as the intensity of ischemia achieved, which bear directly upon the extent of functional recovery.[21]

B. ROLE OF NEUTROPHILS IN THE "NO-REFLOW" PHENOMENON

Neutrophil accumulation in the ischemic myocardium is markedly accelerated during reperfusion.[15] Activation of neutrophils produces cytoskeletal changes that increase their resistance to deformability, impeding their movement through the lumen of the capillary.[16] It has been proposed that this plugging of the microvascular endothelium by activated neutrophils, even after restoration of blood flow through a previously occluded artery,[23] leads to the "no reflow" phenomenon, which perpetuates segmental myocardial ischemic injury.[283] Protrusion of swollen endothelial cells into the capillary lumen or its extrinsic compression by edematous or contracted myocytes also may contribute to microvascular plugging.[10,284] Consistent with this hypothesis, the longer the duration of ischemia and the more extensive the ischemic injury, the more likely it is that areas of the myocardium will fail to be reperfused upon restoration of large vessel blood flow.[23,284] These patchy areas of diminished or absent flow are found only in areas of myocardial necrosis[10,22] and are not as readily demonstrated at the end of ischemia as they are during progressively longer periods of reperfusion.[23] Treatment with SOD at the time of reperfusion markedly diminishes the development of "no-reflow" areas in the postischemic myocardium,[14] suggesting that superoxide anion or one of its metabolites participates in this phenomenon.

C. PROMOTION OF POSTISCHEMIC MYOCARDIAL SALVAGE BY PREVENTING NEUTROPHIL ACCUMULATION

Diverse treatments that decrease circulating neutrophils (Leukopak® filters,[272,285,286] antineutrophil serum,[287,288] or bone marrow-suppressing chemicals like hydroxyurea or mechlorethamine),[288,289] the administration of monoclonal antibodies directed against neutrophil adhesion molecules,[281,290] or drugs that inhibit leukocyte activation (CI-922, perflurochemicals, stable analogs of prostacyclin)[236,291-294] have been reported to reduce postischemic myocardial reperfusion injury and/or ventricular dysfunction *in vivo*. Some studies, however, have not demonstrated benefit from such treatments.[274,275,277,295]

Even though neutrophils may not be the only source of ROS that contributes to the contractile dysfunction produced by short periods of myocardial ischemia,[257,274] ROS produced by activated neutrophils[209,247,296-298] appear capable of contributing to the ventricular dysfunction that accompanies ischemia and reperfusion with necrosis. Normoxic perfusion of hearts *ex vivo* with PMA-activated neutrophils or perfusion of posthypoxic hearts with suspensions of inactivated neutrophils[262,299] results in myocardial contractile dysfunction. The effect is partly reversed by incubation of the activated neutrophils with SOD and catalase prior to perfusion and almost completely reversed by the further addition of thiourea (HO· radical scavenger).[262]

D. ROLE OF THE NEUTROPHIL IN POSTISCHEMIC ENDOTHELIAL CELL INJURY

In the inflammatory response initiated by myocardial ischemia, the interaction of neutrophils with endothelial cells in the microvasculature is of major importance in the capillary plugging that likely contributes to the prevention of an adequate restoration of blood flow to tissue areas crucially susceptible to irreversible injury. In addition, neutrophil interactions with specific endothelial adhesion molecules are necessary for the diapedesis or transendothelial migration of neutrophils into the adjacent myocardium.[300,301] Finally, neutrophil-mediated injury to the endothelium itself may contribute substantially to the overall pathogenesis of myocardial necrosis by increasing capillary permeability, amplifying inflammatory signals, or further compromising blood flow hemodynamics.

Numerous *in vitro* studies with cultured endothelium have implicated ROS as the mediators of neutrophil-induced cytotoxicity. Sacks et al.[302] have addressed the mechanisms of complement activation and granulocyte-mediated pulmonary injury in patients undergoing hemodialysis using monolayers of human umbilical vein endothelial cells (HUVEC). When HUVEC monolayers were incubated with neutrophils and activated serum complement or fractions enriched for the chemoattractant complement component C5a, significant endothelial cell damage occurred as evidenced by ^{51}Cr release from prelabeled HUVEC cultures. Use of the superoxide generating system, xanthine + xanthine oxidase, resulted in an endothelial cytotoxic response similar to that seen with activated neutrophils. A specific role for hydrogen peroxide in this damage was indicated by the fact that both SOD + catalase and catalase alone inhibited endothelial ^{51}Cr release caused by xanthine/xanthine oxidase and by C5a-activated PMNs. Incomplete protection in the latter versus the former case suggests that additional cytotoxic mechanisms may exist with activated neutrophils, which will be explored further below.

Importantly, Sacks et al.[302] found that under conditions where C5a induces an H_2O_2-dependent endothelial injury, little if any myeloperoxidase (MPO) was released from neutrophils, which indicates that the cytotoxic mechanism may not involve an MPO-generated oxidizing hypohalous species. Subsequently, Weiss et al.[303] demonstrated that human neutrophils stimulated with

PMA are destructive for human endothelial cells. In both the studies of Sacks et al.[302] and Weiss et al.[303] endothelial destruction was delayed, with maximum cytotoxicity seen between 4 and 6 h. Cytolysis was prevented by catalase, but not by SOD. Furthermore, addition of azide or cyanide (heme-enzyme inhibitors that inhibit neutrophil MPO activity) actually enhanced ^{51}Cr release from endothelium. Further evidence for H_2O_2-induced endothelial injury was provided by the observation that a glucose-glucose oxidase enzyme system, capable of forming hydrogen peroxide, could replace PMA-stimulated neutrophils in the endothelial injury assay. Whereas addition of exogenous MPO enhanced the endothelial cytotoxicity achieved with low glucose oxidase concentrations, it did not do so with higher enzyme concentrations or with PMA-stimulated neutrophils. The resulting endothelial damage with activated neutrophils also was not affected by addition of amino acids that react rapidly with HOCl derived from H_2O_2 and MPO.

PMA-stimulated human neutrophils also destroy cultured bovine pulmonary endothelial cells by a mechanism that depends upon hydrogen peroxide.[304] Again, this endothelial cytotoxicity, either with stimulated neutrophils or exogenous reagent H_2O_2, is delayed, is not seen at 4 h, but is evident at 8 h and later time points. Interestingly, in the presence of noncytotoxic levels of H_2O_2, injury was seen upon the addition of myeloperoxidase and potassium iodide, indicating susceptibility of these cells to oxidative injury by hypohalous species. This latter effect was markedly increased under acidic conditions (pH 5.8) consistent with the bactericidal effects of the MPO system. This observation may be particularly relevant under pathophysiologic conditions, such as ischemia, where local pH may be reduced.[304] Ultimately, the role of MPO-derived oxidant species in endothelial injury may depend upon availability of the enzyme. For example, PMA, a potent stimulus of the neutrophil respiratory burst, is a weak stimulus for degranulation of neutrophil MPO-containing primary granules.[305,306] However, PMA is apparently a sufficient stimulus for HOCl-mediated (i.e., MPO-dependent) cytolysis of certain tumor cells.[303,307] These results indicate that although endothelial cells may be susceptible to damage caused by hypohalous acids under certain conditions, MPO-dependent formation of these species is not essential for killing of endothelium by activated neutrophils.

In addition to its participation in the myeloperoxidase-catalyzed oxidation of Cl^- into HOCl, H_2O_2 is capable of reacting with transition metals such as Cu^+ and Fe^{2+} to form the highly reactive hydroxyl radical (HO·).[32] Varani et al.[308] have demonstrated the involvement of hydroxyl radicals in neutrophil killing of endothelial cells *in vitro*. Human neutrophils stimulated with immune complexes, opsonized zymosan, or PMA (but not the chemotactic factors, fMLP or PAF) caused marked destruction of cultured bovine pulmonary artery endothelial cells. Endothelial killing correlated with agonist-induced production of H_2O_2, and was inhibited by catalase, dimethylthiourea, the hydroxyl radical scavenger, D-mannitol, and the iron chelator, deferoxamine mesylate. In contrast, SOD, sodium azide, and soybean trypsin inhibitor were without effect. Protection afforded by either catalase, iron chelators, or hydroxyl radical scavengers implicates the hydroxyl radical, formed through the Fenton reaction, in neutrophil-mediated endothelial injury under these conditions *in vitro*. Further studies by the same laboratory have investigated the source of iron in mediating this reaction during neutrophil killing of endothelium.[309] Again, using PMA-stimulated human neutrophils and cultured bovine pulmonary artery endothelial cells, injury was detectable at 2 h and continued to increase through 8 h. Pretreatment of neutrophils with deferoxamine did not affect the amount of reactive oxygen generated upon PMA stimulation nor did it reduce endothelial cytotoxicity. In contrast, pretreatment of the endothelial cells with the iron chelator under conditions that resulted in persistent intracellular association of chelator with the endothelium afforded marked protection against stimulated neutrophil-induced H_2O_2-mediated cytotoxicity.[309]

These data implicate endothelial iron in intracellular formation of hydroxyl radical through a Fenton reaction involving neutrophil-derived H_2O_2. Such a mechanism for endothelial injury would obviously be facilitated by the ability of H_2O_2, in contrast to the negatively charged superoxide anion or the highly reactive hydroxyl radical, to diffuse through cellular membranes.[32] Hydroxyl radicals, even if they were released from activated neutrophils *in vivo*, are so highly reactive that they would likely interact with circulating proteins or lipid macromolecules in plasma long before they could reach the surface of another cell. In contrast, the diffusion of H_2O_2 from one cell to another *in vivo* is made more likely by its relative stability and by the lack of specific enzyme systems in plasma, such as catalase and glutathione peroxidase, to hasten its elimination. However, since the H_2O_2 formed by the activated neutrophil *in vivo* is in close proximity to numerous circulating cells such as erythrocytes, its specific entry into endothelial or other target cells would be greatly facilitated by close apposition of the neutrophil with the target cell as occurs during adhesion.

Endothelial cell iron available to participate in the formation of damaging hydroxyl radicals may be provided by a low molecular mass, intracellular iron pool composed of iron loosely bound to various cellular components such as membranes, nucleic acids, organic acids, and phosphate complexes[32] or released from the iron storage protein ferritin by reducing species such as ascorbate or superoxide anion.[32] Recent evidence for the participation of endothelial-derived superoxide anion in this process is the demonstration that supplementation of endothelial intracellular SOD (by prolonged incubation with high levels of the enzyme) inhibits injury of rat pulmonary artery endothelial cells induced by H_2O_2 or stimulated neutrophils.[249] A possible source of endothelial-derived superoxide is the enzyme xanthine oxidase (XO), derived from xanthine dehydrogenase, which as indicated above, is present in endothelial cells in substantial quantities in certain species.[40] Xanthine dehydrogenase (XD) normally utilizes NAD^+ as the electron acceptor during the oxidation of the purines xanthine and hypoxanthine[310] (Figure 1). The role of thie XD to XO conversion in the generation of superoxide during reoxygenation of ischemic tissues has been described above. Recently, the conversion of XD to XO in rat pulmonary artery endothelial cells as well as the release of XO from these cells by certain inflammatory mediators, such as C5a, TNF-α, and fMLP has been described.[311,312] In addition, XD to XO conversion in these same cells can be achieved by exposure to PMA-stimulated human neutrophils.[42] This cell-mediated conversion is not prevented by SOD or catalase, is not induced by exogenous H_2O_2, and is effected by neutrophils from patients with CGD, which are incapable of generating ROS, thus excluding a role for superoxide anion or H_2O_2 in mediating the increase in endothelial XO activity, despite their importance in cytotoxicity, as described above. Moreoever, supernatants from activated neutrophils also failed to induce XD to XO conversion, suggesting a possible requirement for cell-cell adhesion and an endothelial response as a consequence of such adhesion in this process.

A fundamental role for neutrophil adhesion in reactive oxygen-mediated killing of target cells has recently been elucidated in studies aimed at examining more physiologic mechanisms of neutrophil activation that may operate *in vivo*. A majority of the cytotoxicity studies, such as those described above, have utilized phorbol esters, such as PMA, to activate neutrophils. PMA is a potent activator of neutrophil reactive oxygen secretion, but is obviously nonphysiologic. In contrast to agents like PMA and OAG, chemotactic factors, such as fMLP, LTB$_4$, and C5a, for which neutrophils have specific membrane receptors, elicit little reactive oxygen from neutrophils stimulated in suspension. Furthermore, the respiratory burst that is achieved is very brief compared to the sustained production of superoxide that results from stimulation with phorbol esters. However, exposure of neutrophils to adhesive surfaces, such as polystyrene coated with extracellular matrix proteins or to cultured endothelium, results in production of large quantities of hydrogen peroxide (e.g., 50 to 200 nmol/10^6 PMNs) in response to levels of chemotactic factors, such as fMLP, or cytokines, such as TNF-α which, although capable of promoting adhesion, are insufficient to promote detectable production of reactive oxygen from neutrophils stimulated in suspension.[209] We and others have now extended these observations to human, canine, and rabbit neutrophils using a variety of stimuli and substrates.[247,296,297,313,314] In all cases, adherence-dependent production of hydrogen peroxide (in contrast to that elicited by phorbol esters) is delayed, with a lag phase of 30 to 60 min depending upon the substrate, and prolonged in duration (typically greater than 60 min). Recently, evidence has been generated for the role of the neutrophil CD11/CD18 adhesion complex (β_2 integrin family) in mediating this adhesion-dependent respiratory burst.

E. NEUTROPHIL AND ENDOTHELIAL ADHESION MOLECULES IN INFLAMMATION

The neutrophil CD11/CD18 adhesion molecules are a family of heterodimeric proteins composed of a noncovalently associated common β chain (designated CD18) and immunologically distinct alpha subunits designated CD11a, CD11b, and CD11c for LFA-1 (lymphocyte function-associated antigen 1), Mac-1 (Mo1 or complement receptor type 3), and p150,95, respectively.[315-317] These molecules are members of the integrin supergene family of adhesion molecules,[317-319] which are important in a variety of cell-matrix and cell-cell interactions and which are currently divided into three classes based on their β-subunits. The β_2 (CD18) integrins are restricted in their distribution to leukocytes and mediate a variety of important adhesion reactions in immune and inflammatory processes.[315,316,320,321] In addition to their localization in neutrophil plasma membranes, Mac-1 and p150,95 (but not LFA-1) are also stored in intracellular vesicles, a subset of neutrophil secondary granules or "tertiary" granules,[316,322-325] so that neutrophil stimulation by a variety of chemotactic factors or phorbol esters results in the increased surface expression of these proteins.[316,322,324]

The role of these heterodimeric glycoproteins in neutrophil or lymphocyte adhesion reactions has been elucidated by studies using subunit-specific monoclonal antibodies (MAbs).[315-317,321] Mac-1 (CD11b/CD18), which functions as the receptor for the iC3b component of complement and, thus, participates in phagocytosis of appropriately opsonized particles or microbes,[326-328] also appears to mediate the adhesion of stimulated neutrophils to certain protein-coated surfaces and to unstimulated endothelial monolayers.[316,320,329-332] p150,95 may share similar adhesive properties with Mac-1, but is quantitatively minor in comparison. LFA-1 binds to intercellular adhesion molecule 1 (ICAM-1), a transmembrane protein of the immunoglobulin supergene family that is expressed on a variety of cell types, including cytokine-stimulated endothelium.[333-337] This ligand pair is important in certain immune cell interactions,[317,321,333] but also appears to mediate the adhesion of unstimulated neutrophils to cytokine- or endotoxin (LPS)-stimulated endothelium by a process that results in transendothelial migration.[300,301,337] Stimulation of neutrophils by chemotactic factors enhances adhesion to and migration through cytokine- or LPS-stimulated endothelium, a process that appears to involve a cooperative interaction of both neutrophil Mac-1 and LFA-1 with endothelial ICAM-1.[301,337] The extreme importance of CD11/CD18 integrins in the inflammatory processes is evidenced by the profound inflammatory defects observed in patients with leukocyte adhesion deficiency (LAD), an inherited, autosomal recessive, severe and life-threatening condition characterized by a complete absence or marked reduction of CD11/CD18 molecules on leukocyte cell surfaces.[316] In addition, monoclonal antibodies to CD11b or CD18 have been observed to have marked anti-inflammatory effects in a variety of models *in vivo*, including ischemia-reflow injury in intestine[338] and heart.[281,282]

Recent studies by our laboratory, addressing the role of leukocyte CD18 integrins, specifically Mac-1, in mediating the adhesion-dependent respiratory burst in neutrophils, have used chemotactic factor-stimulated CD18-deficient neutrophils from patients with LAD. These adhesion-deficient neutrophils, in contrast to their essentially normal production of reactive oxygen in response to PMA in suspension, do not produce H_2O_2 on the CD11b/CD18 substrate keyhole limpet hemocyanin (KLH) or on cultured HUVEC monolayers.[247] In addition, H_2O_2 production by fMLP-stimulated normal human neutrophils or by PAF-stimulated canine neutrophils on KLH or endothelium is blocked by monoclonal antibodies to CD11b (MAb 904) and CD18 (MAb R15.7).[247] Nathan et al.[296] have recently shown that TNF-α-stimulated CD18-deficient (LAD) neutrophils do not produce H_2O_2 on serum, fibronectin, vitronectin, fibrinogen, thrombospondin, or laminin. In addition, prolonged exposure to supersaturating levels of the anti-CD18 MAb, IB4, prevented H_2O_2 production by TNF-stimulated normal PMNs on serum, fibrinogen, thrombospondin, and laminin.[296]

F. ROLE OF PROTEOLYTIC ENZYMES IN NEUTROPHIL-MEDIATED INJURY

In addition to the oxidative mechanisms described above, whereby stimulated neutrophils have been shown to destroy cultured endothelium *in vitro*, neutrophils also can adversely affect endothelial function through nonoxidative processes. Harlan et al.[339] demonstrated that although fMLP-stimulated neutrophils failed to induce endothelial cytolysis under the conditions examined, they produced marked disruption in bovine aortic or pulmonary artery endothelial monolayer integrity by a process that was not affected by SOD or catalase and was normal with neutrophils from patients with CGD, but appeared to require neutrophil-endothelial contact. Furthermore, PMA-stimulated neutrophils were demonstrated to cause a reactive oxygen-independent detachment of HUVEC that was absent with neutrophils from a patient with CD18 deficiency[340] and blocked by an anti-CD18 monoclonal antibody that inhibited neutrophil-endothelial adhesion.

In addition to the membrane NADPH oxidase system, neutrophils contain in their primary and secondary storage granules a variety of proteolytic enzymes capable, in theory, of contributing to tissue injury during inflammation *in vivo*.[341] These enzymes, including serine proteases, such as elastase, and metalloproteases, such as collagenase, are released into phagosomes during the phagocytosis and destruction of microorganisms and might also be released into the extracellular environment under certain conditions.[341,342] Elastase has been implicated in the nonlytic detachment of endothelial cells caused by stimulated neutrophils.[343] In contrast, elastase was recently shown by Smedly et al.[344] to be essential for the lytic injury of human microvascular endothelial cells by fMLP- or C5a-stimulated neutrophils following low-level LPS pretreatment. The nonoxidative nature of this injury vs the numerous examples cited above may be due to differences in the endothelial cell preparations employed and in their relative susceptibility to oxidative destruction,[344] due perhaps to differences in content of antioxidant enzymes, such as catalase.[345] In addition, the specific conditions for stimulating neutrophils in these experiments *in vitro* may determine the mechanisms by which the neutrophil exerts its injurious effects on the target cell. For example, PMA, frequently employed in such studies, is a potent stimulator of neutrophil reactive oxygen secretion, but a weak stimulus for neutrophil degranulation.[305,306] The ability of CD18-dependent adhesion, described above as a strikingly potent stimulus for prolonged and massive release of H_2O_2, to induce secretion of primary and secondary granule contents is currently under investigation in our laboratory.

Despite the potential destructive capacity of neutrophil proteolytic enzymes, focus on neutrophil-mediated injury *in vivo* has been largely on the production of ROS. This focus has provided numerous examples of the antiinflammatory effects of various antioxidant interventions in models of neutrophil-mediated tissue damage.[298] Furthermore, neutrophil collagenase is secreted in an inactive form,[346] and tissues are protected *in vivo* from serine proteases, such as elastase, by high levels of antiproteases, particularly the serine protease inhibitor, α_1-antitrypsin.[298,347] However, recent studies have indicated potential mechanisms whereby neutrophil ROS can unmask the destructive mechanism of secreted proteolytic enzymes. Latent, inactive collagenase can be activated by neutrophil-oxidizing species, specifically HOCl derived from the myeloperoxidase system,[298,348] and oxidation of α_1-antitrypsin by HOCl renders it ineffective as an inhibitor of elastase.[298] α_1-Antitrypsin, present in high concentrations in plasma, is a small glycoprotein that can readily diffuse throughout interstitial fluid and, hence, afford essentially all tissues protection from the highly destructive neutrophil elastase.[347]

A synergistic interaction of reactive oxygen products and proteases has also been recently shown in the stimulated neutrophil killing of rat pulmonary artery endothelial cells.[349] Whereas, endothelial injury by PMA-stimulated PMNs at 4 h was iron-dependent and inhibited by catalase, it was not significantly reduced by soybean trypsin inhibitor (STI). In contrast, the progressive injury seen at later time points (18 h) was only partially inhibited by catalase or STI, but these two reagents together provided synergistic protection.[349] Hence, neutrophil-mediated endothelial injury, under certain conditions, might involve complex interactions between ROS and proteolytic enzymes.

G. OTHER PROINFLAMMATORY EFFECTS OF REACTIVE OXYGEN SPECIES

The above comments have focused on the cytotoxic effects of ROS, either by direct endothelial lytic injury or by interactions with processes affecting the actions of proteolytic enzymes. However, oxidants might also function in the initiation or amplification of inflammatory injury, perhaps by modulating neutrophil chemotaxis or adhesion. McCord and co-workers[350] demonstrated that exposure of plasma to xanthine and xanthine oxidase resulted in the formation of chemotactic activity for neutrophils that was inhibited by prior addition of SOD but not catalase. In contrast, this plasma factor did not stimulate neutrophil superoxide production or degranulation. Superoxide-treated plasma was able to elicit neutrophil emigration when injected intradermally in rats, and intradermal injection of xanthine + xanthine oxidase also resulted in a smaller, but significant, neutrophil influx that was abolished by coinjection of SOD but not catalase.[350] In either case, the neutrophil emigration was not accompanied by gross evidence of erythema or edema. Further evidence for a proinflammatory neutrophil chemotactic response due to superoxide anion formation is the report that administration of a ficoll derivative of SOD was anti-inflammatory and markedly reduced neutrophil influx in reverse-passive Arthus and carrageenan-induced foot edema models.[350] Preliminary characterization revealed that this superoxide-dependent plasma chemotactic activity was a lipid component bound to albumin.[350]

Several other examples exist of the formation of known or newly demonstrated chemotactic substances upon nonenzymatic oxidation of fatty acids or the peroxidation of lipids in plasma-derived lipoproteins. Perez et al.[351] demonstrated that the exposure of arachidonic acid to a superoxide-generating system of xanthine oxidase plus acetaldehyde resulted in the formation of a product with neutrophil chemotactic activity that had chemical characteristics consistent with a lipid hydroperoxide. Formation of this product from arachidonate, xanthine oxidase, and acetaldehyde was significantly inhibited by SOD, catalase, the hydroxyl radical scavenger, mannitol, and the singlet oxygen scavengers, histidine, uric acid, and 2,5-dimethylfuran.[351]

Esterbauer and co-workers[352] have identified a number of lipid aldehydes formed during lipid peroxidation of isolated low-density lipoprotein (LDL) *in vitro*. These products, which are potentially involved in the covalent modification of the apoB protein moiety of LDL that may be etiologically relevant to proatherogenic properties of oxidized LDL,[353] accounted for only about 2% of the polyunsaturated fatty acids consumed during LDL oxidation. However, two of the products formed, 4-hydroxyoctenal and 4-hydroxynonenal, have been demonstrated to possess potent chemotactic activity for neutrophils.[354,355] Lenz and co-workers[356] in our laboratory have recently shown that oxidation of LDL *in vitro* results in the formation of much larger quantities of free and esterified hydroperoxy and hydroxy derivatives of linoleic (18:2) and arachidonic (20:4) acid, accounting for approximately 70 and 25% of the 18:2 and 20:4 consumed, respectively, in a 24-h oxidation. Hydroxyeicosatetraenote (HETE) derivatives of arachidonate are chemotactic for neutrophils.[357] Although these metabolites are less potent than LTB_4 in promoting neutrophil chemotaxis, such oxidized lipids might serve as proinflammatory mediators if formed in sufficiently large quantities *in vivo*. In addition to oxidized arachidonic metabolites that may stimulate neutrophils in such a manner, McIntyre and co-workers have recently shown that free radical-initiated oxidation of 1-*O*-palmitoyl-2-arachidonyl-*sn*-glycero-3-phosphocholine results in fragmentation of the *sn*-2 unsaturated lipid to short-chain residues (e.g., propionyl) that are sufficient to allow cross-reactivity and stimulation of neutrophils through their PAF (1-*O*-alkyl-2-acetyl-*sn*-glycero-3-phosphocholine) receptors. Hence, multiple mechanisms exist whereby oxidative stress of sufficient magnitude might generate chemotactic lipid derivatives from plasma lipoproteins or cellular membranes that are capable of activation of neutrophils.

In addition to the nonenzymatic formation of chemoattractant oxidized lipids, ROS, such as H_2O_2, may stimulate endothelium or other target cells in a manner promoting neutrophil adhesion. For example, Lewis et al.[162] have recently demonstrated that bovine pulmonary artery endothelium and HUVEC produce PAF in response to H_2O_2, a glucose-glucose oxidase incubation system, or PMA- or ionophore-stimulated neutrophils. PAF is a potent proinflammatory mediator whose endothelial synthesis is tightly coupled to endothelial-dependent neutrophil adhesion in response to a variety of endothelial agonists.[162,358-360] The synthesis of PAF by endothelium in response to H_2O_2 is rapid, maximal at 20 to 30 min, but still present at 50 to 60 min.[162] The accumulation of PAF induced by H_2O_2 is concentration-dependent over an H_2O_2 concentration range of 1 to 20 mM, with maximal response at 5 to 10 mM. Under these conditions, PAF formation is temporally dissociated from any evidence of endothelial injury. Endothelial synthesis of PAF in response to H_2O_2 is dependent upon extracellular calcium, and H_2O_2 causes an increase in the permeability of endothelium for calcium.[162] That PAF generation may thus occur as a consequence of the activation of a calcium-dependent phospholipase A_2 is evidenced by the coincident endothelial metabolism of arachidonic acid to prostacyclin and the inhibition of PAF production (and endothelial adhesion) by pretreatment of the endothelium with the phospholipase inhibitor, *p*-bromophenacyl bromide.[162]

As mentioned above, endothelial cell-dependent neutrophil adhesion in response to a variety of calcium-mobilizing endothelial agonists is accompanied by the tightly correlated formation of PAF.[162,358-360] Treatment of endothelium with H_2O_2 also induces adhesion of otherwise unstimulated neutrophils.[162,361] This endothelial-dependent neutrophil adhesion is correlated, under every condition examined, with the endothelial synthesis of PAF. Although PAF synthesized in response to H_2O_2 remains associated with the endothelial cell and is not released into the culture medium, it may be positioned in the cell membrane in such a manner as to interact with specific neutrophil PAF receptors.[360] Studies aimed at specific neutrophil receptor desensitization suggested that H_2O_2-induced endothelial-dependent neutrophil adhesion is mediated via PAF interactions with its receptor on neutrophils; these studies demonstrated a substantially decreased adhesion of PMNs to H_2O_2-treated endothelium by prior PMN exposure to PAF, with no effect on adhesion induced by other neutrophil stimuli.[162] The phenomenon has been confirmed more recently by studies employing specific PAF receptor antagonists.[360]

The role of specific adhesion molecules in H_2O_2-induced neutrophil adhesion has been investigated recently in our laboratories employing a system of isolated-perfused canine vessels, radiolabeled isolated canine neutrophils, and a visual adherence assay[300,301] with cultured canine endothelium. As shown

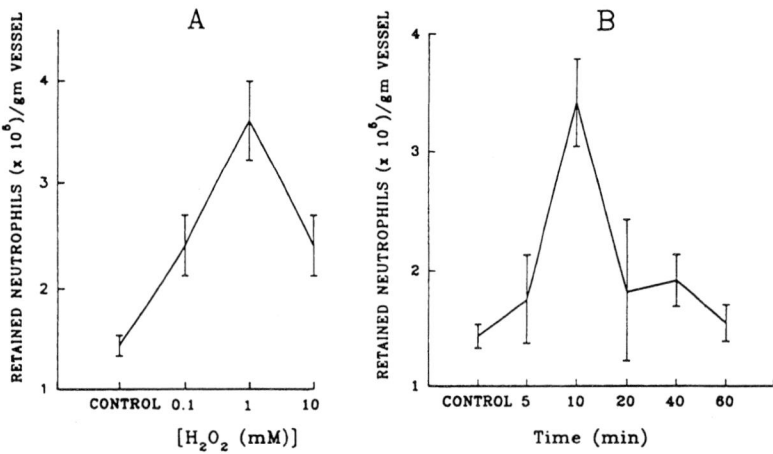

FIGURE 4.   (A) Pretreatment of canine carotid arteries for 20 min with increasing concentrations of H_2O_2 produces a dose-dependent increase in neutrophil retention that is maximal at 1 mM; n = 4, values are mean ± SEM. (B) Neutrophil retention by H_2O_2 (1 mM) preexposed arteries is both time-dependent and transient. Maximal effect is observed 10 min post-H_2O_2 exposure and declined to control values by 60 min; n = 4.

in Figure 4, perfusion of isolated canine carotid arteries or jugular veins *ex vivo* with H_2O_2 induced a dose-dependent retention of canine neutrophils.[361-363] This response was maximal with 1 mM H_2O_2, and transient (with respect to the time between vessel stimulation and neutrophil addition). PMN adhesion was maximal at 10 min postperfusion and had returned to control levels at 60 min, indicating a transient response reminiscent of cultured endothelial PAF production in response to similar levels of H_2O_2 exposure.[162] Repeat H_2O_2 exposure resulted in similar PMN adhesion (as the first exposure).[363] Endothelial injury was not evident as a consequence of H_2O_2 exposure alone, but with subsequent infusion of otherwise unstimulated neutrophils, transmission electron microscopy revealed transendothelial migration of PMNs and swelling and loss of plasma membrane integrity of endothelial cells underlying adherent PMNs.[363] Studies with arteries denuded of endothelium revealed that this H_2O_2-induced adhesion is endothelial-dependent[361-363] and, as shown in Figure 5, this adhesion is similarly reduced to control levels by monoclonal antibodies that react either with canine neutrophil CD18 or with canine endothelial intercellular adhesion molecule 1 (ICAM-1).[335,364] Similar results were obtained with the stimulated canine neutrophil adhesion to cultured canine jugular vein endothelial cells following 10-min exposure to 20 mM H_2O_2.[363] We have previously observed that PAF is a potent stimulus of increased surface expression of CD11b/CD18 on canine neutrophils and induces a CD18-dependent adhesion of canine neutrophils to protein-coated surfaces,[247] cultured endothelium,[313] and cytokine-stimulated isolated canine myocytes (see below).[313] Therefore, our results are consistent with the induction of a CD18-dependent adhesion as a consequence of sublethal exposure

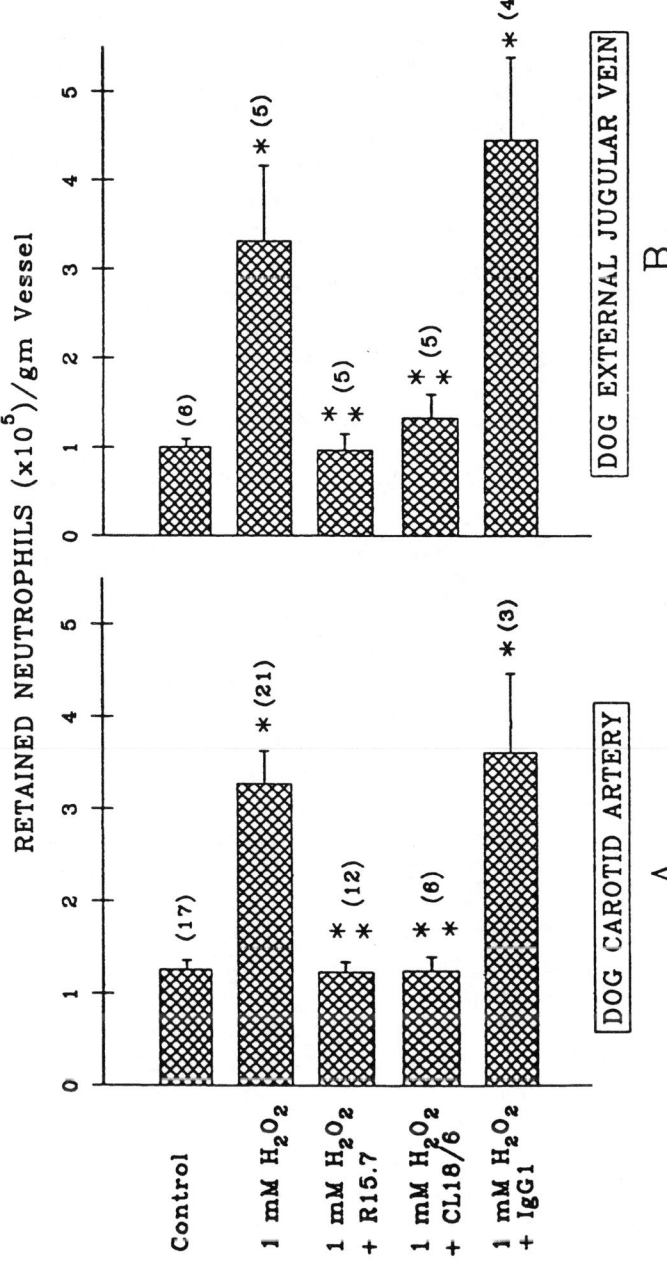

FIGURE 5.   Hydrogen peroxide-induced PMN retention by *ex vivo* perfused canine carotid arteries (A) and external jugular veins (B) is completely inhibited by incubation of the neutrophils with monoclonal antibodies (MAbs) against CD18 (R15.7), but not with an isotype-matched nonbinding control (IgG1). Similarly, in vessels perfused with MAb against canine endothelial ICAM-1 (CL18/6) 5 min before PMN instillation, the increased PMN retention produced by H_2O_2 pretreatment is completely inhibited. Values are mean ± SEM () = number of vessels. *$p < 0.05$ compared with control; **$p < 0.05$ compared to H_2O_2-treated vessels.

to H_2O_2 through the stimulated formation of endothelial PAF. This mechanism has been further substantiated by our recent observation that the neutrophil adhesion occurring in perfused canine vessels exposed to H_2O_2 is also blocked by the PAF receptor antagonist WEB-2086.[362,363]

Another adhesion protein may also be involved in the rapidly induced endothelial-dependent adhesion of neutrophils that occurs in response to the stimuli described above, including H_2O_2, which are capable of inducing endothelial production of PAF. Granule membrane protein-140 (GMP-140) is a membrane glycoprotein located in secretory granules of platelets and endothelium.[365,366] Agonists, such as histamine and thrombin, which promote rapid neutrophil adhesion to endothelium, cause rapid and transient expression of GMP-140 on the endothelial surface.[365,366] GMP-140 is a member of a newly characterized family of adhesion molecules, termed selectins, which, like the other two molecules in this class, ELAM-1 and Mel-14, are composed of an N-terminal lectin domain, a region homologous to epidermal growth factor, and a number of repeated domains homologous to those found in complement regulatory proteins.[367] ELAM-1 (endothelial leukocyte adhesion molecule 1) is expressed on endothelial cells in a protein synthesis-dependent manner, reaching a maximum at 2 to 4 h following stimulation of endothelium with certain cytokines.[368] Like endothelial ICAM-1, ELAM-1 mediates adhesion of neutrophils to cytokine-stimulated endothelium.[368,369] The murine Mel-14 antigen is a lymphocyte homing receptor,[370-373] the human homolog of which is also expressed on neutrophils[374] and contributes to a CD18-independent adhesion under certain circumstances.[375]

Direct evidence for the ability of GMP-140 to support rapid neutrophil adhesion has recently been obtained. COS cells transfected with a cDNA for GMP-140 support adhesion of isolated human neutrophils, and this adhesion, as well as neutrophil adhesion to microtiter wells coated with purified GMP-140, is blocked by certain monoclonal antibodies to GMP-140.[376] Furthermore, the rapid neutrophil adhesion to endothelium stimulated with phorbol esters or histamine is blocked by monoclonal antibodies to GMP-140.[376] Although studies were not performed with H_2O_2 as a stimulus, the similarities in the time course, divalent cation requirements, and simultaneous PAF production in response to stimuli, such as thrombin and histamine and millimolar levels of H_2O_2, would indicate that endothelial GMP-140 might also be involved in the adhesion of neutrophils to H_2O_2-treated endothelium.

A more delayed neutrophil adhesion to endothelium stimulated with low doses of H_2O_2 or t-butyl hydroperoxide (tBHP) has recently been described.[377] This delayed adhesion appears to be mediated by endothelial GMP-140, but is unrelated to PAF formation.[377] Exposure of endothelium to 250 μM H_2O_2 or tBHP induced a PMN adherence that was evident after 1 h and was maximal at 4 h. Endothelial PAF was not detected at these time points, and adhesion was not affected by PAF receptor antagonists, anti-ELAM antibodies, or protein synthesis inhibitors. However, the delayed adhesion was blocked by various antioxidant interventions (e.g., butylated hydroxyanisole, nordihydroguaiaretic acid, ascorbate, and deferoxamine) and by anti-GMP-140 MAbs.[377,377a] This suggests an oxidative mechanism for the delayed surface expression of preformed endothelial GMP-140, and, like the expression of PAF and the induction of a CD18-dependent adhesion, provides a molecular mechanism whereby ROS can promote neutrophil-endothelial interactions. Although the neutrophil ligand for endothelial GMP-140 has not yet been identified, and the functional significance of neutrophil adhesion mediated through this mechanism is not yet known, CD18-ICAM-1 interactions clearly participate in the transendothelial migration of neutrophils,[300,301] and CD18-dependent adhesion markedly amplifies neutrophil production of ROS.[247,296,313] Thus, mechanisms exist that might allow for sublethal concentrations of ROS to initiate and amplify neutrophil responses leading to cytotoxic cell-cell interactions. These might include direct endothelial injury and damage to vascular integrity by processes outlined above, migration into adjacent myocardium, and, ultimately, to myocyte necrosis through specific neutrophil-myocyte interactions, which are described below.

Evidence for the contribution of ROS to increased neutrophil adhesion to the vasculature *in vivo* has been generated by Granger and co-workers in a feline model of intestinal ischemia-reflow injury. Suzuki et al. reported that following 60 min of reduced (20% of control) mesenteric and intestinal blood flow and 60 min of reperfusion, intravenous administration of human recombinant superoxide dismutase (hSOD) or anti-CD18 monoclonal antibody, IB4, attenuated the reperfusion-induced neutrophil adherence as assessed with intravital microscopy 10 min following administration.[338] Hypoxia reoxygenation-induced adhesion of feline neutrophils to cultured bovine microvascular endothelium was also blocked by hSOD or IB4, whereas H_2O_2-inactivated hSOD was ineffective *in vivo* or *in vitro*.[338] A role for H_2O_2 and endothelial PAF-neutrophil CD18 interactions in neutrophil adhesion to venular endothelium is supported by preliminary reports from these investigators that H_2O_2 promotes leukocyte adhesion to cat mesenteric venules *in vivo*[378] and that the reperfusion-induced leukocyte adhesion to cat mesenteric venules is blocked by PAF receptor antagonists[378a] as well as by anti-CD18 MAb.[338] There is also evidence supporting a role for PAF in the increased cardiac permeability of isolated perfused rat hearts following 15 min of global ischemia, a condition inhibited by the PAF receptor antagonist, CV-6209.[379]

H. MECHANISMS OF NEUTROPHIL/REACTIVE OXYGEN-INDUCED MYOCYTE INJURY

As described above, substantial data implicate ROS in the pathogenesis of both the regional myocardial dysfunction (stunning) that follows brief periods of ischemia and the extension of tissue necrosis that occurs with reperfusion following longer periods of coronary artery occlusion. The potential sources of such ROS have been detailed previously. Although the role of neutrophil activation in the etiology of contractile dysfunction following brief ischemic episodes is uncertain, overwhelming evidence, reviewed above, argues for the participation of neutrophils in the cellular necrosis attributed to reperfusion of the myocardium. Multiple mechanisms exist whereby this damage can be mediated, including prolongation of ischemic injury per se by occlusion of the microvasculature, by direct endothelial injury, and/or by changes in vascular resistance that may contribute to contractile dysfunction.[380-384] In addition, the early infiltration of neutrophils into the myocardial tissue upon reperfusion[15] and the potential for myocyte injury by ROS generated by neutrophils make it likely that direct cytotoxic neutrophil-myocyte interactions are etiologically important in the extension of myocardial necrosis. The actual biochemical derangements responsible for oxidant injury of myocytes may be qualitatively similar in both contractile dysfunction and overt cell necrosis. Alternatively, due to the localization of neutrophil-generated reactive oxygen at the neutrophil-myocyte interface, injury in overt cell necrosis may be dictated by sarcolemmal membrane localization of lipid or protein oxidative events or by intracellular diffusion of H_2O_2. The discussion below (summarized in Figure 6) regarding myocyte cellular pathophysiologic consequences of ROS is not intended to be a complete or definitive analysis, but simply to provide a relatively mechanistic approach to understanding how neutrophils may irreversibly injure myocytes. The molecular basis for an adhesive interaction between neutrophils and myocytes has recently been elucidated and will be considered in detail here with regard to neutrophil oxidative injury to myocytes.

The biochemical alterations leading to actual myocyte necrosis involve cellular processes similar to those that have been studied in detail in ischemic or xenobiotic injury of myocytes, hepatocytes, and a variety of other cell types. These changes include decreases in cellular ATP levels, decreased activity of membrane ATPases responsible for maintaining intracellular ion concentrations, and membrane permeability changes as a result of oxidative processes.[385] Sodium (Na^+) accumulation as a result of increased membrane permeability and decreased Na^+/K^+-ATPase activity is accompanied by increases in intracellular Ca^{2+} due to a decline in Na^+-Ca^{2+} exchange. The calcium increase is confounded by impaired activity of Ca^{2+}-ATPases in the plasma membrane or in intracellular organelles, such as mitochondria or endoplasmic reticulum (sarcoplasmic reticulum in myocytes), that are responsible for buffering intracellular cytoplasmic calcium levels. Intracellular calcium accumulation appears to represent a common final pathway in cellular necrosis.[386] This increased intracellular Ca^{2+} activates phospholipases and proteases and produces cytoskeletal alterations that ultimately result in the morphologic characteristics of irreversible cellular injury and overt necrosis (Figure 6).

Many of the cellular changes typical of injury that lead to cell death are initiated by ischemia per se, and if of sufficient duration, will lead to irreversible injury.[1,11-13] With restoration of blood flow, altered ion homeostasis and further mitochondrial injury can be caused by chemical reactions initiated

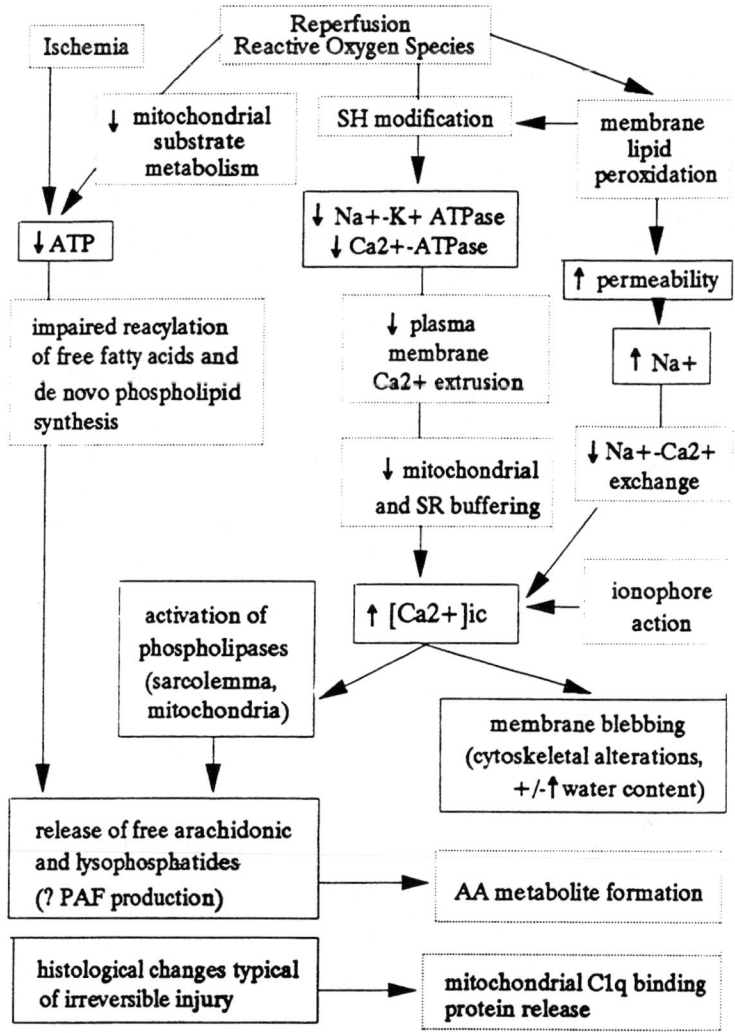

FIGURE 6. Ischemia and reperfusion induce several biochemical changes (shown in dashed boxes) that lead to cellular structural or functional abnormalities (shown in solid boxes). See text for details of these processes.

by ROS, including those derived from neutrophils that infiltrate areas of myocardium that are still viable. Studies with myocytes or subcellular or- ganelles *in vitro* and reactive oxygen-generating systems or reagent peroxides have indicated that ROS-mediated injury can be due to adverse effects on multiple cellular processes. Independent effects of reactive oxygen on mem- brane lipids, plasma membrane, and sarcoplasmic reticular ATPases, and mitochondrial metabolism, when combined, can cause massive accumulation of intracellular calcium. Neutrophil-derived oxidants during neutrophil-myocyte interactions would likely similarly affect multiple cell targets, as indicated in Figure 6.

As also indicated in Figure 6, perturbation of membrane ATPase function can occur as a secondary result of ATP depletion or membrane lipid peroxidation or as a direct consequence of oxidative modification of sulfhydryl groups crucial to enzyme function. Exposure of the isolated-perfused rat heart, heart tissue slices, or isolated myocytes to HOCl or H_2O_2 results in significant oxidation of methionine and cysteine residues in cellular proteins.[387] Overall, heart tissue was observed to be more susceptible than lung tissue, and HOCl was more potent than H_2O_2.[387] In a study addressing the role of reactive oxygen species in the reduction of sarcolemmal Na^+/K^+-ATPase activity upon reperfusion, 2 h of ischemia in the isolated guinea pig heart following by 1 h of reperfusion was found to result in reduced Na^+/K^+-ATPase activity, decreased specific ^3H-labeled oubain binding to this enzyme in ventricular homogenates, and a concomitant increase in tissue malondialdehyde (MDA) concentration.[388] The adverse effects of ischemia and reperfusion on Na^+/K^+-ATPase activity were inhibited to various degrees by administration of SOD, catalase, DMSO, histidine, and vitamin E;[388] the extent of protection of ATPase activity was found to correlate with the inhibition of MDA production,[388] which indicates that ROS-induced lipid peroxidation of sarcolemmal membranes may adversely affect the Na^+/K^+-ATPase. In support of the hypothesis that sarcolemmal membrane enzymes may be inhibited secondary to lipid peroxidation, Kramer et al.[389] found that adult canine myocytes exposed to a free radical-generating system (dihyroxyfumarate and Fe^{3+}-ADP) showed loss of enzyme activities of both sarcolemmal and microsomal membranes along with associated increases in lipid peroxidation as measured by MDA formation. A positive correlation existed between MDA production and the percentage inhibition of sarcolemmal Na^+/K^+-ATPase activity and microsomal NADH cytochrome c reductase activity, but the rates of lipid peroxidation and enzyme inhibition were much greater for the sarcolemma membrane,[389] suggesting a greater sensitivity of this membrane to peroxidative injury. In contrast, Scherer et al.[390] found that sarcoplasmic reticulum (SR) microsomes from lobster abdominal muscle that were oxidized by exposure to peroxy disulfate, H_2O_2, or iron/ascorbate had a decline in Ca^{2+}-ATPase activity and an increase in Ca^{2+} permeability that did not correlate with the amount of MDA formed and that was not prevented by the lipid antioxidants α-tocopherol or butylated hydroxyanisole. Instead, the detected decline in sulfhydryl content of the oxidized SR microsomes and the ability of the reducing agents dithiothreitol, glutathione, and β-mercaptoethanol to prevent the decline in Ca^{2+}-ATPase activity indicated that oxidation of SH groups might be responsible for the adverse affect of SR oxidation.[390] This is consistent with the known biochemistry of the Ca^{2+}-ATPase, the activity of which is dependent on certain essential −SH groups and can be inhibited by SH-binding agents.[391] Hence, redox regulation of Ca^{2+}-ATPases in the SR and plasma membrane, which are essential for maintaining intracellular Ca^{2+}

homeostasis, provides a possible mechanism whereby an intracellular or extracellular oxidative stress might adversely affect cell viability.

Rowe et al.[392] found that PMA-stimulated human peripheral leukocytes inhibited Ca^{2+} uptake and Ca^{2+}-stimulated Mg^{2+}-dependent ATPase activity in vesicles prepared from canine heart SR. This inhibition was blocked by a combination of SOD plus either catalase or 100 μM sodium azide, indicating that the responsible agent was a reactive oxygen species, perhaps derived from myeloperoxidase. In contrast, exogenous H_2O_2 uncoupled Ca^{2+} uptake activity from ATP hydrolysis.[392] Furthermore, the cyclooxygenase inhibitors indomethacin and ibuprofen prevented ATPase inhibition by stimulated leukocytes but did not prevent the depression of Ca^{2+} accumulation. However, indomethacin + catalase blocked both effects of stimulated leukocytes.[392] Based on these results, the authors suggested that H_2O_2 was responsible for depressing Ca^{2+} uptake but that hydroxyl radical, perhaps generated by cyclooxygenase metabolism of arachidonic acid, was responsible for ATPase enzyme inhibition. The leukocyte preparation employed in these studies included 10% monocytes, a cell type with the capacity to produce cyclooxygenase products. Although the specific results obtained may have reflected the experimental design used in these studies, the data indicate that ROS can suppress Ca^{2+} sequestration by myocyte SR. Similar results were recently obtained by Kaminishi et al.[393] in an examination of the sensitivity of mitochondrial and SR Ca^{2+} uptake to H_2O_2 and HOCl. In saponin-permeabilized rat cardiac myocytes, HOCl inhibited Ca^{2+} uptake by mitochondria and SR at 180 nM (similar to diastole) intracellular Ca^{2+} concentration. At higher intracellular Ca^{2+} concentrations (750 nM), SR Ca^{2+} uptake was still sensitive to HOCl, but the mitochondrial uptake, perhaps representing a biochemically distinct system with lower affinity but higher capacity, was relatively resistant. In contrast, both mitochondrial and SR Ca^{2+} uptakes were less sensitive to H_2O_2.[393]

Although studies using isolated organelles or permeabilized cells may provide some insight into intracellular actions of ROS, their relevance to the situation occurring with intact cells that possess a variety of antioxidant defense mechanisms is uncertain. Although neutrophil-generated H_2O_2 could easily diffuse into adjacent myocytes, HOCl, implicated as important in ATPase inhibition in these studies, perhaps by $-SH$ modification,[387] is highly reactive and likely short-lived. HOCl can react with nitrogen-containing compounds to form chloramines that retain oxidizing activity, some of which are relatively long-lived.[174] Reaction with taurine, which is present in high concentrations in neutrophils, can form a relatively stable chloramine.[174] *N*-Chlorotaurine is a hydrophilic molecule capable of oxidizing sulfhydryl groups,[174] and hence, like HOCl, might be expected to react with plasma membrane enzymes. In contrast, reaction of HOCl with other molecules might form more lipophilic products capable of diffusing into cells. Reaction of HOCl with ammonium ion forms monochloroamine, a lipophilic product with potent oxidizing properties.[174] The importance of such species in myocyte injury will require more detailed investigations, particularly with neutrophils and intact myocytes. Preliminary studies with intact myocytes and plasma membrane enzymes, however, have indicated that H_2O_2 adversely affects activities of the sarcolemmal Na^+/K^+-ATPase and Na^+-Ca^{2+} exchange, and that HOCl can indeed impair intracellular calcium homeostasis.[393-395]

In addition to the potential adverse effect on membrane-localized enzymes, such as the sarcolemmal membrane Na^+/K^+-ATPase and plasma membrane, mitochondrial, and sarcoplasmic reticular Ca^{2+}-ATPases, the peroxidative destruction of cellular membranes will compromise cellular function through a general increase in permeability to ions, such as sodium, and ultimately to cytoplasmic constituents including macromolecules due to loss of integrity of the membrane phospholipid bilayer. Studies demonstrating the susceptibility of isolated plasma membranes or membranous organelles to peroxidative destruction must be interpreted with caution when addressing mechanisms of injury to intact cells, which possess significant capacities to detoxify peroxides.[396] Few studies exist, as indicated above, that adequately address the occurrence of peroxidative membrane destruction as a consequence of myocardial reperfusion. However, even if the amounts of reactive oxygen that are generated by target tissues (myocytes, endothelium) immediately upon reperfusion are insufficient to overwhelm antioxidant defenses, membrane peroxidation might still occur as a consequence of postischemic inflammatory neutrophil infiltration and activation that begins soon after reperfusion[15,231] and appears to continue over the first several hours following restoration of blood flow.[24,283] Neutrophils adherent to endothelium or stimulated with PMA release massive amounts of ROS (> 100 nmol $H_2O_2/10^6$ cells) over a period of 1 h or more.[209,247] Under very specific conditions of stimulation described later, neutrophils also will adhere to myocytes and undergo a sustained respiratory burst. Myocyte antioxidant defenses against membrane peroxidation, particularly after prolonged ischemia, may be sufficiently overwhelmed in such circumstances and allow initiation and propagation of membrane lipid peroxidation. The cellular consequences of such a plasma membrane attack will be similar to those already described for ATPase inhibition, including an increase in intracellular Na^+ and ultimately Ca^{2+} concentration, as summarized in Figure 6.

Induction of lipid peroxidation, documented by measurement of conjugated dienes, following treatment of isolated beating adult rat myocytes with either cumene hydroperoxide (100 μM) or the GSH-depleting reagent diamide (200 μM) resulted in severe myocyte ultrastructural changes.[397] Blebbing of plasma membranes was evident within 30 min and occurred more rapidly with higher concentrations of either reagent. Examination of cells by transmission electron microscopy revealed a loss of normal transverse striations, with cells appearing shortened and contracted and having randomly distributed plasma membrane blebs that occasionally appeared to be released from cells. Mitochondria contained inclusions consisting of electron-dense material between the cristae membranes and amorphous matrix densities. As expected, these myocytes demonstrated profound increases in membrane permeability. Cells treated with concentrations of either diamide or cumene hydroperoxide that resulted in these morphological changes had profound decreases in ATP content that could be partially prevented by pretreatment with the antioxidant, butylated hydroxytoluene. In addition, myocytes exposed to lipid peroxides generated with liver microsome preparations underwent a slow decrease in ATP concentrations.

Similar severe plasma membrane and mitochondrial injury can be induced in isolated myocytes by addition of hydrogen peroxide.[397,398] ATP decreases under such conditions may result from increased consumption as membrane ATPases attempt to maintain intracellular ion concentrations and from decreased production, due to impaired oxidative phosphorylation as a consequence of mitochondrial injury. Although −SH groups critical to function of membrane ATPases may be oxidized directly by a high concentration of diamide, the fact that peroxide treatment also causes inhibition indicates that reduced ATPase activity may be due to loss of GSH required for maintaining these enzymes in a functional reduced state or to secondary protein alterations resulting from membrane peroxidative damage. The mitochondrial changes seen in peroxide-treated isolated myocytes[397] are characteristic of those that occur in irreversible cell injury.[10,11,24,27] Inhibition of mitochondrial oxidative phosphorylation may be due to membrane lipid changes and/or the accumulation of Ca^{2+}. A contribution of altered energy production to myocyte injury and cell death is evidenced by the fact that plasma membrane morphological changes similar to those described above can be induced by treatment of myocytes with mitochondrial uncoupling agents.[397,399]

As indicated in Figure 6, peroxidative injury to myocyte sarcolemmal membranes may be potentiated by lipid amphophilic molecules, such as lysophosphatidylcholine, that may accumulate during myocardial ischemia.[400] In sarcolemmal membranes prepared from canine cardiac myocytes and exposed to a free radical-generating system, lipid peroxidation, assessed by MDA formation, was accelerated by pretreatment of the sarcolemma with palmitoyl-CoA, palmitoylcarnitine, or lysophosphatidylcholine.[400] This increased rate of membrane lipid peroxidation was accompanied by a greater loss in the activity of the Na^+/K^+-ATPase. Although demonstrated with isolated membrane preparations and, hence, subject to the cautions mentioned above, the results suggest that these amphophilic lipids, which have been demonstrated to accumulate in some studies during ischemia as a result of altered lipid metabolism,[5,401-405] may potentiate membrane injury when ROS are present, such as occurs during neutrophil-myocyte interactions.

Decrease in cellular ATP content is an important biochemical consequence of myocardial ischemia[2] that can be expected to have a deleterious effect on a variety of cellular metabolic processes and contribute, through incompletely delineated mechanisms, to sarcolemmal phospholipid degradation and arachidonate release with the development of irreversible cell death.[108,406] ROS may adversely affect myocyte function so as to lead to further declines in

ATP or to prevent restoration of ATP levels despite a return to normoxic conditions. Neutrophil plugging of the microvasculature would be expected to perpetuate the decline in high-energy phosphate levels initiated by ischemia per se. In addition, ROS derived from neutrophils or generated intracellularly may result in continued depletion of ATP by leading to alterations in intracellular ion concentrations and subsequent increases in ion ATPase activities and by mitochondrial structural injury impeding normal oxidative phosphorylation and high-energy phosphate production. ROS may have more subtle effects on substrate oxidation by myocytes, perhaps without overt signs of myocyte injury. Exposure of isolated adult rat heart cells to xanthine + xanthine oxidase resulted in marked inhibition of the oxidation of exogenous glucose, lactate, and octanoate with an accompanying decline in high-energy phosphate levels.[407] Under the conditions examined, cell viability and morphology were not visibly affected. The metabolic consequences of exposure to reactive oxygen exposure were prevented by SOD and catalase.[407] Similarly, hypoxia-reoxygenation adversely affected myocyte glucose oxidation.[408]

Oxidants cause prominent changes in myocyte morphology. Addition of hydrogen peroxide (0.5 mM) to isolated beating (electrically stimulated) rat myocytes resulted in a marked decrease in the percentage of rod-shaped cells that contracted.[399] This effect was more pronounced at higher extracellular Ca^{2+} concentrations, and was partially reversed by dimethyl sulfoxide. Although in this study depletion of ATP by exposure of the same myocytes to the mitochondrial uncoupling agent carbonyl cyanide *m*-chlorophenylhydrazone resulted in Ca^{2+}-dependent morphologic changes, ATP levels with H_2O_2 treatment were not reported.[399] As described above, exposure of isolated adult rat myocytes to concentrations of diamide or cumene hydroperoxide resulting in morphological changes was accompanied by a severe decline in intracellular ATP that was partially prevented by prior addition of butylated hydroxyanisole.[397] These results indicate a possible role of membrane lipid peroxidation in the reduction of high-energy phosphate levels, but certainly do not exclude the contribution of other pathologic processes. Another report[398] demonstrated that the addition of 300 μM H_2O_2 to cultured adult rat myocytes resulted in blebbing and a loss of cell viability (trypan blue staining) after 30 to 60 min that was accompanied by a significant increase in MDA formation and depletion of ATP. However, pretreatment of cells with N,N'-diphenylenediamine or deferoxamine prevented MDA formation, membrane blebbing, and cell death, but had no observable effect on ATP preservation,[398] indicating a possible mechanism of ATP reduction besides that resulting from membrane peroxidation.

Studies with patch-clamped guinea pig ventricular myocytes demonstrated that exposure to xanthine + xanthine oxidase or H_2O_2 resulted in the activation of ATP-sensitive potassium channels, a decrease in calcium current, and cell contracture due to irreversible inhibition of glycolytic and oxidative metabolic processes.[409] The cellular changes seen upon exposure to ROS-generating systems were similar to those observed with combined administration of mitochondrial and glycolytic inhibitors. Exposure of myocytes to the mitochondrial uncoupler, carbonyl cyanide-*p*-trifluoromethoxyphenylhydrazone (FCCP), and the glycolytic inhibitor, 2-deoxyglucose (2-DG), caused a progressive shortening in action potential duration, a rapid decrease in the amplitude of the voltage-dependent inward Ca^{2+} current, and changes in current-voltage relations consistent with the activation of the ATP-sensitive K^+ current (efflux of K^+ as a consequence of ATP depletion).[409] Arterial perfusion of isolated rabbit ventricular septum or exposure of isolated cardiac myocytes to H_2O_2 or xanthine + xanthine oxidase caused potassium efflux followed by progressive shortening of action potential duration, decreased Ca^{2+} current, changes in current-voltage relations, decrease in developed tension, and increases in resting tension, changes markedly attenuated with SOD and catalase.[409]

The metabolic basis of the activation of ATP-sensitive K^+ channels mediated by ROS was investigated in isolated myocytes permeabilized at one end with saponin. Such channels can be activated by removing substrates sufficient for endogenous ATP generation via glycolysis or oxidative metabolism (and suppressed by readdition of substrates or addition of exogenous ATP).[409] In addition, activation of ATP-sensitive K^+ channels by metabolic inhibitors is accompanied by myocyte shortening. In the presence of either xanthine + xanthine oxidase or H_2O_2, metabolic substrates were ineffective at suppressing the ATP-sensitive K^+ channels following activation of these channels by ATP removal.[409] Cell shortening was also observed with H_2O_2, and was not reversed by addition of substrates for endogenous ATP production. However, both K^+ channel activation and cell shortening were reversed following H_2O_2 exposure upon addition of exogenous ATP. The inhibition of endogenous ATP generation by ROS was demonstrated to be irreversible and separate substrate addition studies revealed that both oxidative phosphorylation and glycolysis were inhibited.[409] Mitochondrial inhibition in isolated cell studies was not due to calcium overload. Inhibition of oxidative metabolism may be due to subtle injury to mitochondria, and inhibition of glycolytic enzymes might result from injury to the nearby sarcolemma. However, these data provide evidence that in addition to more overt direct ROS-mediated damage to sarcolemmal and organelle lipids and proteins (with deleterious effects on intracellular ion homeostasis), biochemical and functional changes similar to those observed in the reperfused postischemic myocardium may also be mediated in part by inhibitory effects of H_2O_2 on myocyte metabolism.

Hence, as summarized in Figure 6, multiple interacting mechanisms exist whereby ROS such as H_2O_2 can cause myocyte dysfunction and ultimately necrosis. Neutrophils, in particular, can generate large quantities of hydrogen peroxide for sustained periods under appropriate stimulatory conditions described previously. The rapid infiltration of neutrophils into the myocardium upon reperfusion,[15,410] mediated at least in part by a CD18-dependent adhesion to and migration through endothelium,[410] would suggest that interactions of myocytes with activated neutrophils may contribute to myocyte injury. Neutrophil-derived oxidants are obvious candidates for the mediators of myocyte injury. Recently, the molecular mechanisms whereby neutrophils specifically adhere to isolated myocytes have been elucidated.[313] Canine neutrophils will adhere to isolated adult canine cardiac myocytes if the neutrophils are stimulated with agents, such as phorbol esters or chemotactic factors (zymosan-activated serum or PAF), that increase the surface expression and/or activate the CD18 complex.[313] Such adhesion also requires the stimulation of myocytes, either with phorbol esters or cytokines, such as interleukin-1 (IL-1), tumor necrosis factor-alpha (TNF-α), or interleukin-6 (IL-6).[313,411] The myocyte response requires 1 to 4 h of incubation with stimulus and requires protein synthesis (i.e., is blocked by treatment with cycloheximide or actinomycin D).[313] Canine cardiac lymph obtained from myocardial regions at risk during ischemia-reflow has been demonstrated to stimulate isolated canine myocytes to support neutrophil adhesion, with activity observable in lymph obtained early in reperfusion and peaking at approximately 90 min into reperfusion.[411a]

The adhesion of stimulated neutrophils to isolated, cytokine- or lymph-stimulated myocytes is specifically blocked by a monoclonal antibody (R15.7) directed against canine neutrophil CD18.[313] As described previously for neutrophil-endothelial[247] and neutrophil-extracellular matrix protein[296] interactions, we have demonstrated that this CD18-dependent adhesion of neutrophils to myocytes results in a delayed and marked activation of the neutrophil respiratory burst.[313] This phenomenon is illustrated in Figure 7, using PAF and IL-1 as stimuli. Significant H_2O_2 production was typically seen only under conditions that promote adhesion (e.g., neutrophils stimulated with chemotactic factors and myocytes stimulated with cytokines).[313] The adhesion-dependent production of H_2O_2 under such conditions is blocked by the anti-CD18 MAb R15.7,[313] as shown in Figure 7.

Myocyte ICAM may be the relevant ligand for neutrophil CD18 in neutrophil-myocyte interactions, as indicated by recent studies with a MAb that recognizes canine ICAM-1.[412,412a] This MAb blocks the adhesion of stimulated neutrophils to cytokine-stimulated myocytes.[412] In addition, Northern blot analyses indicated that the cytokines IL-1 and IL-6 increase the expression in myocytes of an mRNA species that hybridizes with a partial cDNA for canine ICAM-1.[413] This probe recognizes a transcript of the same size in cytokine-stimulated canine endothelium, a cell type well known to express ICAM-1,[301] as indicated above. Hence, during the myocardial inflammatory response initiated by ischemia-reflow, cytokine induction of ICAM-1 expression on the surface of endothelial cells and myocytes may represent an important regulatory mechanism allowing for transendothelial migration of neutrophils into the myocardium and cytotoxic neutrophil-endothelial and neutrophil-myocyte interactions. Studies by Youker et al.[414] employing neutralizing antibodies to cytokines such as IL-1, TNF-α, and IL-6 have shown that IL-6, a potent myocyte stimulus for neutrophil adhesion, may be the

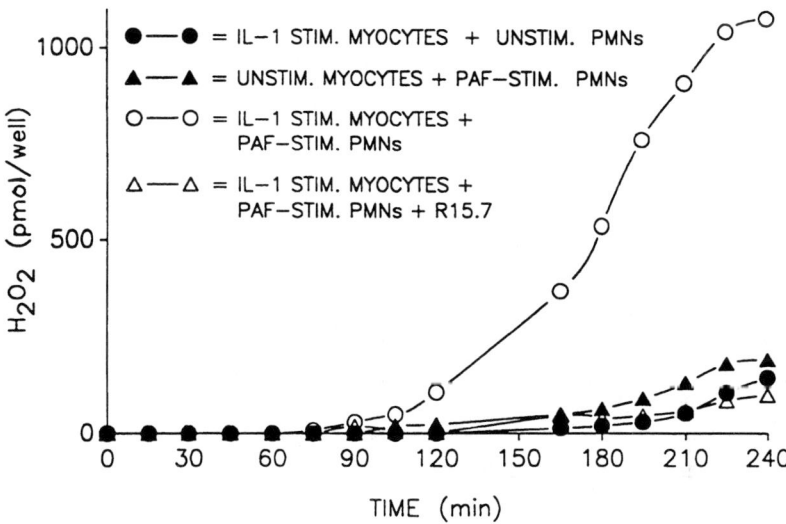

FIGURE 7. H_2O_2 production by canine neutrophils adherent to isolated canine myocytes. Neutrophils and myocytes were coincubated in collagen I-coated 96-well microtiter plates and H_2O_2 production was determined by the oxidation of scopoletin as described.[313] As explained in the text, adhesion requires chemotactic stimulation of neutrophils and cytokine stimulation of myocytes and is mediated by neutrophil CD18 adhesion glycoproteins. This adhesion results in a delayed and sustained neutrophil generation of H_2O_2. H_2O_2 production was not observed with stimulation of either cell type alone (e.g., 200 ng/mL PAF or pretreatment of myocytes with 3 U/mL IL-1 for 3 h at 37°C). The PAF and IL-1 induced neutrophil respiratory burst was blocked by the anti-CD18 MAb R15.7, which also blocks adhesion under the same conditions. From Entman, M. L. et al., *J. Clin. Invest.*, 85, 1497, 1990. With permission.

major factor present in reperfusion cardiac lymph capable of stimulating myocytes to support neutrophil adhesion *ex vivo*. Hence, IL-6 production (by resident macrophages, endothelial cells, or other relevant cell types in the heart) may be an important component regulating the deleterious interaction of infiltrating neutrophils with myocardial cells during the extension of necrosis upon reperfusion.

In addition, recent studies have indicated that, in a manner analogous to cultured endothelium and to perfused arteries and veins, the sublethal (short duration) exposure of isolated canine myocytes to H_2O_2 results in a CD18-dependent adhesion of unstimulated (no exogenous chemotactic factor) neutrophils.[411] This adhesion is blocked by the PAF receptor antagonist WEB-2086, indicating that the mechanism whereby sublethal H_2O_2 exposure results in a myocyte-dependent neutrophil adhesion may be identical to that shown for endothelial-dependent adhesion. As H_2O_2-mediated endothelial and myocyte cytotoxicity requires prolonged incubation *in vitro*, this phenomenon of H_2O_2-induced adhesion (now illustrated for the two most relevant cell types in reperfusion injury) demonstrates one possible mechanism whereby sublethal formation of reactive oxygen (such as that seen immediately upon reperfusion or following initial neutrophil influx and activation) might initiate the sustained specific adhesion of neutrophils that is likely required for ROS-mediated cell killing.

I. ROLE OF LIPID-DERIVED INFLAMMATORY MEDIATORS IN MYOCARDIAL ISCHEMIA-REFLOW INJURY

The lipid-derived chemoattractants that have received the most attention thus far in myocardial inflammation are the lipoxygenase metabolites of arachidonic acid. The most potent of these mediators is LTB_4, which promotes neutrophil chemotaxis and adhesion in the 1 to 10 nM range.[128,415] As mentioned above, experiments demonstrating reduction of neutrophil accumulation and/or infarct size with the administration of agents capable of inhibiting 5-lipoxygenase (5-LOX) has been taken as indirect evidence for the role of LTB_4 in promoting neutrophil accumulation during myocardial reperfusion. However, as also mentioned previously, interpretation of the protective mechanism of such agents has largely been complicated by the fact that most 5-LOX inhibitors are also potent antioxidants,[242] so that such agents might offer protection by scavenging ROS, in a manner analogous to SOD, catalase, or other antioxidants. As described earlier, the 5-LOX enzyme has a nonheme iron at the active site, and inhibitors may function by reducing a required ferric species to ferrous iron. The ability of agents to inhibit 5-LOX by such a redox mechanism does not correlate completely with their antioxidant potencies, indicating that other structural requirements exist (e.g., active site access). Hence, certain compounds, such as REV-5901, are reasonably potent 5-LOX inhibitors (comparable in effectiveness to agents like NDGA), while possessing little or no capacity to scavenge ROS.

Some 5-LOX inhibitors also have been reported to inhibit numerous neutrophil functions elicited by a variety of stimuli *in vitro*.[238,239,241] The mechanism(s) by which such agents inhibited neutrophil function were not clear, but were presumed to be due to the inhibition of LOX. However, the ability of these agents to inhibit neutrophil functions like superoxide production or directed migration did not correlate well with their ability to inhibit leukotriene formation. Furthermore, such agents appeared capable of inhibiting neutrophil responses to stimuli under conditions that do not appear to result in the formation of detectable LOX metabolites.[240,416-419] Since these 5-LOX inhibitors could, however, function by relatively nonspecific mechanisms, they might be capable of preventing neutrophil infiltration and neutrophil-mediated myocardial injury *in vivo* regardless of the factor or factors actually promoting neutrophil adhesion and migration.

We have recently demonstrated that inhibition of chemotactic stimulus-induced neutrophil motility, adhesion, and secretory functions is not a general property of 5-LOX inhibitors.[240] The antioxidant NDGA exhibited a dose-dependent inhibition of fMLP and recombinant C5a-induced neutrophil bipolarization, fMLP-induced Mac-1 (CD11b/CD18) upregulation and aggregation, and neutrophil adherence to and migration through IL-1-stimulated endothelial monolayers; however, the nonantioxidant 5-LOX inhibitor REV-5901 and the nonantioxidant dual COX/LOX inhibitor SK&F 86002 had no effect on these functions or on neutrophil oxidative metabolism at concentrations significantly higher than those required to inhibit 5-LOX.[240] Importantly, both of these agents have been observed to reduce myocardial ischemia-reflow injury in various animal models,[419a] and the inability of these agents to scavenge ROS and to directly inhibit neutrophil function suggests that their protective mechanism may indeed be due to the reduced formation of chemotactic LOX metabolites (especially LTB_4), which work at specific neutrophil receptors. More recently, the latter hypothesis has been reexamined by the use of specific LTB_4 receptor antagonists.

The development of specific LTB_4 receptor antagonists has proved to be a difficult endeavor, perhaps due to the general similarity of structure of LTB_4 to other mono and dihydroxy derivatives of arachidonate, the relatively simple structure overall for this ligand, and difficulty in removing partial agonist properties of candidate agents. However, apparently highly specific antagonists[420-422] have recently become available that are adequate for research purposes. The pharmacokinetic properties of these compounds has thus far made continuous infusion necessary in order to achieve active levels of these highly protein-bound drugs in blood of animals during ischemia-reflow protocols. Hahn et al.[423] recently reported that administration of the LTB_4 antagonist LY 255283 to dogs undergoing 1 h of circumflex artery occlusion followed by 5 h of reperfusion resulted in an infarct size (infarct area/risk area or AI/AR) of $32 \pm 5\%$ vs $43 \pm 5\%$ in controls, which was not a statistically significant difference with eight animals in each group. Furthermore, MPO content, as an index of neutrophil accumulation, in infarcted tissues and tissue at risk was not different between control and treated dogs.[423] In a similar number of rabbits, using the LTB_4 antagonist LY 223982, which specifically antagonizes LTB_4-induced human neutrophil activation *in vitro*,[420] we have observed a small but significant reduction of infarct size (AI/AR = $57.7 \pm 6.6\%$ in treated vs $74.8 \pm 3.7\%$ in control), following 45 min of left circumflex occlusion followed by 3 h of reflow, along with a slight but significant reduction in MPO activity.[424]

The reason for the positive response in only one of these studies with LTB_4 antagonists may reflect slight differences in agents, protocols, or species employed. However, even in our study, the amount of tissue salvage was much less than in previous investigations employing neutrophil depletion or LOX inhibitors. Whether this indicates that LTB_4 is itself only responsible for a minor component of neutrophil activation and that LOX inhibitors have protected via other nondefined mechanisms or that the pharmacologic properties of the antagonists have rendered the protocols employed thus far as yielding submaximal receptor antagonism will require continued analysis.

Recently a class of indole-derived compounds that are extremely potent and thus far are highly specific for inhibition of leukotriene formation have been identified and developed. Such agents, which have been termed leukotriene biosynthesis inhibitors[425,426] should be highly useful in further clarifying the role of LOX metabolites in ischemia-reflow injury and other inflammatory conditions as well. These compounds, the prototype of which is designated MK-886, both prevent and reverse the membrane translocation of 5-lipoxygenase,[426] previously demonstrated to occur in ionophore-stimulated neutrophils and postulated to be a key step in enzyme activation.[166] These agents have no effect on 5-lipoxygenase activity in cell-free systems, demonstrating further that their mechanism of inhibition of leukotriene formation in intact cells is indeed by preventing association of the enzyme with cellular membranes upon increases in intracellular Ca^{2+}. This novel mechanism results in these agents being about three orders of magnitude more potent than previous redox acting agents in inhibiting leukotriene production by intact neutrophils,[425] and MK-886, which is orally active, has been shown to have potent anti-inflammatory properties in a variety of *in vivo* inflammatory models,[425] including antigen-induced bronchoconstriction in monkeys[425] and guinea pig anaphylaxis.[427] Furthermore, these agents have made possible the identification of a specific membrane protein of 18 kDa to which MK-886 binds and which is necessary for leukotriene production in intact cells.[428,429] The cDNA for the rat and human MK-886 binding protein indicates a protein whose sequence is consistent with three transmembrane domains and which may represent a specific anchoring protein (termed 5-lipoxygenase anchoring protein or FLAP) for 5-LOX translocation.[429] Preliminary reports indicate that the FLAP binding MK-886 is specific in its anti-inflammatory mechanism and does not directly inhibit rat neutrophil adhesive or aggregatory responses,[425] indicating that it may indeed be a useful compound for further delineating the role for chemotactic LOX metabolites in myocardial reperfusion injury.

V. SUMMARY AND FUTURE DIRECTIONS

The goal of current therapy in myocardial infarction is to restore blood flow as quickly as possible (with thrombolytic agents or angioplasty) so as to reduce the amount of myocardium irreversibly injured by ischemia. Since reperfusion itself contributes to the extent of myocardial necrosis, adjunctive therapy to minimize reperfusion injury administered at the time of blood flow restoration might allow for a further reduction in infarct size, which is the single most important prognostic factor in patient survival following acute myocardial infarction.[430] The demonstration of myocardial salvage in animal models of ischemia-reflow injury by prior neutrophil depletion or by treatment with pharmacologic agents that limit neutrophil infiltration indicates the important role for neutrophil activation in reperfusion injury and suggests that rational therapy aimed at preventing neutrophil-mediated necrosis also might be effective in humans. This chapter has examined the importance of neutrophil adhesion molecules and neutrophil-generated ROS in the adhesive and cytotoxic interactions of neutrophils with the vascular endothelium and myocardial tissue. Therapy aimed at reducing neutrophil-mediated injury thus might be directed at multiple steps in the process of neutrophil activation, including: (1) the neutrophil adhesion molecules themselves or their target tissue ligands; (2) the inflammatory mediators responsible for neutrophil activation, adhesion to the vasculature, and migration into adjacent tissue; and (3) the stimulation of secretory responses during neutrophil-target cell interactions, including the generation of ROS that occurs as a consequence of neutrophil activation.

Studies performed *in vivo* have already indicated that inhibition of neutrophil adhesion molecules, specifically CD11/CD18 integrins, may offer one feasible approach to myocardial salvage. Reduction of infarct size in the dog following 6 or 72 h of reperfusion by anti-CD11b monoclonal antibody (MAb) 904 IgG[281] or F(ab')$_2$ fragments,[277] respectively, may be related not only to inhibition of myocardial neutrophil accumulation, but also to inhibition of adhesion-dependent processes, such as directed migration,[320] neutrophil aggregation,[320] and adhesion-dependent secretion of ROS.[247] Other studies have been performed with MAbs directed against the common beta subunit (CD18) of leukocyte integrins. Reimer et al.[431] reported that the anti-CD18 MAb, IB4, reduced neutrophil accumulation in the canine heart following 90 min of circumflex artery occlusion and 3 h of reperfusion, but no infarct size reduction was evident at this early time point. In studies addressing both the time course and transmural distribution of neutrophil accumulation in the canine myocardium during reperfusion, Dreyer et al.[410] observed that neutrophil accumulation was greatest in the first hour of reperfusion, gradually declining in subsequent hours. Neutrophil accumulation at 1 h of reperfusion occurred preferentially at the subendocardial surface, and was inversely related to regional myocardial blood flow. Compared to 1-h reperfusion controls, animals treated with the anti-CD18 MAb R15.7 (1 mg/kg iv) prior to occlusion showed significant attenuation of neutrophil localization in the subendocardial region.[431a] Administration of an anti-CD18 MAb (R3.3) to rabbits 15 min before a 60-min occlusion of the left anterior descending coronary artery followed by 5 h of reperfusion resulted in a reduction of infarct size without affecting circulating leukocyte counts.[290] Use of a MAb directed against CD11a (the alpha subunit of LFA-1) also reduced infarct size in the same protocol, but was not effective if given only during reperfusion.[290] The use of an antibody (R6.5) directed against a potential target for CD18-dependent adhesion, ICAM-1, reduced infarct size but also resulted in marked decreases in leukocyte counts during the treatment period, raising concerns about mechanism of protection.[290] Similar studies with F(ab')$_2$ preparations of anti-ICAM-1 antibodies may be particularly informative.

Results of such studies using antibodies against different CD11/CD18 subunits or different MAbs (perhaps recognizing distinct epitopes) to the same subunits as well as MAbs against potential endothelial or other cell type ligands will likely be quite useful in identifying targets for therapeutic interventions aimed at reducing infarct size. Complementary studies *in vitro* to further delineate crucial functional domains of adhesion molecules (employing specific epitope-directed antibodies, site-directed mutagenesis to specifically alter regions of adhesion molecules, and synthetic peptides based on primary or predicted tertiary structures of adhesion molecules) may ultimately allow for development of specific lower molecular weight therapeutic agents more suitable for routine human administration.

We have emphasized the importance of neutrophil adhesion, specifically, CD18-mediated adhesion, to extracellular matrix proteins, endothelium, and myocytes themselves as an important stimulus for activation of neutrophil reactive oxygen production. Given the large quantities of superoxide, H_2O_2, and metabolites, such as hypohalous acids or hydroxyl radical, that might be generated under such conditions, antioxidant interventions might be capable of minimizing the extent of myocardial necrosis that is a direct result of neutrophil activation. Numerous examples for effectiveness in myocardial salvage of SOD and/or catalase and other strategies aimed at reducing reactive oxygen damage have been cited here. Recent studies in our laboratories into the nature of neutrophil-myocyte oxidative interactions have indicated that such processes might not be accessible to SOD and catalase. Oxidation of myocyte intracellular fluorescent probes as a result of neutrophil CD18-dependent adhesion and respiratory burst activation is not inhibited by SOD and catalase.[431b] The potential localization of neutrophil NADPH oxidase activation at the site of cell-cell contact[298] perhaps with direct diffusion of species such as H_2O_2 into myocytes and the potential protection of such a microenvironment from large molecular weight molecules in the extracellular fluid may limit the ability of antioxidant enzymes to intervene. As such, reduction of necrosis by SOD and catalase *in vivo* may not occur through the direct scavenging of neutrophil-generated oxidants. One possible alternative is that other sources (such as xanthine oxidase, fatty acid oxygenases, or mitochondria after reperfusion) generate a sufficient magnitude of ROS to produce myocardial necrosis and such ROS have an extracellular "existence" sufficient to render them susceptible to antioxidant enzymes. A related alternative is that such interventions achieve an "anti-inflammatory effect" by elimination of extracellular ROS that contribute to subsequent neutrophil-mediated injury.

Recent attention to the pathologic significance of the quantities and sources of ROS generated at the onset of reperfusion[253,259,260] and ongoing elucidation of the molecular mechanisms whereby sublethal quantities of ROS might exert more subtle proinflammatory effects[162,411,432] have suggested numerous mechanisms whereby antioxidant enzymes or other antioxidant interventions might result in tissue salvage. Any reduction in neutrophil accumulation and neutrophil-mediated injury with antioxidant interventions *in vivo* would not necessarily imply a causative role of ROS in the molecular mechanisms of neutrophil adhesion and chemotaxis. Reduction in tissue necrosis, per se, via inhibition of oxidative injury might result in a secondary decrease in neutrophil accumulation through the decreased formation of chemotactic factors, such as generated via activation of the complement cascade. However, direct demonstration with intravital microscopy of reduced neutrophil adhesion to venular endothelium following administration of SOD during postischemic reperfusion in the feline intestine[338] and the studies *in vitro* demonstrating H_2O_2-induced endothelial-dependent neutrophil adhesion in our laboratory and by others[162] indicate potential mechanisms whereby SOD- and catalase-inhibitable transformation might initiate or potentiate neutrophil-mediated cytotoxic interactions. A possible role for PAF and CD18-dependent adhesion in these processes has been detailed above. Continued elucidation of the precise mechanisms whereby sublethal quantities of ROS might generate and amplify neutrophil-mediated inflammatory responses and the relative importance of such processes *in vivo* will allow for more specifically designed therapeutic interventions, both in terms of the agents employed and the times at which their administration might be most beneficial.

The chemotactic factors responsible for activating neutrophils during reperfusion injury are not clearly established. Neutrophils possess receptors for a number of clearly defined chemotactic ligands, including fMLP,[433] the C5a fragment of the fifth component of complement,[326,434] LTB_4,[435] peptide chemoattractants such as IL-8 (also described previously as neutrophil-activating factor or NAF),[436,437] and PAF.[438,439] Although subtle differences may exist in the cellular responses to these agents, including some species dependencies, the biochemical changes and, hence, the neutrophil activation responses initiated by these various agents are, by and large, essentially the same. The neutrophil receptors for chemoattractants appear to be coupled by a G-protein (similar if not identical to G_i) to a phospholipase C, such that cellular responses are mediated by phospholipid hydrolysis and subsequent intracellular responses, such as Ca^{2+} mobilization, initiated by the second messengers formed (e.g., inositol trisphosphate). These chemoattractants promote directed migration and adhesion at very low concentrations (e.g., 1 to 10 nM for LTB_4) with weak secretion of reactive oxygen and granule contents achieved only at much higher concentrations (e.g., 1 μM). In addition, neutrophils respond to a variety of proteins that may be present during inflammation, derived from monocytes, endothelium, or other cells, including TNF, granulocyte-macrophage colony-stimulating factor (GM-CSF), and platelet-derived growth factor.

It is obviously very likely that multiple soluble mediators contribute to neutrophil activation during myocardial inflammation. The same agents that promote initial neutrophil contact with and migration through endothelium are also capable of promoting directed migration in inflamed tissue, a process requiring adhesive contacts with appropriate cellular or extracellular matrix supports. Neutrophil receptors for chemoattractants appear to undergo rapid homologous desensitization,[438] such that continued migration may require increased concentrations of the acting ligand, a phenomenon in keeping with the capacity of neutrophils to migrate in an increasing gradient of chemotactic stimulus. In addition, continued cell activation, adhesion-contact formation, and cellular contraction leading to movement might occur in the presence of different (heterologous) stimulating ligands. The multiple mediators contributing to neutrophil stimulation during the evolution of myocardial inflammation may be generated differentially, both temporally and spatially (e.g., intravascular, endothelial, myocardium, or accumulated neutrophils). Interfering at any level with such a complicated interactive process may provide substantial amelioration of the inflammatory response.

REFERENCES

1. **Jennings, R. B. and Reimer, K. A.,** Lethal myocardial ischemic injury, *Am. J. Pathol.,* 102, 241, 1981.
2. **Jennings, R. B., Reimer, K. A., Hill, M. L., and Mayer, S. E.,** Total ischemia in dog hearts, *in vitro,* comparison of high energy phosphate production, utilization, and depletion, and of adenine nucleotide catabolism in total ischemia *in vitro* vs. severe ischemia *in vivo, Circ. Res.,* 49, 892, 1981.
3. **Schaefer, S., Camacho, S. A., Gober, J., Obregon, R. G., DeGroot, M. A., Botvinick, E. H., Massie, B., and Weiner, M. W.,** Response of myocardial metabolites to graded regional ischemia: ^{31}P NMR spectroscopy of porcine myocardium *in vivo, Circ. Res.,* 64, 968, 1989.
4. **Siegmund, B., Koop, A., Klietz, T., Schwartz, P., and Piper, H. M.,** Sarcolemmal integrity and metabolic competence of cardiomyocytes under anoxia-reoxygenation, *Am. J. Physiol.,* 258, H285, 1990.
5. **Chien, K. R., Reeves, J. P., Buja, L. M., Bonte, F., Parkey, R. W., and Willerson, J. T.,** Phospholipid alterations in canine ischemic myocardium: temporal and topographical correlations with Tc-99m-PPi accumulation and an *in vitro* sarcolemmal Ca^{2+} permeability defect, *Circ. Res.,* 48, 711, 1981.
6. **Lamers, J. M. J., Post, J. A., Verkleij, A. J., Ten Cate, F. J., van der Giessen, W. J., and Verdouw, P. D.,** Loss of functional and structural integrity of the sarcolemma: an early indicator of irreversible injury of myocardium? *Biomed. Biochim. Acta,* 46, S517, 1987.
7. **Duan, J. and Karmazyn, M.,** Relationship between oxidative phosphorylation and adenine nucleotide translocase activity of two populations of cardiac mitochondria and mechanical recovery of ischemic hearts following reperfusion, *Can. J. Physiol. Pharmacol.,* 67, 704, 1989.
8. **Shin, G., Sugiyama, M., Shoji, T., Kagiyama, A., Sato, H., and Ogura, R.,** Detection of mitochondrial membrane damages in myocardial ischemia with ESR spin labeling technique, *J. Mol. Cell. Cardiol.,* 21, 1029, 1989.
9. **Kloner, R. A., Ellis, S. G., Lange, R., and Braunwald, E.,** Studies of experimental coronary artery reperfusion: effect on infarct size, myocardial function, biochemistry, ultrastructural and microvascular damage, *Circulation,* 68 (Suppl. I), I-8, 1983.
10. **Kloner, R. A., Rude, R. E., Carlson, N., Maroko, P. R., DeBoer, L. W. V., and Braunwald, E.,** Ultrastructural evidence of microvascular damage and myocardial cell injury after coronary artery occlusion: Which comes first? *Circulation,* 62, 945, 1980.
11. **Jennings, R. B., Sommers, H. M., Smyth, G. A., Flack, H. A., and Linn, H.,** Myocardial necrosis induced by temporary occlusion of a coronary artery in the dog, *AMA Arch. Pathol.,* 70, 68, 1960.
12. **Jennings, R. B., Ganote, C. E., and Reimer, K. A.,** Ischemic tissue injury, *Am. J. Pathol.,* 81, 179, 1975.

13. **Schaper, J., Mulch, J., Winkler, B., and Schaper, W.**, Ultrastructural, functional, and biochemical criteria for estimation of reversibility of ischemic injury: a study on the effects of global ischemia on the isolated dog heart, *J. Mol. Cell. Cardiol.*, 11, 521, 1979.

14. **Reimer, K. A., Lowe, J. E., Rasmussen, M. M., and Jennings, R. B.**, The wavefront phenomenon of ischemic cell death, myocardial infarct size vs duration of coronary occlusion in dogs, *Circulation*, 56, 786, 1977.

15. **Engler, R. L., Schmid-Schonbein, G. W., and Pavelec, R. S.**, Leukocyte capillary plugging in myocardial ischemia and reperfusion in the dog, *Am. J. Pathol.*, 111, 98, 1983.

16. **Schmid-Schonbein, G. W.**, Capillary plugging by granulocytes and the no-reflow phenomenon in the microcirculation, *Fed. Proc. Fed. Am. Soc. Exp. Biol.*, 46, 2397, 1987.

17. **Schaper, W., Gorge, G., Winkler, B., and Schaper, J.**, The collateral circulation of the heart, *Prog. Cardiovasc. Dis.*, 31, 57, 1988.

18. **Heyndrickx, G. R., Millard, R. W., and McRitchie, R. J.**, Regional myocardial function and electrophysiological alterations after brief coronary artery occlusion in conscious dogs, *J. Clin. Invest.*, 56, 978, 1975.

19. **Weiner, J. M., Apstein, C. S., and Arthur, J. H.**, Persistence of myocardial injury following brief periods of coronary occlusion, *Cardiovasc. Res.*, 10, 678, 1976.

20. **Braunwald, E. and Kloner, R. A.**, The stunned myocardium: prolonged, postischemic ventricular dysfunction, *Circulation*, 66, 1146, 1982.

21. **Bolli, R., Zhu, W.-X., Thornby, J. I., O'Neill, P. G., and Roberts, R.**, Time course and determinants of recovery of function after reversible ischemia in conscious dogs, *Am. J. Physiol.*, 254, H102, 1988.

22. **Ambrosio, G., Becker, L. C., Hutchins, G. M., Weisman, H. F., and Weisfeldt, M. L.**, Reduction in experimental infarct size by recombinant human superoxide dismutase: insights into the pathophysiology of reperfusion injury, *Circulation*, 74, 1424, 1986.

23. **Ambrosio, G., Weisman, H. F., and Becker, L. C.**, The "no-reflow" phenomenon: a misnomer? *Circulation*, 74, II-260, 1986.

24. **Kloner, R. A., Ellis, S. G., Lange, R., and Braunwald, E.**, Studies of experimental coronary artery reperfusion: effects on infarct size, myocardial function, biochemistry, ultrastructure, and microvascular damage, *Circulation*, 58 (Suppl. I), I-1, 1985.

25. **The I.S.A.M. Study Group**, A prospective trial of intravenous streptokinase in acute myocardial infarction (I.S.A.M.), *N. Engl. J. Med.*, 314, 1465, 1986.

26. **Koren, G., Weiss, A. T., Hasin, Y., Appelbaum, D., Welber, S., Rozenman, Y., Lotan, C., Mosseri, M., Sapoznikov, D., Luria, M. H., and Gotsman, M. S.**, Prevention of myocardial damage in acute myocardial ischemia by early treatment with intravenous streptokinase, *N. Engl. J. Med.*, 313, 1384, 1985.

27. **Hearse, D. J., Humphrey, S. M., Nayler, W. G., Slade, A., and Border, D.**, Ultrastructural damage associated with reoxygenation of the anoxic myocardium, *J. Mol. Cell. Cardiol.*, 7, 315, 1975.

28. **Hearse, D. J., Humphrey, S. M., and Chain, E. B.**, Abrupt reoxygenation of the anoxic potassium-arrested perfused rat heart: a study of myocardial enzyme release, *J. Mol. Cell. Cardiol.*, 5, 395, 1973.

29. **Mullane, K. M.**, Oxygen-derived free radicals and reperfusion injury of the heart, in *New Frontiers in Cardiovascular Therapy: Focus on ACE Inhibition*, Sonnenblick, Laragh, and Lesch, Eds., Excerpta Medica, Princeton, NJ, 1989, 234.

30. **Halliwell, B. and Gutteridge, J. M. C.**, Lipid peroxidation: a radical chain reaction, in *Free Radicals in Biology and Medicine*, Halliwell, B. and Gutteridge, J. M. C., Eds., Clarendon Press, Oxford, 1989, 188.

31. **Gutteridge, J. M. C. and Halliwell, B.**, The measurement and mechanism of lipid peroxidation in biological systems, *TIBS*, 15, 129, 1990.

32. **Halliwell, B. and Gutteridge, J. M. C.**, Oxygen free radicals and iron in relation to biology and medicine: some problems and concepts, *Arch. Biochem. Biophys.*, 246, 501, 1986.

33. **Boveris, A. and Chance, B.**, The mitochondrial generation of hydrogen peroxide, *Biochem. J.*, 134, 707, 1973.

34. **Vandeplassche, G., Hermans, C., Thoné, F., and Borgers, M.**, Mitochondrial hydrogen peroxide generation in NADH-oxidase activity following regional myocardial ischemia in the dog, *J. Mol. Cell. Cardiol.*, 21, 383, 1989.

35. **Vandeplassche, G., Thoné, F., and Borgers, M.**, Cytochemical evidence of NADH-oxidase activity in the isolated working rabbit heart subjected to normothermic global ischaemia, *Histochem. J.*, 22, 11, 1990.

36. **Turrens, J. F. and Boveris, A.**, Generation of superoxide anion by the NADH dehydrogenase of bovine heart mitochondria, *Biochem. J.*, 191, 421, 1980.

37. **Jarasch, E.-D., Bruder, G., and Heid, H. W.**, Significance of xanthine oxidase in capillary endothelial cells, *Acta Physiol. Scand. Suppl.*, 548, 39, 1986.

38. **Jarasch, E.-D., Grund, C., Bruder, G., Heid, H. W., Keenan, T. W., and Franke, W. W.**, Localization of xanthine oxidase in mammary-gland epithelium and capillary endothelium, *Cell*, 25, 67, 1981.

39. **McCord, J. M.**, Oxygen-derived free radicals in post-ischemic tissue injury, *N. Engl. J. Med.*, 312, 159, 1985.

40. **Downey, J. M., Hearse, D. J., and Yellon, D. M.**, The role of xanthine oxidase during myocardial ischemia in several species including man, *J. Mol. Cell. Cardiol.*, 20 (Suppl. II), 55, 1988.

41. **McCord, J. M.**, Free radicals and myocardial ischemia: overview and outlook, *Free Radical Biol. Med.*, 4, 9, 1988.

42. **Phan, S. H., Gannon, D. E., Varani, J., Ryan, U. S., and Ward, P. A.**, Xanthine oxidase activity in rat pulmonary artery endothelial cells and its alteration by activated neutrophils, *Am. J. Pathol.*, 134, 1201, 1989.

43. **Pilloud, M.-C., Doussiere, J., and Vignais, P. V.**, The $O_2^-\cdot$ generating oxidase activation of bovine neutrophils: evidence for synergism of multiple cytosolic factors in a cell-free system, *FEBS Lett.*, 257, 167, 1989.

44. **Quinn, M. T., Parkos, C. A., and Jesaitis, A. J.**, The lateral organization of components of the membrane skeleton and superoxide generation in the plasma membrane of stimulated human neutrophils, *Biochim. Biophys. Acta*, 987, 83, 1989.

45. **Fujimoto, S., Smith, R. M., Curnutte, J. T., and Babior, B. M.**, Evidence that activation of the respiratory burst oxidase in a cell-free system from human neutrophils is accomplished in part through an alteration of the oxidase-related 67-kDa cytosolic protein, *J. Biol. Chem.*, 264, 21629, 1989.

46. **Della Bianca, V., Grzeskowiak, M., and Rossi, F.**, Studies on molecular regulation of phagocytosis and activation of the NADPH oxidase in neutrophils: IgG- and C3b-mediated ingestion and associated respiratory burst independent of phospholipid turnover and Ca^{2+} transients, *J. Immunol.*, 144, 1411, 1990.

47. **Doussière, J., Pilloud, M.-C., and Vignais, P. V.**, Cytosolic factors in bovine neutrophil oxidase activation. Partial purification and demonstration of translocation to a membrane fraction, *Biochemistry*, 29, 2225, 1990.

48. **Singal, P. K., Kapur, N., Dhillon, K. S., Beamish, R. E., and Dhalla, N. S.**, Role of free radicals in catecholamine-induced cardiomyopathy, *Can. J. Physiol. Pharmacol.*, 60, 1390, 1982.

49. **Newman, W. H., Mathur, P. P., and Walton, R. P.**, Catecholamines and local rebound in left ventricular contractile force after release of coronary artery occlusion, *Cardiovasc. Res.*, 5, 81, 1971.

50. **Grisham, M. B.**, Myoglobin-catalyzed hydrogen peroxide-dependent arachidonic acid peroxidation, *Free Radical Biol. Med.*, 1, 227, 1985.

51. **Mitsos, S. E., Kim, D., Lucchesi, B. R., and Fantone, J. C.,** Modulation of myoglobin-H_2O_2-mediated peroxidation reactions by sulfhydryl compounds, *Lab. Invest.*, 59, 824, 1988.
52. **Mullane, K. M.,** Eicosanoids in myocardial ischemia/reperfusion injury, in *Advanced Inflammation Research*, Lewis, A., Ed., Raven Press, New York, 1988, 191.
53. **Mullane, K. M., Westlin, W., and Kraemer, R.,** Activated neutrophils release mediators that may contribute to myocardial injury and dysfunction associated with ischemia and reperfusion, *Ann. N.Y. Acad. Sci.*, 524, 103, 1988.
54. **Kukreja, R. C., Kontos, H. A., Hess, M. L., and Ellis, E. F.,** PGH synthase and lipoxygenase generate superoxide in the presence of NADH or NADPH, *Circ. Res.*, 59, 612, 1986.
55. **Egan, R. W., Gale, P. H., Baptista, E. M., Kennicott, K. L., VandenHeuvel, W. J. A., Walker, R. W., Fagerness, P. E., and Kuehl, F. A., Jr.,** Oxidation reactions by prostaglandin cyclooxygenase-hydroperoxidase, *J. Biol. Chem.*, 256, 7352, 1981.
56. **Hemler, M. E. and Lands, W. E. M.,** Evidence for a peroxide-initiated free radical mechanism of prostaglandin biosynthesis, *J. Biol. Chem.*, 255, 6253, 1980.
57. **Simpson, P. J. and Lucchesi, B. R.,** Free radicals and myocardial ischemia and reperfusion injury, *J. Lab. Clin. Med.*, 110, 13, 1987.
58. **Kloner, R. A., Przyklenk, K., and Whittaker, P.,** Deleterious effects of oxygen radicals in ischemia/reperfusion: resolved and unresolved issues, *Circulation*, 80, 1115, 1989.
59. **Rasmussen, U. F.,** The oxidation of added NADH by intact heart mitochondria, *FEBS Lett.*, 2, 157, 1969.
60. **Rasmussen, U. F. and Rasmussen, H. N.,** The NADH oxidase system (external) of muscle mitochondria and its role in the oxidation of cytoplasmic NADH, *Biochem. J.*, 229, 631, 1985.
61. **van der Giessen, W. J., Verdouw, P. D., Ten Cate, F. J., Essed, C. E., Rijsterborgh, H., and Lamers, J. M. J.,** *In vitro* cyclic AMP induced phosphorylation of phospholamban: an early marker of long-term recovery of function following reperfusion of ischaemic myocardium? *Cardiovasc. Res.*, 22, 714, 1988.
62. **Ferrari, R., Ceconi, C., Curello, S., Guarnieri, C., Caldarera, C. M., Albertini, A., and Visioli, O.,** Oxygen-mediated myocardial damage during ischemia and reperfusion: role of the cellular defenses against oxygen toxicity, *J. Mol. Cell. Cardiol.*, 17, 937, 1985.
63. **Corte, E. D. and Stirpe, F.,** The regulation of rat liver xanthine oxidase. Involvement of thiol groups in the conversion of the enzyme activity from dehydrogenase (type D) into oxidase (type O) and purification of the enzyme, *Biochem. J.*, 126, 739, 1972.
64. **Waud, W. R. and Rajagopalan, K. V.,** Purification and properties of the NAD^+-dependent (type D) and O_2-dependent (type O) forms of rat liver xanthine dehydrogenase, *Arch. Biochem. Biophys.*, 172, 354, 1976.
65. **Engerson, T. D., McKelvey, T. G., Rhyne, D. B., Boggio, E. B., Snyder, S. J., and Jones, H. P.,** Conversion of xanthine dehydrogenase to oxidase in ischemic rat tissues, *J. Clin. Invest.*, 79, 1564, 1987.
66. **Halliwell, B.,** Free radicals, reactive oxygen species and human disease: a critical evaluation with special reference to atherosclerosis, *Br. J. Exp. Pathol.*, 70, 737, 1989.
67. **Halliwell, B.,** Protection against tissue damage *in vivo* by desferrioxamine: what is its mechanism of action, *Free Radical Biol. Med.*, 7, 645, 1989.
68. **Terada, L. S., Beehler, C. J., Banerjee, A., Brown, J. M., Grosso, M. A., Harken, A. H., McCord, J. M., and Repine, J. E.,** Hyperoxia and self- or neutrophil-generated O_2 metabolites inactivate xanthine oxidase, *J. Appl. Physiol.*, 65, 2349, 1988.
69. **Werns, S. W., Shea, M. J., Mitsos, S. E., Dysko, R. C., Fantone, J. C., Schork, M. A., Abrams, G. D., Pitt, B., and Lucchesi, B. R.,** Reduction of the size of infarction by allopurinol in the ischemic-reperfused canine heart, *Circulation*, 73, 518, 1986.
70. **Charlat, M. L., O'Neill, P. G., Egan, J. M., Abernethy, D. R., Michael, L. H., Myers, M. L., Roberts, R., and Bolli, R.,** Evidence for a pathogenetic role of xanthine oxidase in the "stunned" myocardial, *Am. J. Physiol.*, 252, H566, 1987.

71. **Puett, D. W., Forman, M. B., Cates, C. U., Wilson, B. H., Hande, K. R., Friesinger, G. C., and Virmani, R.**, Oxypurinol limits myocardial stunning but does not reduce infarct size after reperfusion, *Circulation*, 76, 678, 1987.

72. **Akizuki, S., Yoshida, S., Chambers, D. E., Eddy, L. J., Parmley, L. F., Yellon, D. M., and Downey, J. M.**, Infarct size limitation by the xanthine oxidase inhibitor, allopurinol, in closed-chest dogs with small infarcts, *Cardiovasc. Res.*, 19, 686, 1985.

73. **Kingma, J. G., Jr., Denniss, A. R., Hearse, D. J., Downey, J. M., and Yellon, D. M.**, Limitation of infarct size for 24 hours by combined treatment with allopurinol plus verapamil during acute myocardial infarction in the dog, *Circulation*, 75 (Suppl. V), V-25, 1987.

74. **Badylak, S. F., Voorhees, W. D., Babbs, C. F., and Simmons, A.**, The effectiveness of postischemic oxypurinol administration upon myocardial function in the isolated rat heart, *Resuscitation*, 16, 31, 1988.

75. **Chambers, D. J., Braimbridge, M. V., and Hearse, D. J.**, Free radicals and cardioplegia: allopurinol and oxypurinol reduce myocardial injury following ischemic arrest, *Ann. Thorac. Surg.*, 44, 291, 1987.

76. **Brown, J. M., Terada, L. S., Grosso, M. A., Whitmann, G. J., Velasco, S. E., Patt, A., Harken, A. H., and Repine, J. E.**, Xanthine oxidase produces hydrogen peroxide which contributes to reperfusion injury of ischemic, isolated perfused rat hearts, *J. Clin Invest.*, 81, 1297, 1988.

77. **Brown, J. M., Terada, L. S., Grosso, M. A., Whitman, G. J., Velasco, S. E., Patt, A., Harken, A. H., and Repine, J. E.**, Hydrogen peroxide mediates reperfusion injury in the isolated rat heart, *Mol. Cell. Biochem.*, 84, 173, 1988.

78. **Thompson-Gorman, S. L. and Zweier, J. L.**, Evaluation of the role of xanthine oxidase in myocardial reperfusion injury, *J. Biol. Chem.*, 265, 6656, 1990.

79. **Muxfeldt, M. and Schaper, W.**, The activity of xanthine oxidase in hearts of pigs, guinea pigs, rabbits, rats, and humans, *Basic Res. Cardiol.*, 82, 486, 1987.

80. **Downey, J. M., Miura, T., Eddy, L. J., Chambers, D. E., Mellert, T., Hearse, D. J., and Yellon, D. M.**, Xanthine oxidase is not a source of free radicals in the ischemic rabbit heart, *J. Mol. Cell. Cardiol.*, 19, 1053, 1987.

81. **Grum, C. M., Ragsdale, R. A., Ketai, L. H., and Shlafer, M.**, Absence of xanthine oxidase or xanthine dehydrogenase in the rabbit myocardium, *Biochem. Biophys. Res. Commun.*, 141, 1104, 1986.

82. **Grum, C. M., Ketai, L. H., Myers, C. L., and Shlafer, M.**, Purine efflux after cardiac ischemia: relevance to allopurinol cardioprotection, *Am. J. Physiol.*, 252, H368, 1987.

83. **Grum, C. M., Gallagher, K. P., Kirsh, M. M., and Shlafer, M.**, Absence of detectable xanthine oxidase in human myocardium, *J. Mol. Cell. Cardiol.*, 21, 263, 1989.

84. **Eddy, L. J., Stewart, J. R., Jones, H. P., Engerson, T. D., McCord, J. M., and Downey, J. M.**, Free radical-producing enzyme, xanthine oxidase, is undetectable in human hearts, *Am. J. Physiol.*, 253, H709, 1987.

85. **Wajner, M. and Harkness, R. A.**, Distribution of xanthine dehydrogenase and oxidase activities in human and rabbit tissues, *Biochim. Biophys. Acta*, 991, 79, 1989.

86. **Lasley, R. D., Ely, S. W., Berne, R. M., and Mentzer, R. M., Jr.**, Allopurinol enhanced adenine nucleotide repletion after myocardial ischemia in the isolated rat heart, *J. Clin. Invest.*, 81, 16, 1988.

87. **Moorhouse, P. C., Grootveld, M., Halliwell, B., Quinlan, J. G., and Gutteridge, J. M. C.**, Allopurinol and oxypurinol and hydroxyl radical scavengers, *FEBS Lett.*, 213, 23, 1987.

88. **Das, D. K., Engelman, R. M., Clement, R., Otani, H., Prasad, M. R., and Rao, P. S.**, Role of xanthine oxidase inhibitor as free radical scavenger: a novel mechanism of action of allopurinol and oxypurinol in myocardial salvage, *Biochem. Biophys. Res. Commun.*, 148, 314, 1987.

89. **Zimmerman, B. J., Parks, D. A., Grisham, M. B., and Granger, D. N.**, Allopurinol does not enhance antioxidant properties of extracellular fluid, *Am. J. Physiol.*, 255, H2O2, 1988.

90. **Myers, C. L., Weiss, S. J., Kirsh, M. M., Shepard, B. M., and Shlafer, M.**, Effects of supplementing hypothermic crystalloid cardioplegic solution with catalase, superoxide dismutase, allopurinol, or deferoxamine on functional recovery of globally ischemic and reperfused isolated hearts, *J. Thorac. Cardiovasc. Surg.*, 91, 281, 1986.

91. **Sedor, J. R.**, Editorial. Free radicals and prostanoid synthesis, *J. Lab. Clin. Med.*, 108, 521, 1986.

92. **Needleman, P., Turk, J., Jakschik, B. A., Morrison, A. R., and Lefkowith, J. B.**, Arachidonic acid metabolism, *Annu. Rev. Biochem.*, 55, 69, 1986.

93. **Marnett, L. J., Wlodawer, P., and Samuelsson, B.**, Co-oxygenation of organic substrates by the prostaglandin synthetase of sheep vesicular gland, *J. Biol. Chem.*, 250, 8510, 1975.

94. **Lambeir, A. M., Markey, C. M., Dunford, H. B., and Marnett, L. J.**, Spectral properties of the higher oxidation states of prostaglandin H synthase, *J. Biol. Chem.*, 260, 14894, 1985.

95. **Reed, G. A., Brooks, E. A., and Eling, T. E.**, Phenylbutazone-dependent epoxidation of 7,8-dihydroxy-7,8-dihydrobenzo(a)pyrene, *J. Biol. Chem.*, 259, 5591, 1984.

96. **Ohki, S., Ogino, N., Yamamoto, S., Hayaishi, O.**, Prostaglandin hydroperoxidase, an integral part of prostaglandin endoperoxide synthetase from bovine vesicular gland microsomes, *J. Biol. Chem.*, 254, 829, 1979.

97. **Bosia, A. and Pescarmona, G. P.**, Role of glutathione in blood platelet function, in Coenzymes and Cofactors, Vol. III,B, Dolphin, D. R., Poulson, R., and Avramovic, O., Eds., John Wiley & Sons, New York, 1989, 235.

98. **O'Brien, P. J. and Hawco, F. J.**, Hydroxyl-radical formation during prostaglandin formation catalysed by prostaglandin cyclo-oxygenase, *Biochem. Soc. Trans.*, 6, 1169, 1978.

99. **Egan, R. W., Paxton, J., and Kuehl, F. A., Jr.**, Mechanism for irreversible self-deactivation of prostaglandin synthetase, *J. Biol. Chem.*, 251, 7329, 1976.

100. **Spector, A. A., Gordon, J. A., and Moore, S. A.**, Hydroxyeicosatetraenoic acids (HETEs), *Prog. Lipid Res.*, 27, 271, 1988.

101. **Samuelsson, B. and Funk, C. D.**, Enzymes involved in the biosynthesis of leukotriene B$_4$, *J. Biol. Chem.*, 264, 19469, 1989.

102. **Hammarstrom, S.**, Leukotrienes, *Annu. Rev. Biochem.*, 52, 355, 1983.

103. **Bryant, R. W. and Bailey, J. M.**, Role of selenium-dependent glutathionine peroxidase in platelet lipoxygenase metabolism, *Prog. Lipid Res.*, 20, 189, 1982.

104. **Shimizu, T., Izumi, T., Seyama, Y., Tadokoro, K., Radmark, O., and Samuelsson, B.**, Characterization of leukotriene A$_4$ synthase from murine mast cells: evidence for its identity to arachidonate 5-lipoxygenase, *Proc. Natl. Acad. Sci. U.S.A.*, 83, 4175, 1986.

105. **Rouzer, C. A., Rands, E., Kargman, S., Jones, R. E., Register, R. B., and Dixon, R. A.**, Characterization of cloned human leukocyte 5-lipoxygenase expressed in mammalian cells, *J. Biol. Chem.*, 263, 10135, 1988.

106. **Rouzer, C. A., Matsumoto, T., and Samuelsson, B.**, Single protein from human leukocytes possesses 5-lipoxygenase and leukotriene A$_4$ synthase activities, *Proc. Natl. Acad. Sci. U.S.A.*, 83, 857, 1986.

107. **Schulam, P. G. and Shearer, W. T.**, Evidence for 5-lipoxygenase activity in human B cell lines: a possible role for arachidonic acid metabolites during B cell signal transduction, *J. Immunol.*, 144, 2696, 1990.

108. **Chien, K. R., Sen, A., Reynolds, R., Chang, A., Kim, Y., Gunn, M. D., Buja, L. M., and Willerson, J. T.**, Release of arachidonate from membrane phospholipids in cultured neonatal rat myocardial cells during adenosine triphosphate depletion, *J. Clin. Invest.*, 75, 1770, 1985.

109. **Gerritsen, M. E. and Printz, M. P.**, Sites of prostaglandin synthesis in the bovine heart and isolated bovine coronary microvessels, *Circ. Res.*, 49, 1152, 1981.

110. **Hsueh, W. and Needleman, P.**, Sites of lipase activation and prostaglandin synthesis in isolated, perfused rabbit hearts and hydronephrotic kidneys, *Prostaglandins*, 16, 661, 1978.

111. **Needleman, P., Bronson, S. D., Wyche, A., Sivakoff, M., and Nicolaou, K. C.,** Cardiac and renal prostaglandin I_2. Biosynthesis and biological effects in isolated perfused rabbit tissues, *J. Clin. Invest.,* 61, 839, 1978.

112. **Escoubet, B., Griffaton, G., and Lechat, P.,** Verapamil depresses the synthesis of lipoxygenase products by hypoxic cardiac rat fibroblasts in culture, *Biochem. Pharmacol.,* 35, 1879, 1986.

113. **Shappell, S. B., Hughes, H., Youker, K., Entman, M. L., Wong, A., Mong, S., and Taylor, A. A.,** Detection of 5-lipoxygenase in isolated canine cardiac myocytes, *Fed. Am. Soc. Exp. Biol. J.,* 4, A2077, 1990.

114. **Nicosia, S. and Patrono, C.,** Eicosanoid biosynthesis and action: novel opportunities for pharmacological intervention, *Fed. Am. Soc. Exp. Biol. J.,* 3, 1941, 1989.

115. **Mullane, K. M. and Pinto, A.,** Endothelium, arachidonic acid, and coronary vascular tone, *Fed. Proc. Fed. Am. Soc. Exp. Biol.,* 46, 54, 1987.

116. **Mullane, K., Barst, S., and McGiff, J. C.,** Leukotrienes in myocardial ischemia, *Prog. Clin. Biol. Res.,* 199, 127, 1985.

117. **Coller, B. S.,** Platelets and thrombolytic therapy, *N. Engl. J. Med.,* 322, 33, 1990.

118. **Mullane, K. M. and McGiff, J. C.,** Platelet depletion and infarct size in an occlusion-reperfusion model of myocardial ischemia in anesthetized dogs, *J. Cardiovasc. Pharmacol.,* 7, 733, 1985.

119. **Gerritsen, M. E.,** Eicosanoid production by the coronary microvascular endothelium, *Fed. Proc. Fed. Am. Soc. Exp. Biol.,* 46, 47, 1987.

120. **Boxer, L. A., Allen, J. M., Schmidt, M., Yoder, M., and Baehner, R. L.,** Inhibition of polymorphonuclear leukocyte adherence by prostacyclin, *J. Lab. Clin. Med.,* 95, 672, 1980.

121. **Fantone, J. C. and Kinnes, D. A.,** Prostaglandin E_1 and prostaglandin I_2 modulation of superoxide production by human neutrophils, *Biochem. Biophys. Res. Commun.,* 113, 506, 1983.

122. **Ham, E. A., Soderman, D. D., Zanetti, M. E., Dougherty, H. W., MCCauley, E., and Kuehl, F. A., Jr.,** Inhibition by prostaglandins of leukotriene B_4 release from activated neutrophils, *Proc. Natl. Acad. Sci. U.S.A.,* 80, 4349, 1983.

123. **Lucchesi, B. R., Mickelson, J. K., Homeister, J. W., and Jackson, C. V.,** Interaction of the formed elements of blood with the coronary vasculature *in vivo, Fed. Proc. Fed. Am. Soc. Exp. Biol.,* 46, 63, 1987.

124. **Botting, R. and Vane, J. R.,** Mediators and the anti-thrombotic properties of the vascular endothelium, *Ann. Med. (Hagerstown Md.),* 21, 31, 1989.

125. **Gryglewski, R. J., Botting, R. M., and Vane, J. R.,** Mediators produced by the endothelial cell, *Hypertension,* 12, 530, 1988.

126. **Levi, R., Genovese, A., and Pinckard, R. N.,** Alkyl chain homologs of platelet-activating factor and their effects on the mammalian heart, *Biochem. Biophys. Res. Commun.,* 161, 1341, 1989.

127. **Sasaki, K., Ueno, A., Katori, M., and Kikawada, R.,** Detection of leukotriene B_4 in cardiac tissue and its role in infarct extension through leucocyte migration, *Cardiovasc. Res.,* 22, 142, 1988.

128. **Parker, C. W.,** 5-Lipoxygenase, leukotrienes, and regulation of inflammatory responses, *Drug. Dev. Res.,* 10, 277, 1987.

129. **Mullane, K. M., Salmon, J. A., and Kraemer, R.,** Leukocyte-derived metabolites of arachidonic acid in ischemia-induced myocardial injury, *Fed. Proc. Fed. Am. Soc. Exp. Biol.,* 46, 2422, 1987.

130. **Lucchesi, B. R. and Mullane, K. M.,** Leukocytes and ischemia-induced myocardial injury, *Annu. Rev. Pharmacol. Toxicol.,* 26, 201, 1986.

131. **Ezra, D., Boyd, L. M., Feuerstein. G., and Goldstein, R. E.,** Coronary constriction by leukotriene C_4, D_4, and E_4 in the intact pig heart, *Am. J. Cardiol.,* 51, 1451, 1983.

132. **Michelassi, F., Landa, L., Hill, R. D., Lowenstein, E., Watkins, W. D., Petkau, A. J., and Zapol, W. M.,** Leukotriene D_4: a potent coronary artery vasoconstrictor associated with impaired ventricular contraction, *Science,* 217, 841, 1982.

133. **Ito, B. R., Roth, D. M., and Engler, R. L.**, Thromboxane A_2 and peptidoleukotrienes contribute to the myocardial ischemia and contractile dysfunction in response to intracoronary infusion of complement C5a in pigs, *Circ. Res.*, 66, 596, 1990.

134. **Keller, A. M., Clancey, R. M., Barr, M. L., Marboe, C. C., and Cannon, P. J.**, Acute reoxygenation injury in the isolated rat heart: role of resident cardiac mast cells, *Circ. Res.*, 63, 1044, 1988.

135. **Del Balzo, U. H., Levi, R., and Polley, M. J.**, Cardiac dysfunction caused by purified human C3a anaphylatoxin, *Proc. Natl. Acad. Sci. U.S.A.*, 82, 886, 1985.

136. **Piomelli, D., Feinmark, S. J., and Cannon, P. J.**, Leukotriene biosynthesis by canine and human coronary arteries, *J. Pharmacol. Exp. Ther.*, 241, 763, 1987.

137. **Feinmark, S. J. and Cannon, P. J.**, Endothelial cell leukotriene C_4 synthesis results from intercellular transfer of leukotriene A_4 synthesized by polymorphonuclear leukocytes, *J. Biol. Chem.*, 261, 16466, 1986.

138. **Maclouf, J. A. and Murphy, R. C.**, Transcellular metabolism of neutrophil-derived leukotriene A_4 by human platelets, *J. Biol. Chem.*, 263, 174, 1988.

139. **Fradin, A., Zirrolli, J. A., Maclouf, J., Vausbinder, L., Henson, P. M., and Murphy, R. C.**, Platelet-activating factor and leukotriene biosynthesis in whole blood: a model for the study of transcellular arachidonate metabolism, *J. Immunol.*, 143, 3680, 1989.

140. **Maclouf, J., Murphy, R. C., and Henson, P. M.**, Transcellular sulfidopeptide leukotriene biosynthetic capacity of vascular cells, *Blood*, 74, 703, 1989.

141. **Grimminger, F., Kreusler, B., Schneider, U., Becker, G., and Seeger, W.**, Influence of microvascular adherence on neutrophil leukotriene generation: evidence for cooperative eicosanoid synthesis, *J. Immunol.*, 144, 1866, 1990.

142. **Hemler, M. E., Cook, H. W., Lands, W. E. M.**, Prostaglandin biosynthesis can be triggered by lipid peroxides, *Arch. Biochem. Biophys.*, 193, 340, 1979.

143. **Kulmacz, R. J. and Lands, W. E. M.**, Requirements for hydroperoxide by the cyclooxygenase and peroxidase activities of prostaglandin H synthase, *Prostaglandins*, 25, 531, 1983.

144. **Lands, W., Lee, R., and Smith, W.**, Factors regulating the biosynthesis of various prostaglandins, *Ann. N.Y. Acad. Sci.*, 180, 107, 1971.

145. **Smith, W. L. and Lands, W. E. M.**, Oxygenation of polyunsaturated fatty acids during prostaglandin biosynthesis by sheep vesicular gland, *Biochemistry*, 11, 3276, 1972.

146. **Smith, W. L. and Lands, W. E. M.**, Oxygenation of unsaturated fatty acids by soybean lipoxygenase, *J. Biol. Chem.*, 247, 1038, 1972.

147. **Cook, H. W. and Lands, W. E. M.**, Mechanism for suppression of cellular biosynthesis of prostaglandins, *Nature (London)*, 260, 630, 1976.

148. **Rouzer, C. A. and Samuelsson, B.**, The importance of hydroperoxide activation for the detection and assay of mammalian 5-lipoxygenase, *FEBS Lett.*, 204, 293, 1986.

149. **Bryant, R. W., She, H. S., Ng, K. J., and Siegel, M. I.**, Modulation of the 5-lipoxygenase activity of MC-9 mast cells: activation by hydroperoxides, *Prostaglandins*, 32, 615, 1986.

150. **Egan, R. W., Tischler, A. N., Baptista, E. M., Ham, E. A., Soderman, D. D., and Gale, P. H.**, Specific inhibition and oxidative regulation of 5-lipoxygenase, *Adv. Prostaglandin Thromboxane Leukotriene Res.*, 11, 151, 1983.

151. **Marshall, P. J. and Lands, W. E. M.**, In vitro formation of activators for prostaglandin synthesis by neutrophils and macrophages from humans and guinea pigs, *J. Lab. Clin. Med.*, 108, 525, 1986.

152. **Hemler, M. E., Graff, G., and Lands, W. E. M.**, Accelerative autoactivation of prostaglandin biosynthesis by PGG_2, *Biochem. Biophys. Res. Commun.*, 85, 1325, 1978.

153. **Siegel, M. I., McConnell, R. T., Abrahams, S. L., Porter, N. A., and Cuatrecasas, P.**, Regulation of arachidonate metabolism via lipoxygenase and cyclo-oxygenase by 12-HPETE, the product of human platelet lipoxygenase, *Biochem. Biophys. Res. Commun.*, 89, 1273, 1979.

154. **Lands, W.**, Biological consequences of fatty acid oxygenase reaction mechanisms, *Prostaglandin Leukotriene Med.*, 13, 35, 1984.
155. **Marshall, P. J., Kulmacz, R. J., and Lands, W. E. M.**, Constraints on prostaglandin biosynthesis in tissues, *J. Biol. Chem.*, 262, 3510, 1987.
156. **Harlan, J. M. and Callahan, K. S.**, Role of hydrogen peroxide in the neutrophil-mediated release of prostacyclin from cultured endothelial cells, *J. Clin. Invest.*, 74, 442, 1984.
157. **Whorton, A. R., Montgomery, M. E., and Kent, R. S.**, Effect of hydrogen peroxide on prostaglandin production and cellular integrity in cultured porcine aortic endothelial cells, *J. Clin. Invest.*, 76, 295, 1985.
158. **Moncada, S., Gryglewski, R. J., Bunting, S., and Vane, J. R.**, A lipid peroxide inhibits the enzyme in blood vessel microsomes that generates from prostaglandin endoperoxides the substance (prostaglandin X) which prevents platelet aggregation, *Prostaglandins*, 12, 715, 1976.
159. **Weiss, S. J., Turk, J., and Needleman, P.**, A mechanism for the hydroperoxide-mediated inactivation of prostacyclin synthetase, *Blood*, 53, 1191, 1979.
160. **Ham, E. A., Egan, R. W., Soderman, D. D., Gale, P. H., and Kuehl, F. A., Jr.**, Peroxidase-dependent deactivation of prostacyclin synthetase, *J. Biol. Chem.*, 254, 2191, 1979.
161. **Kent, R. S., Diedrich, S. L., and Whorton, A. R.**, Regulation of vascular prostaglandin synthesis by metabolites of arachidonic acid in perfused rabbit aorta, *J. Clin. Invest.*, 72, 455, 1983.
162. **Lewis, M. S., Whatley, R. E., Cain, P., McIntyre, T. M., Prescott, S. M., and Zimmerman, G. A.**, Hydrogen peroxide stimulates the synthesis of platelet-activating factor by endothelium and induces endothelial cell-dependent neutrophil adhesion, *J. Clin. Invest.*, 82, 2045, 1988.
163. **Elliott, S. J., Eskin, S. G., and Schilling, W. P.**, Effect of t-butyl-hydroperoxide on bradykinin-stimulated changes in cytosolic calcium in vascular endothelial cells, *J. Biol. Chem.*, 264, 3806, 1989.
164. **Elliott, S. J. and Schilling, W. P.**, Carmustine augments the effects of tert-butyl hydroperoxide on calcium signaling in cultured pulmonary artery endothelial cells, *J. Biol. Chem.*, 265, 103, 1990.
165. **Rouzer, C. A., Shimizu, T., and Samuelsson, B.**, On the nature of the 5-lipoxygenase reaction in human leukocytes: characterization of a membrane-associated stimulatory factor, *Proc. Natl. Acad. Sci. U.S.A.*, 82, 7505, 1985.
166. **Rouzer, C. A. and Kargman, S.**, Translocation of 5-lipoxygenase to the membrane in human leukocytes challenged with ionophore A23187, *J. Biol. Chem.*, 263, 10980, 1988.
167. **Wong, A., Hwang, S. M., Cook, M. N., Hogaboom, G. K., and Crooke, S. T.**, Interactions of 5-lipoxygenase with membranes: studies on the association of soluble enzyme with membranes and alterations in enzyme activity, *Biochemistry*, 27, 6763, 1988.
168. **Sporn, P. H. S. and Peters-Golden, M.**, Hydrogen peroxide inhibits alveolar macrophage 5-lipoxygenase metabolism in association with depletion of ATP, *J. Biol. Chem.*, 263, 14776, 1988.
169. **Henderson, W. R., Jorg, A., and Klebanoff, S. J.**, Eosinophil peroxidase-mediated inactivation of leukotrienes B_4, C_4, and D_4, *J. Immunol.*, 128, 2609, 1982.
170. **Goetzl, E. J.**, The conversion of leukotriene C_4 to isomers of leukotriene B_4 by human eosinophil peroxidase, *Biochem. Biophys. Res. Commun.*, 106, 270, 1982.
171. **Lee, C. W., Lewis, R. A., Corey, E. J., Barton, A., Oh, H., Tauber, A. I., and Austen, K. F.**, Oxidative inactivation of leukotriene C_4 by stimulated human polymorphonuclear leukocytes, *Proc. Natl. Acad. Sci. U.S.A.*, 79, 4166, 1982.
172. **Henderson, W. R. and Klebanoff, S. J.**, Leukotriene B_4, C_4, D_4 and E_4 inactivation by hydroxyl radicals, *Biochem. Biophys. Res. Commun.*, 110, 266, 1983.

173. **Henderson, W. R. and Klebanoff, S. J.**, Leukotriene production and inactivation by normal, chronic granulomatous disease and myeloperoxidase-deficient neutrophils, *J. Biol. Chem.*, 258, 13522, 1983.

174. **Klebanoff, S. J.**, Phagocytic cells: products of oxygen metabolism, in *Inflammation: Basic Principles and Clinical Correlates*, Gallin, J. I., Goldstein, I. M., and Snyderman, R., Eds., Raven Press, New York, 1988, 391.

175. **Ward, P. A., Warren, J. S., and Johnson, K. J.**, Oxygen radicals, inflammation, and tissue injury, *Free Radical Biol. Med.*, 5, 403, 1988.

176. **Markert, M., Glass, G. A., Babior, B. M.**, Respiratory burst oxidase from human neutrophils: purification and some properties, *Proc. Natl. Acad. Sci. U.S.A.*, 82, 3144, 1985.

177. **Umei, T., Takeshige, K., and Minakami, S.**, NADPH binding component of neutrophil superoxide-generating oxidase, *J. Biol. Chem.*, 261, 5229, 1986.

178. **Cross, A. R., Jones, O. T. G., Garcia, R., and Segal, A. W.**, The association of FAD with the cytochrome b-245 of human neutrophils, *Biochem. J.*, 208, 759, 1982.

179. **Borregaard, N. and Tauber, A. I.**, Subcellular localization of the human neutrophil NADPH oxidase: b-cytochrome and associated flavoprotein, *J. Biol. Chem.*, 259, 47, 1984.

180. **Borregaard, N., Heiple, J. M., Simons, E. R., and Clark, R. A.**, Subcellular localization of the b-cytochrome component of the human neutrophil microbiocidal oxidase — translocation during activation, *J. Cell Biol.*, 97, 52, 1983.

181. **Garcia, R. C. and Segal, A. W.**, Changes in the subcellular distribution of the cytochrome b-245 on stimulation of human neutrophils, *Biochem. J.*, 219, 233, 1984.

182. **Higson, F. K., Durbin, L., Pavlotsky, N., and Tauber, A. I.**, Studies of cytochrome B-245 translocation in the PMA stimulation of the human neutrophil NADPH-oxidase, *J. Immunol.*, 135, 519, 1985.

183. **Ohno, Y., Seligmann, B. E., and Gallin, J. I.**, Cytochrome b translocation to human neutrophil plasma membranes and superoxide release: differential effects of N-formylmethionylleucylphenylalanine, phorbol myristate acetate and A23187, *J. Biol. Chem.*, 260, 2409, 1985.

184. **Ambruso, D. R., Bolscher, B. G. J. M., Stokman, P. M., Verhoeven, A. J., and Roos, D.**, Assembly and activation of the $NADPH:O_2$ oxidoreductase in human neutrophils after stimulation with phorbol myristate acetate, *J. Biol. Chem.*, 265, 924, 1990.

185. **Bromberg, Y. and Pick, E.**, Activation of NADPH-dependent superoxide production in a cell-free system by sodium dodecyl sulfate, *J. Biol. Chem.*, 260, 13539, 1985.

186. **Curnutte, J. T., Kuver, R., and Scott, P. J.**, Activation of neutrophil NADPH oxidase in a cell-free system, *J. Biol. Chem.*, 262, 5563, 1987.

187. **Curnutte, J. T., Kuver, R., and Babior, B. M.**, Activation of the respiratory burst oxidase in a fully soluble system from human neutrophils, *J. Biol. Chem.*, 262, 6450, 1987.

188. **Curnutte, J. T., Scott, P. J., and Mayo, L. A.**, Cytosolic components of the respiratory burst oxidase: resolution of four components, two of which are missing in complementing types of chronic granulomatous disease, *Proc. Natl. Acad. Sci. U.S.A.*, 86, 825, 1989.

189. **Clark, R. A., Volpp, B. D., Leidal, K. G., and Nauseef, W. M.**, Two cytosolic components of the human neutrophil respiratory burst oxidase translocate to the plasma membrane during cell activation, *J. Clin. Invest.*, 85, 714, 1990.

190. **Lomax, K. J., Leto, T. L., Nunoi, H., Gallin, J. I., and Malech, H. I.**, Recombinant 47-kilodalton cytosol factor restores NADPH oxidase in chronic granulomatous disease, *Science*, 245, 409, 1989.

191. **Volpp, B. D., Nauseef, W. M., and Clark, R. A.**, Two cytosolic neutrophil oxidase components absent in autosomal chronic granulomatous disease, *Science*, 242, 1295, 1988.

192. **Clark, R. A., Malech, H. L., Gallin, J. I., Nunoi, H., Volpp, B. D., Pearson, D. W., Nauseef, W. M., and Curnutte, J. T.**, Genetic variants of chronic granulomatous disease: prevalence of deficiencies of two cytosolic components of the NADPH oxidase system, *N. Engl. J. Med.*, 321, 647, 1989.

193. **Gabig, T. G., English, D., Akard, L. P., and Schell, M. J.**, Regulation of neutrophil NADPH oxidase activation in a cell-free system by guanine nucleotides and fluoride, *J. Biol. Chem.*, 262, 1685, 1987.

194. **Seifert, R. and Schultz, G.**, Fatty-acid-induced activation of NADPH oxidase in plasma membranes of human neutrophils depends on neutrophil cytosol and is potentiated by stable guanine nucleotides, *Eur. J. Biochem.*, 162, 563, 1987.

195. **Gabig, T. G., Eklund, E. A., Potter, G. B., and Dykes, J. R., III**, A neutrophil GTP-binding protein that regulates cell-free NADPH oxidase activation is located in the cytosolic fraction, *J. Immunol.*, 145, 945, 1990.

196. **Volpp, B. D., Nauseef, W. M., Donelson, J. E., Moser, D. R., and Clark, R. A.**, Cloning of the cDNA and functional expression of the 47-kilodalton cytosolic component of human neutrophil respiratory burst oxidase, *Proc. Natl. Acad. Sci. U.S.A.*, 86, 7195, 1989.

197. **Quinn, M. T., Parkos, C. A., Walker, L., Orkin, S. H., Dinauer, M. C., and Jesaitis, A. J.**, Association of a Ras-related protein with cytochrome b of human neutrophils, *Nature (London)*, 342, 198, 1989.

198. **Tauber, A. I.**, Review: protein kinase C and the activation of the human neutrophil NADPH-oxidase, *Blood*, 69, 711, 1987.

199. **Nishizuka, Y.**, Studies and perspectives of protein kinase C, *Science*, 233, 305, 1986.

200. **Heyworth, P. G. and Segal, A. W.**, Further evidence for the involvement of a phosphoprotein in the respiratory burst oxidase of human neutrophils, *Biochem. J.*, 239, 723, 1986.

201. **Berridge, M. J.**, Inositol trisphosphate and diacylglycerol: two interacting second messengers, *Annu. Rev. Biochem.*, 56, 159, 1987.

202. **Kikkawa, U., Kishimoto, A., and Nishizuka, Y.**, The protein kinase C family: heterogeneity and its implications, *Annu. Rev. Biochem.*, 58, 31, 1989.

203. **Rando, R. R.**, Regulation of protein kinase C activity by lipids, *Fed. Am. Soc. Exp. Biol. J.*, 2, 2348, 1988.

204. **Nishizuka, Y.**, Review article. The molecular heterogeneity of protein kinase C and its implications for cellular regulation, *Nature (London)*, 334, 661, 1988.

205. **Abdel-Latif, A. A.**, Calcium-mobilizing receptors, polyphosphoinositides, and the generation of second messengers, *Pharmacol. Rev.*, 38, 227, 1986.

206. **Berkow, R. L., Dodson, R. W., and Kraft, A. S.**, The effect of a protein kinase C inhibitor, H-7, on human neutrophil oxidative burst and degranulation, *J. Leukocyte Biol.*, 41, 441, 1987.

207. **Gerard, C., McPhail, L. C., Marfat, A., Stimler-Gerard, N. P., Bass, D. A., and McCall, C. E.**, Role of protein kinases in stimulation of human polymorphonuclear leukocyte oxidative metabolism by various agonists, *J. Clin. Invest.*, 77, 61, 1986.

208. **Snyderman, R. and Pike, M. C.**, Chemoattractant receptors on phagocytic cells, *Annu. Rev. Immunol.*, 2, 257, 1984.

209. **Nathan, C. F.**, Neutrophil activation on biological surfaces: massive secretion of hydrogen peroxide in response to products of macrophages and lymphocytes, *J. Clin. Invest.*, 80, 1550, 1987.

210. **Snyderman, R. and Uhing, R.**, Phagocytic cells: stimulus-response coupling mechanisms, in *Inflammation: Basic Principles and Clinical Correlates*, Gallin, J. I., Goldstein, I. M., and Snyderman, R., Eds., Raven Press, New York, 1988.

211. **Heinsohn, C., Polgar, P., Fishman, J., and Taylor, L.**, The effect of bovine serum albumin on the synthesis of prostaglandin and incorporation of [^{3}H]acetate into platelet-activating factor, *Arch. Biochem. Biophys.*, 257, 251, 1987.

212. **Honeycutt, P. J. and Niedel, J. E.**, Cytochalasin B enhancement of the diacylglycerol response in formyl peptide-stimulated neutrophils, *J. Biol. Chem.*, 261, 15900, 1986.

213. **Burton, K. P.**, Superoxide dismutase enhances recovery following myocardial ischemia, *Am. J. Physiol.*, 248, H637, 1985.

214. **Werns, S. W., Shea, M. J., Driscoll, E. M., Cohen, C., Abrams, G. D., Pitt, B., and Lucchesi, B. R.**, The independent effects of oxygen radical scavengers on canine infarct size: reduction by superoxide dismutase but not catalase, *Circ. Res.*, 56, 895, 1985.

215. **Jolly, S. R., Kane, W. J., Bailie, M. B., Abrams, G. D., and Lucchesi, B. R.**, Canine myocardial reperfusion injury: its reduction by the combined administration of superoxide dismutase and catalase, *Circ. Res.*, 54, 277, 1984.

216. **Ambrosio, G., Weisfeldt, M. L., Jacobus, W. E., and Flaherty, J. T.**, Evidence for a reversible oxygen radical-mediated component of reperfusion injury: reduction by recombinant human superoxide dismutase administered at the time of reflow, *Circulation*, 75, 282, 1987.

217. **Chi, L., Tamura, Y., Hoff, P. T., Macha, M., Gallagher, K. P., Schork, M. A., and Lucchesi, B. R.**, Effect of superoxide dismutase on myocardial infarct size in the canine heart after 6 hours of regional ischemia and reperfusion: a demonstration of myocardial salvage, *Circ. Res.*, 64, 665, 1989.

218. **Bando, K., Teramoto, S., Tago, M., Seno, S., Murakami, T., Nawa, S., and Senoo, Y.**, Oxygenated perfluorocarbon, recombinant human superoxide dismutase, and catalase ameliorate free radical induced myocardial injury during heart preservation and transplantation, *J. Thorac. Cardiovasc. Surg.*, 96, 930, 1988.

219. **Massey, K. D. and Burton, K. P.**, α-Tocopherol attenuates myocardial membrane-related alterations resulting from ischemia and reperfusion, *Am. J. Physiol.*, 256, H1192, 1989.

220. **Ferreira, R., Burgos, M., Llesuy, S., Molteni, L., Milei, J., Gonzalez Flecha, B., and Boveris, A.**, Reduction of reperfusion injury with mannitol cardioplegia, *Ann. Thorac. Surg.*, 48, 77, 1989.

221. **Mitsos, S. E., Fantone, J. C., Gallagher, K. P., Walden, K. M., Simpson, P. J., Abrams, G. D., Schork, M. A., and Lucchesi, B. R.**, Myocardial reperfusion injury: protection by a free radical scavenger, N-2-mercaptopropionyl glycine, *J. Cardiovasc. Pharmacol.*, 8, 978, 1986.

222. **Mitsos, S. E., Askew, T. E., Fantone, J. C., Kunkel, S. L., Abrams, G. D., Schock, A., and Lucchesi, B. R.**, Protective effects of N-2-mercaptopropionyl glycine against myocardial reperfusion injury after neutrophil depletion in the dog: evidence for the role of intracellular-derived free radicals, *Circulation*, 73, 1077, 1986.

223. **Bolli, R., Zhu, W., Hartley, C. J., Michael, L. H., Repine, J. E., Hess, M. L., Kukreja, R. C., and Roberts, R.**, Attenuation of dysfunction in the postischemic 'stunned' myocardium by dimethylthiourea, *Circulation*, 76, 458, 1987.

224. **Portz, S. J., Lesnefsky, E. J., VanBenthuysen, K. M., Repine, J. E., Parker, N. B., McMurtry, I. F., and Horwitz, L. D.**, Dimethylthiourea, but not dimethylsulfoxide, reduces canine myocardial infarct size, *Free Radical Biol. Med.*, 7, 53, 1989.

225. **Godin, D. V. and Bhimji, S.**, Effects of allopurinol on myocardial ischemic injury induced by coronary artery ligation and reperfusion, *Biochem. Pharmacol.*, 36, 2101, 1987.

226. **Ambrosio, G., Zweier, J. L., Jacobus, W. E., Weisfeldt, M. L., and Flaherty, J. T.**, Improvement of postischemic myocardial function and metabolism induced by administration of deferoxamine at the time of reflow: the role of iron in the pathogenesis of reperfusion injury, *Circulation*, 76, 906, 1987.

227. **Reddy, B. R., Kloner, R. A., and Przyklenk, K.**, Early treatment with deferoxamine limits myocardial ischemic/reperfusion injury, *Free Radical Biol. Med.*, 7, 45, 1989.

228. **Bolli, R., Patel, B. S., Zhu, W., O'Neill, P. G., Hartley, C. J., Charlat, M. L., and Roberts, R.**, The iron chelator desferrioxamine attenuates postischemic ventricular dysfunction, *Am. J. Physiol.*, 253, H1372, 1987.
229. **Repine, J. E., Eaton, J. W., Anders, M. W., Hoidal, J. R., and Fox, R. B.**, Generation of hydroxyl radical by enzymes, chemicals, and human phagocytes, *in vitro*, *J. Clin. Invest.*, 64, 1642, 1979.
230. **Wasil, M., Halliwell, B., Grootveld, M., Moorhouse, C. P., Hutchison, D. C. S., and Baum, H.**, The specificity of thiourea, dimethylthiourea, and dimethyl sulfoxide as scavengers of hydroxyl radicals, *Biochem. J.*, 243, 867, 1987.
231. **Engler, R. L., Dahlgren, M. D., Peterson, M. A., Dobbs, A., and Schmid-Schonbein, G. W.**, Accumulation of polymorphonuclear leukocytes during 3-h experimental myo-cardial ischemia, *Am. J. Physiol.*, 251, H93, 1986.
232. **Warren, J. S., Yabroff, K. R., Mandel, D. M., Johnson, K. J., and Ward, P. A.**, Role of O_2^- in neutrophil recruitment into sites of dermal and pulmonary vasculitis, *Free Radical Biol. Med.*, 8, 163, 1990.
233. **Zweier, J. L., Duke, S. S., Kuppusamy, P., Sylvester, J. T., and Gabrielson, E. W.**, Electron paramagnetic resonance evidence that cellular oxygen toxicity is caused by the generation of superoxide and hydroxyl free radicals, *FEBS Lett.*, 252, 12, 1989.
234. **Nohl, H.**, A novel superoxide radical generator in heart mitochondria, *FEBS Lett.*, 214, 269, 1987.
235. **Babbitt, D. G., Forman, M. B., Jones, R., Bajaj, A. K., and Hoover, R. L.**, Prevention of neutrophil-mediated injury to endothelial cells by perfluorochemical, *Am. J. Pathol.*, 136, 451, 1990.
236. **Bajaj, A. K., Cobb, M. A., Virmani, R., Gay, J. C., Light, R. T., and Forman, M. B.**, Limitation of myocardial reperfusion injury by intravenous perfluorochemicals: role of neutrophil activation, *Circulation*, 79, 645, 1989.
237. **Mullane, K. M. and Moncada, S.**, The salvage of ischaemic myocardium by BW755C in anesthetized dogs, *Prostaglandins*, 24, 255, 1982.
238. **Bednar, M., Smith, B., Pinto, A., and Mullane, K. M.**, Nafazatrom-induced salvage of ischemic myocardium in anesthetized dogs is mediated through inhibition of neutrophil function, *Circ. Res.*, 57, 131, 1985.
239. **Ozaki, Y., Ohashi, T., and Niwa, Y.**, A comparative study on the effects of inhibitors of the lipoxygenase pathway on neutrophil function: inhibitory effects on neutrophil function may not be attributed to inhibition of the lipoxygenase pathway, *Biochem. Pharmacol.*, 35, 3481, 1986.
240. **Shappell, S. B., Taylor, A. A., Hughes, H., Mitchell, J. R., Anderson, D. C., and Smith, C. W.**, Comparison of antioxidant and non-antioxidant lipoxygenase inhibitors on neutrophil function. Implications for pathogenesis of myocardial reperfusion injury, *J. Pharmacol. Exp. Ther.*, 252, 531, 1990.
241. **Showell, H. J., Naccache, P. H., Sha'afi, R. I., and Becker, E. L.**, Inhibition of rabbit neutrophil lysosomal enzyme secretion, non-stimulated and chemotactic factor stimulated locomotion by nordihydroguaiaretic acid, *Life Sci.*, 27, 421, 1980.
242. **Egan, R. W. and Gale, P. W.**, Comparative biochemistry of lipoxygenase inhibitors, in *Prostaglandins, Leukotrienes, and Lipoxins*, Bailey, I. and Martyn, J., Eds., Plenum Press, New York, 1985, 593.
243. **Chambers, D. E., Parks, D. A., Patterson, G., Roy, R., McCord, J. M., Yoshida, S., Parmley, L. F., and Downey, J. M.**, Xanthine oxidase as a source of free radical damage in the myocardial ischemia, *J. Mol. Cell. Cardiol.*, 17, 145, 1985.
244. **Tamura, Y., Chi, L., Driscoll, E. M., Jr., Hoff, P. T., Freeman, B. A., Gallagher, K. P., and Lucchesi, B. R.**, Superoxide dismutase conjugated to polyethylene glycol provides sustained protection against myocardial ischemia/reperfusion injury in canine heart, *Circ. Res.*, 63, 944, 1988.

245. Nejima, J., Knight, D. R., Fallon, J. T., Uemura, N., Manders, W. T., Canfield, D. R., Cohen, M. V., and Vatner, S. F., Superoxide dismutase reduces reperfusion arrhythmias but fails to salvage regional function or myocardium at risk in conscious dogs, *Circulation,* 79, 143, 1989.

246. Przyklenk, K. and Kloner, R. A., "Reperfusion injury" by oxygen-derived free radicals? Effect of superoxide dismutase plus catalase, given at the time of reperfusion, on myocardial infarct size, contractile function, coronary microvasculature, and regional myocardial blood flow, *Circ. Res.,* 64, 86, 1989.

247. Shappell, S. B., Toman, C., Anderson, D. C., Taylor, A. A., Entman, M. L., and Smith, C. W., Mac-1 (CD11b/CD18) mediates adherence-dependent hydrogen peroxide production by human and canine neutrophils, *J. Immunol.,* 144, 2702, 1990.

248. Hamby, R., Location of exogenous superoxide dismutase in myocardial cells, *Fed. Am. Soc. Exp. Biol. J.,* 4, A, 1990.

249. Markey, B. A., Phan, S. H., Varani, J., Ryan, U. S., and Ward, P. A., Inhibition of H_2O_2 and neutrophil-induced endothelial cell cytotoxicity by intracellular superoxide dismutase supplementation, *Fed. Am. Soc. Exp. Biol. J.,* 4, A4372, 1990.

250. Vander Heide, R. S., Sobotka, P. A., and Ganote, C. E., Effects of the free radical scavenger DMTU and mannitol on the oxygen paradox in perfused rat hearts, *J. Mol. Cell. Cardiol.,* 19, 615, 1987.

251. Ytrehus, K., Gunnes, S., Myklebust, R., and Mjos, O. D., Protection by superoxide dismutase and catalase in the isolated rat heart reperfused after prolonged cardioplegia: a combined study of metabolic, functional, and morphometric ultrastructural variables, *Cardiovasc. Res.,* 21, 492, 1987.

252. Otani, H., Umemoto, M., Kagawa, K., Nakamura, Y., Omoto, K., Tanaka, K., Sato, T., Nonoyama, A., and Kagawa, T., Protection against oxygen-induced reperfusion injury of the isolated canine heart by superoxide dismutase and catalase, *J. Surg., Res.,* 41, 126, 1986.

253. Baker, J. E., Felix, C. C., Olinger, G. N., Kalyanaraman, B., Myocardial ischemia and reperfusion: direct evidence for free radical generation by electron spin resonance spectroscopy, *Proc. Natl. Acad. Sci. U.S.A.,* 85, 2786, 1988.

254. Garlick, P. B., Davies, M. J., Hearse, D. J., and Slater, T. F., Direct detection of free radicals in the reperfused rat heart using electron spin resonance spectroscopy, *Circ. Res.,* 61, 757, 1987.

255. Zweier, J. L., Flaherty, J. T., and Weisfeldt, M. L., Direct measurement of free radical generation following reperfusion of ischemic myocardium, *Proc. Natl. Acad. Sci. U.S.A.,* 84, 1404, 1987.

256. Zweier, J. L., Measurement of superoxide-derived free radicals in the reperfused heart, *J. Biol. Chem.,* 25, 1353, 1988.

257. Zweier, J. L., Rayburn, B. K., Flaherty, J. T., and Weisfeldt, M. L., Recombinant superoxide dismutase reduces oxygen free radical concentrations in reperfused myocardium, *J. Clin. Invest.,* 80, 1728, 1987.

258. Kramer, J. H., Arroyo, C. M., Dickens, B. F., and Weglicki, W. B., Spin-trapping evidence that graded myocardial ischemia alters post-ischemic superoxide production, *Free Rad. Biol. Med.,* 3, 153, 1987.

259. Culcasi, M., Pietri, S., and Cozzone, P. J., Use of 3,3,5,5-tetramethyl-1-pyrroline-1-oxide spin trap for the continuous flow ESR monitoring of hydroxyl radical generation in the ischemic and reperfused myocardium, *Biochem. Biophys. Res. Commun.,* 164, 1274, 1989.

260. Bolli, R., Patel, B. S., Jeroudi, M. O., Lai, E. K., and McCay, P. B., Demonstration of free radical generation in "stunned" myocardium of intact dogs with the use of the spin trap alpha-phenyl-N-tert-butyl nitrone, *J. Clin. Invest.,* 82, 476, 1988.

261. Tosaki, A., Blasig, I. E., Pali, T., and Ebert, B., Heart protection and radical trapping by DMPO during reperfusion in isolated working rat hearts, *Free Radical Biol. Med.,* 8, 363, 1990.

262. **Semb, A. G., Ytrehus, K., Vaage, J., Myklebust, R., and Mjos, O. D.**, Functional impairment in isolated rat hearts induced by activated leukocytes: protective effect of oxygen free radical scavengers, *J. Mol. Cell. Cardiol.*, 21, 877, 1989.
263. **Baker, J. E. and Kalyanaraman, B.**, Ischemia-induced changes in myocardial paramagnetic metabolites: implications for intracellular oxy-radical generation, *FEBS Lett.*, 244, 311, 1989.
264. **Nakazawa, H., Ichimori, K., Shinozaki, Y., Okino, H., and Hori, S.**, Is superoxide demonstration by electron-spin resonance spectroscopy really superoxide? *Am. J. Physiol.*, 255, H213, 1988.
265. **Kramer, J. H., Liang, J.-H., and Weglicki, W. B.**, A non-invasive approach to ESR spin trapping in post-ischemic rat hearts, *Fed. Am. Soc. Exp. Biol. J.*, 4, A2084, 1990.
266. **Myers, M. L., Bolli, R., Lekich, R. F., Hartley, C. J., and Roberts, R.**, Enhancement of recovery of myocardial function by oxygen free-radical scavengers after reversible regional ischemia, *Circulation*, 72, 915, 1985.
267. **Gross, G. J., Faber, N. E., Hardman, H. F., and Warltier, D. C.**, Beneficial actions of superoxide dismutase and catalase in stunned myocardium of dogs, *Am. J. Physiol.*, 250, H372, 1986.
268. **Przyklenk, K. and Kloner, R. A.**, Beneficial actions of superoxide dismutase and catalase in stunned myocardium of dogs, *Circ. Res.*, 58, 148, 1986.
269. **Westlin, W. and Mullane, K.**, Does captopril attenuate reperfusion-induced myocardial dysfunction by scavenging free radicals, *Circulation*, 77, 130, 1988.
270. **Myers, M. L., Bolli, R., Leikich, R. F., Hartley, C. J., and Roberts, R.**, N-2-mercaptopropionylglycine improves recovery of myocardial function after reversible regional ischemia, *J. Am. Coll. Cardiol.*, 8, 1161, 1986.
271. **Farber, N. E., Vercellotti, G. M., Jacob, H. S., Pieper, G. M., and Gross, G. J.**, Evidence for a role of iron-catalyzed oxidants in functional and metabolic stunning in the canine heart, *Circ. Res.*, 63, 351, 1988.
272. **Engler, R. and Covell, J. W.**, Granulocytes cause reperfusion ventricular dysfunction after 15-minute ischemia in the dog, *Circ. Res.*, 61, 20, 1987.
273. **Westlin, W. and Mullane, K. M.**, Alleviation of myocardial stunning by leukocyte and platelet depletion, *Circulation*, 80, 1828, 1989.
274. **Jeremy, R. W. and Becker, L. C.**, Neutrophil depletion does not prevent myocardial dysfunction after brief coronary occlusion, *J. Am. Coll. Cardiol.*, 13, 1155, 1989.
275. **O'Neill, P. G., Charlat, M. L., Michael, L. H., Roberts, R., and Bolli, R.**, Influence of neutrophil depletion on myocardial function and flow after reversible ischemia, *Am. J. Physiol.*, 256, H341, 1989.
276. **Go, L. O., Murry, C. E., Richard, V. J., Weischedel, G. R., Jennings, R. B., and Reimer, K. A.**, Myocardial neutrophil accumulation during reperfusion after reversible or irreversible ischemic injury, *Am. J. Physiol.*, 255, H1188, 1988.
277. **Schott, R. J., Nao, B. S., McClanahan, T. B., Simpson, P. J., Stirling, M. C., Todd, R. F., III, and Gallagher, K. P.**, F(ab')$_2$ fragments of anti-Mo1 (904) monoclonal antibodies do not prevent myocardial stunning, *Circ. Res.*, 65, 1112, 1989.
278. **Fiedler, V. B.**, Reduction of acute myocardial ischemia in rabbit hearts by nafazatrom, *J. Cardiovasc. Pharmacol.*, 6, 318, 1984.
279. **Shea, M. J., Murtagh, J. J., Jolly, S. R., Abrams, G. D., Pitt, B., and Lucchesi, B. R.**, Beneficial effects of nafazatrom on ischemic reperfused myocardium, *Eur. J. Pharmacol.*, 102, 63, 1984.
280. **O'Neill, P. G., Charlat, M. L., Kim, H.-S., Pocius, J., Michael, L. H., Harley, C. J., Zhu, W.-X., Roberts, R., and Bolli, R.**, Lipoxygenase inhibitor nafazatrom fails to attenuate postischemic ventricular dysfunction, *Cardiovasc. Res.*, 21, 755, 1987.
281. **Simpson, P. J., Todd, R. F., III, Fantone, J. C., Mickelson, J. K., Griffin, J. D., and Lucchesi, B. R.**, Reduction of experimental canine myocardial reperfusion injury by a monoclonal antibody (anti-Mo1, anti-CD11b) that inhibits leukocyte adhesion, *J. Clin. Invest.*, 81, 624, 1988.

282. **Simpson, P. J., Todd, R. F., III, Mickelson, J. K., Fantone, J. C., Gallagher, K. P., Lee, K. A., Tamura, Y., Cronin, M., and Lucchesi, B. R.**, Sustained limitation of myocardial reperfusion injury by a monoclonal antibody that alters leukocyte function, *Circulation,* 81, 226, 1990.

283. **Ambrosio, G., Weisman, H. F., Mannisi, J. A., and Becker, L. C.**, Progressive impairment of regional myocardial perfusion after initial restoration of postischemic blood flow, *Circulation,* 80, 1846, 1989.

284. **Engler, R. L., Dahlgren, M. D., Morris, D. D., Peterson, M. A., and Schmid-Schonbein, G. W.**, Role of leukocytes in response to acute myocardial ischemia and reflow in dogs, *Am. J. Physiol.,* 251, H314, 1986.

285. **Breda, M. A., Drinkwater, D. C., Laks, H., Bhuta, S., Corno, A. F., Davtyan, H. G., and Chang, P.**, Prevention of reperfusion injury in the neonatal heart with leukocyte-depleted blood, *J. Thorac. Cardiovasc. Surg.,* 97, 654, 1989.

286. **Litt, M. R., Jeremy, R. W., Weisman, H. F., Winkelstein, J. A., and Becker, L. C.**, Neutrophil depletion limited to reperfusion reduces myocardial infarct size after 90 minutes of ischemia: evidence for neutrophil-mediated reperfusion injury, *Circulation,* 80, 1816, 1989.

287. **Romson, J. L., Hook, B. G., Kunkel, S. L., Abrams, G. D., Schork, M. A., and Lucchesi, B. R.**, Reduction of the extent of ischemic myocardial injury by neutrophil depletion in the dog, *Circulation,* 67, 1016, 1983.

288. **de Lorgeril, M. A., Basmadjian, A., Lavallée, M., Clément, R., Millette, D., Rousseau, G., and Latour, J.-G.**, Influence of leukopenia on collateral flow, reperfusion flow, reflow ventricular fibrillation, and infarct size in dogs, *Am. Heart J.,* 117, 523, 1989.

289. **Mullane, K. M., Read, N., Salmon, J. A., and Moncada, S.**, Role of leukocytes in acute myocardial infarction in anesthetized dogs:relationship to myocardial salvage by anti-inflammatory drugs, *J. Pharmacol. Exp. Ther.,* 228, 510, 1984.

290. **Seewaldt-Becker, E., Rothlein, R., and Dammgen, J. W.**, CDw18 dependent adhesion of leukocytes to endothelium and its relevance for cardiac reperfusion, in *Leukocyte Adhesion Molecules: Structure, Function, and Regulation,* Springer-Verlag, New York, 1989, 138.

291. **Werns, S. B., Eller, B. T., Shea, M. J., Simpson, P. J., Dysko, R. C., Abrams, G. D., and Lucchesi, B. R.**, Protection of reperfused ischemic canine myocardium by CI-922, a new inhibitor of leukocyte activation, *J. Cardiovasc. Pharmacol.,* 12, 608, 1988.

292. **Ferrari, R., Cargnoni, A., Ceconi, C., Curello, S., Belloli, S., Albertini, A., and Visioli, O.**, Protective effect of a prostacyclin-mimetic on the ischaemic-reperfused rabbit myocardium, *J. Mol. Cell. Cardiol.,* 20, 1095, 1988.

293. **Simpson, P. J., Mickelson, J., Fantone, J. C., Gallagher, K. P., and Lucchesi, B. R.**, Ilprost inhibits neutrophil function *in vitro* and *in vivo* and limits experimental infarct size in canine heart, *Circ. Res.,* 60, 666, 1987.

294. **van der Giessen, W. J., Schoutsen, B., Tijssen, J. G. P., and Verdouw, P. D.**, Iloprost (ZK 36374) enhances recovery of regional myocardial function during reperfusion after coronary artery occlusion in the pig, *Br. J. Pharmacol.,* 87, 23, 1986.

295. **Carlson, R. E., Schott, R. J., and Buda, A. J.**, Neutrophil depletion fails to modify myocardial no reflow and functional recovery after coronary reperfusion, *J. Am. Coll. Cardiol.,* 14, 1803, 1989.

296. **Nathan, C. F., Srimal, S., Farber, C., Sanchez, E., Kabbash, L., Asch, A., Gailit, J., and Wright, S. D.**, Cytokine-induced respiratory burst of human neutrophils: dependence on extracellular matrix proteins and CD11/CD18 integrins, *J. Cell Biol.,* 109, 1341, 1989.

297. **Nathan, C. F.**, Respiratory burst in adherent human neutrophils: triggering by colony-stimulating factors CSF-GM and CSF-G, *Blood,* 73, 301, 1989.

298. **Weiss, S. J.,** Tissue destruction by neutrophils, *N. Engl. J. Med.,* 320, 365, 1989.
299. **Kraemer, R. and Mullane, K. M.,** Neutrophils delay functional recovery of the post-hypoxic heart of the rabbit, *J. Pharmacol. Exp. Ther.,* 251, 620, 1989.
300. **Smith, C. W., Rothlein, R., Hughes, B. J., Mariscalco, M. M., Rudloff, H. E., Schmalstieg, F. C., and Anderson, D. C.,** Recognition of an endothelial determinant for CD18-dependent human neutrophil adherence and transendothelial migration, *J. Clin. Invest.,* 82, 1746, 1988.
301. **Smith, C. W., Marlin, S. D., Rothlein, R., Toman, C., and Anderson, D. C.,** Cooperative interactions of LFA-1 and Mac-1 with intercellular adhesion molecular-1 in facilitating adherence and transendothelial migration of human neutrophils *in vitro, J. Clin. Invest.,* 83, 2008, 1989.
302. **Sacks, T., Moldow, C. F., Craddock, P. R., Bowers, T. K., and Jacob, H. S.,** Oxygen radicals mediate endothelial cell damage by complement-stimulated granulocytes, *J. Clin. Invest.,* 21, 1161, 1978.
303. **Weiss, S. J., Young, J., LoBuglio, A. F., and Slivka, A.,** Role of hydrogen peroxide in neutrophil-mediated destruction of cultured endothelial cells, *J. Clin. Invest.,* 68, 714, 1981.
304. **Marin, W. J., II,** Neutrophils kill pulmonary endothelial cells by a hydrogen-peroxide-dependent pathway, *Am. Rev. Respir. Dis.,* 130, 209, 1984.
305. **Wright, D. G., Bralove, D. A., and Gallin, J. I.,** The differential mobilization of human neutrophil granules, *Am. J. Pathol.,* 87, 273, 1977.
306. **White, J. G. and Estensen, R. D.,** Selective labilization of specific granules in poly-morphonuclear leukocytes by phorbol myristate acetate, *Am. J. Pathol.,* 75, 45, 1974.
307. **Slivka, A., LoBuglio, A. F., and Weiss, S. J.,** A potential role for hypochlorous acid in granulocyte-mediated tumor cell cytotoxicity, *Blood,* 55, 347, 1980.
308. **Varani, J., Fligiel, S. E., Till, G. O., Kunkel, R. G., Ryan, U. S., and Ward, P. A.,** Pulmonary endothelial cell killing by human neutrophils. Possible involvement of hydroxyl radical, *Lab. Invest.,* 53, 656, 1985.
309. **Gannon, D. E., Varani, J., Phan, S. H., Ward, J. H., Kaplan, J., Till, G. O., Simon, R. H., Ryan, U. S., and Ward, P. A.,** Source of iron in neutrophil-mediated killing of endothelial cells, *Lab. Invest.,* 57, 37, 1987.
310. **Parks, D. A. and Granger, D. N.,** Xanthine oxidase: biochemistry, distribution and physiology, *Acta Physiol. Scand.,* 548 (Suppl.), 87, 1986.
311. **Friedl, H. P., Till, G. O., Ryan, U. S., and Ward, P. A.,** Mediator-induced activation of xanthine oxidase in endothelial cells, *Fed. Am. Soc. Exp. Biol. J.,* 3, 2512, 1989.
312. **Scovic, J. S., Friedl, H. P., Mahrougui, M., Paul, J. B., Till, G. O., and Ward, P. A.,** Release by inflammatory mediators of xanthine oxidase and histamine from vascular endothelial cells, *Fed. Am. Soc. Exp. Biol. J.,* 4(3), 3619A, 1990.
313. **Entman, M. L., Youker, K., Shappell, S. B., Siegel, C., Rothlein, R., Dreyer, W. J., Schmalstieg, F. C., and Smith, C. W.,** Neutrophil adherence to isolated adult canine myocytes. Evidence for a CD18-dependent mechanism, *J. Clin. Invest.,* 85, 1497, 1990.
314. **Mariscalco, M., Shappell, S., Anderson, D. C., and Smith, C. W.,** Role of CD18 in adherence-dependent hydrogen peroxide (H_2O_2) production by rabbit neutrophils (PMNs), *Fed. Am. Soc. Exp. Biol. J.,* 4, 494a, 1990.
315. **Arnaout, M. A.,** Structure and function of the leukocyte adhesion molecules CD11/CD18, *Blood,* 75, 1037, 1990.
316. **Anderson, D. C. and Springer, T. A.,** Leukocyte adhesion deficiency: an inherited defect in the Mac-1, LFA-1, and p150,95 glycoproteins, *Annu. Rev. Med.,* 38, 175, 1987.
317. **Kishimoto, T. K., Larson, R. S., Corbi, A. L., Dustin, M. L., Staunton, D. E., and Springer, T. A.,** Leukocyte integrins, in *Leukocyte Adhesion Molecules. Structure, Function, and Regulation,* Springer, T. A., Anderson, D. C., Rosenthal, A. S., and Rothlein, R., Eds., Springer-Verlag, New York, 1990, 7.

318. **Hynes, R. O.**, Integrins: a family of cell surface receptors, *Cell*, p. 549, 1990.

319. **Ruoslahti, E. and Pierschbacher, M. D.**, New perspectives in cell adhesion: RGD and integrins, *Science*, 238, 491, 1987.

320. **Anderson, D. C., Miller, L. J., Schmalstieg, F. C., Rothlein, R., and Springer, T. A.**, Contributions of the Mac-1 glycoprotein family to adherence-dependent granulocyte functions: structure-function assessments employing subunit-specific monoclonal antibodies, *J. Immunol.*, 137, 15, 1986.

321. **Springer, T. A.**, Adhesion receptors of the immune system, *Nature (London)*, 346, 425, 1990.

322. **Todd, R. F., III, Arnaout, M. A., Rosin, R. E., Crowley, C. A., Peters, W. A., and Babior, B. M.**, Subcellular localization of the large subunit of Mo1 (Mo1-α; formerly gp 110), a surface glycoprotein associated with neutrophil adhesion, *J. Clin. Invest.*, 74, 1280, 1984.

323. **Jones, D. H., Anderson, D. C., Burr, B. L., Rudloff, H. E., Smith, C. W., Krater, S. S., and Schmalstieg, F. C.**, Quantitation of intracellular Mac-1 (CD11b/CH18) pools in human neutrophils, *J. Leukocyte Biol.*, 44, 535, 1988.

324. **Arnaout, M. A., Spits, H., Terhorst, C., Pitt, J., and Todd, R. F., III**, Deficiency of a leukocyte surface glycoprotein (LFA-1) in two patients with Mo1 deficiency. Effects of cell activation on Mo1/LFA-1 surface expression in normal and deficient leukocytes, *J. Clin. Invest.*, 74, 1291, 1984.

325. **Bainton, D. F., Miller, L. J., Kishimoto, T. K., and Springer, T. A.**, Leukocyte adhesion receptors are stored in peroxidase-negative granules of human neutrophils, *J. Exp. Med.*, 166, 1641, 1987.

326. **Arnaout, M. A., Todd, R. F., III, Dana, N., Melamed, J., Schlossman, S. F., and Colten, H. R.**, Inhibition of phagocytosis of complement C3- or immunoglobulin G-coated particles and of C3bi binding by monoclonal antibodies to a monocyte-granulocyte membrane glycoprotein (Mo1), *J. Clin. Invest.*, 72, 171, 1983.

327. **Wright, S. D., Rao, P. E., Van Voorhis, W. C., Craigmyle, L. S., Iida, K., Talle, M. A., Westberg, E. F., Goldstein, G., and Silverstein, S. C.**, Identification of the C3bi receptor of human monocytes and macrophages by using monoclonal antibodies, *Proc. Natl. Acad. Sci. U.S.A.*, 80, 5699, 1983.

328. **Rothlein, R. and Springer, T. A.**, Complement receptor type three-dependent degradation of opsonized erythrocytes by mouse macrophages, *J. Immunol.*, 135, 2668, 1985.

329. **Tonnesen, M. G., Anderson, D. C., Springer, T. A., Knedler, A., Avdi, N., and Henson, P. M.**, Adherence of neutrophils to cultured human microvascular endothelial cells. Stimulation by chemotactic peptides and lipid mediators and dependence upon the Mac-1, LFA-1, p150,95 glycoprotein family, *J. Clin. Invest.*, 83, 637, 1989.

330. **Pohlman, T. H., Stanness, K. A., Beatty, P. G., Ochs, H. D., and Harlan, J. M.**, An endothelial cell surface factor(s) induced *in vitro* by lipopolysaccharide, interleukin 1, and tumor necrosis factor α increases neutrophil adherence by a CDw18-dependent mechanism, *J. Immunol.*, 136, 4548, 1986.

331. **Zimmerman, G. A. and McIntyre, T. M.**, Neutrophil adherence to human endothelium *in vitro* occurs by CDw18 (Mo1, MAC-1/LFA-1/GP 150,95) glycoprotein-dependent and -independent mechanisms, *J. Clin. Invest.*, 81, 531, 1988.

332. **Gamble, J. K., Harlan, J. M., Klebanoff, S. J., and Vadas, M. A.**, Stimulation of the adherence of neutrophils to umbilical vein endothelium by human recombinant tumor necrosis factor, *Proc. Natl. Acad. Sci. U.S.A.*, 82, 8667, 1985.

333. **Dustin, M. L., Garcia-Aguilar, J., Hibbs, M. L., Larson, R. S., Stacker, S. A., Staunton, D. E., Wardlaw, A. J., and Springer, T. A.**, Structure and regulation of the leukocyte adhesion receptor LFA-1 and its counterreceptors, ICAM-1 and ICAM-2, *Cold Spring Harbor Symp. Quant. Biol.*, 54, 753, 1989.

334. **Rothlein, R., Dustin, M. L., Marlin, S. D., and Springer, T. A.**, A human intercellular adhesion molecule (ICAM-1) distinct from LFA-1, *J. Immunol.*, 137, 1, 1986.

335. **Staunton, D. E., Marlin, S. D., Stratowa, C., Dustin, M. L., and Springer, T. A.,** Primary structure of ICAM-1 demonstrates interaction between members of the immunoglobulin and integrin supergene families, *Cell,* 62, 825, 1988.

336. **Dustin, M. L., Rothlein, R., Bhan, A. K., Dinarello, C. A., and Springer, T. A.,** Induction by IL1 and interferon-gamma: tissue distribution, biochemistry, and function of a natural adherence molecule (ICAM-1), *J. Immunol.,* 137, 245, 1986.

337. **Smith, C. W., Marlin, S. D., Rothlein, R., Lawrence, M. B., McIntire, L. V., and Anderson, D. C.,** Role of ICAM-1 in the adherence of human neutrophils to human endothelial cells *in vitro,* in *Leukocyte Adhesion Molecules: Structure, Function, and Regulation,* Springer, T. A., Anderson, D. C., Rothlein, R., and Rosenthal, A. S., Eds., Springer-Verlag, New York, 1990, 170.

338. **Suzuki, M., Inauen, W., Kvietys, P. R., Grisham, M. B., Meinnger, C., Schelling, M. E., Granger, H. J., and Granger, D. N.,** Superoxide mediates reperfusion-induced leukocyte-endothelial cell interactions, *Am. J. Physiol.,* 257, H1740, 1989.

339. **Harlan, J. M., Schwartz, B. R., Reidy, M. A., Schwartz, S. M., Ochs, H. D., and Harker, L. A.,** Activated neutrophils disrupt endothelial monolayer integrity by an oxygen radical-independent mechanism, *Lab. Invest.,* 52, 141, 1985.

340. **Diener, A. M., Beatty, P. G., Ochs, H. D., and Harlan, J. M.,** The role of neutrophil membrane glycoprotein 150 (Gp-150) in neutrophil-mediated endothelial cell injury *in vitro,* *J. Immunol.,* 135, 537, 1985.

341. **Henson, P. M., Henson, J. E., Fittschen, C., Kimani, G., Bratton, D. L., and Riches, D. W. H.,** Phagocytic cells: degranulation and secretion, in *Inflammation: Basic Principles and Clinical Correlates,* Gallin, J. I., Goldstein, I. M., and Snyderman, R., Eds., Raven Press, New York, 1988, 363.

342. **Henson, P. M. and Johnston, R. B.,** Tissue injury in inflammation. Oxidants, proteinases, and cationic proteins, *J. Clin. Invest.,* 79, 669, 1987.

343. **Harlan, J. M., Killen, P. D., Harker, L. A., Striker, G. E., and Wright, D. G.,** Neutrophil-mediated endothelial injury *in vitro.* Mechanisms of cell detachment, *J. Clin. Invest.,* 68, 1394, 1981.

344. **Smedly, L. A., Tonnesen, M. G., Sandhaus, R. A., Haslett, C., Guthrie, L. A., Johnston, R. B., Jr., Henson, P. M., and Worthen, G. S.,** Neutrophil-mediated injury to endothelial cells. Enhancement by endotoxin and essential role of neutrophil elastase, *J. Clin. Invest.,* 77, 1233, 1986.

345. **Shingu, M., Yoshioka, K., Nobunaga, M., and Yoshida, K.,** Human vascular smooth muscle cells and endothelial cells lack catalase activity and are susceptible to hydrogen peroxide, *Inflammation,* 9, 309, 1985.

346. **Weiss, S. J. and Peppin, G. J.,** Collagenolytic metalloenzymes of the human neutrophil. Characteristics, regulation and potential function *in vivo, Biochem. Pharmacol.,* 35, 3189, 1986.

347. **Carrell, R. W.,** Alpha-1-antitrypsin: molecular pathology, leukocytes, and tissue damage, *J. Clin. Invest.,* 78, 1427, 1986.

348. **Weiss, S. J., Peppin, G., Ortiz, X., Ragsdale, C., and Test, S. T.,** Oxidative autoactivation of latent collagenase by human neutrophils, *Science,* 227, 747, 1985.

349. **Varani, J., Ginsburg, I., Schuger, L., Gibbs, D. F., Bromberg, J., Johnson, K. J., Ryan, U. S., and Ward, P. A.,** Endothelial cell killing by neutrophils: synergistic interactions of oxygen products and proteases, *Am. J. Pathol.,* 135, 435, 1989.

350. **Petrone, W. F., English, D. K., Wong, K., and McCord, J. M.,** Free radicals and inflammation: superoxide-dependent activation of a neutrophil chemotactic factor in plasma, *Proc. Natl. Acad. Sci. U.S.A.,* 77, 1159, 1980.

351. **Perez, H. D., Weksler, B. B., and Goldstein, I. M.,** Generation of a chemotactic lipid from arachidonic acid by exposure to a superoxide-generating system, *Inflammation,* 4, 313, 1980.

352. **Esterbauer, H., Jurgens, G., Quehenberger, O., and Koller, E.,** Autoxidation of human low density lipoprotein: loss of polyunsaturated fatty acids and vitamin E and generation of aldehydes, *J. Lipid Res.,* 28, 495, 1987.

353. **Steinbrecher, U. P., Lougheed, M., Kwan, W.-C., and Dirks, M.**, Recognition of oxidized low density lipoprotein by the scavenger receptor of macrophages results from derivatization of apolipoprotein B by products of fatty acid peroxidation, *J. Biol. Chem.*, 264, 15216, 1989.

354. **Curzio, M., Torrielli, M. V., Giroud, J. P., Esterbauer, H., and Dianzani, M. U.**, Neutrophil chemotactic responses to aldehydes, *Res. Commun. Chem. Pathol. Pharmacol.*, 36, 463, 1985.

355. **Curzio, M., Di Mauro, C., Esterbauer, H., and Dianzani, M. U.**, Chemotactic activity of aldehydes, structural requirements. Role in inflammatory process, *Biomed. Pharmacother.*, 41, 304, 1987.

356. **Lenz, M. L., Hughes, H., Mitchell, J. R., Via, D. P., Guyton, J. R., Taylor, A. A., Gotto, A. M., Jr., and Smith, C. V.**, Lipid hydroperoxy and hydroxy derivatives in copper-catalyzed oxidation of low density lipoprotein, *J. Lipid Res.*, 31, 1043, 1990.

357. **Goetzl, E. J. and Pickett, W. C.**, The human PMN leukocyte chemotactic activity of complex hydroxy-eicosatetraenoic acids (HETEs), *J. Immunol.*, 125, 1789, 1980.

358. **McIntyre, T. M., Zimmerman, G. A., and Prescott, S. M.**, Leukotrienes C4 and D4 stimulate human endothelial cells to synthesize platelet-activating factor and bind neutrophils, *Proc. Natl. Acad. Sci. U.S.A.*, 83, 2004, 1986.

359. **Zimmerman, G. A., McIntyre, T. M., and Prescott, S. M.**, Thrombin stimulates neutrophil adherence by an endothelial cell-dependent mechanism: characterization of the response and relationship to platelet-activating factor synthesis, *Ann. N.Y. Acad. Sci.*, 485, 349, 1986.

360. **Zimmerman, G. A., McIntyre, T. M., Mehra, M., and Prescott, S. M.**, Endothelial cell-associated platelet-activating factor: a novel mechanism for signaling intercellular adhesion, *J. Cell Biol.*, 110, 529, 1990.

361. **Gasic, A. C., McGuire, G. M., Smith, C. W., and Taylor, A. A.**, H_2O_2 pretreatment of vascular endothelium induces neutrophil adhesion by a CD 18-dependent mechanism, *Fed. Am. Soc. Exp. Biol. J.*, 4, A3728, 1990.

362. **Taylor, A. A., Gasic, A. C., McGuire, G. M., Entman, M. L., Krater, S. S., and Smith, C. W.**, Perfusion of canine arteries with H_2O_2 CD18- ICAM-1-, and platelet activating factor-dependent neutrophil adhesion, *Circulation*, 82 (Suppl 4), A456, 1990.

363. **Gasic, A. C., McGuire, G., Krater, S., Farhood, A. I., Goldstein, M. A., Smith, C. W., Entman, M. L., and Taylor, A. A.**, Hydrogen peroxide pretreatment of perfused canine vessels induces ICAM-1 and CD18-dependent neutrophil adherence, *Circulation*, in press, 1991.

364. **Simmons, D., Makgoba, M. W., and Seed, B.**, ICAM, an adhesion ligand of LFA-1, is homologous to the neural cell adhesion molecule NCAM, *Nature (London)*, 331, 624, 1990.

365. **McEver, R. P., Beckstead, J. H., Moore, K. L., Marshall-Carlson, L., and Bainton, D. F.**, GMP-140, a platelet alpha-granule membrane protein, is also synthesized by vascular endothelial cells and is localized in Weibel-Palade bodies, *J. Clin. Invest.*, 84, 92, 1989.

366. **Hattori, R., Hamilton, K. K., Fugates, R. D., McEver, R. P., and Sims, P. J.**, Stimulated secretion of endothelial von Willebrand factor is accompanied by rapid redistribution to the cell surface of the intracellular granule membrane protein GMP-140, *J. Biol. Chem.*, 264, 7768, 1989.

367. **Johnston, G. I., Cook, R. G., and McEver, R. P.**, Cloning of GMP-140, a granule membrane protein of platelets and endothelium: sequence similarity to proteins involved in cell adhesion and inflammation, *Cell*, 56, 1033, 1989.

368. **Bevilacqua, M. P., Stengelin, S., Gimbrone, M. A., Jr., and Seed, B.**, Endothelial leukocyte adhesion molecule 1: An inducible receptor for neutrophils related to complement regulatory protein and lectins, *Science*, 243, 1160, 1989.

369. **Bevilacqua, M. P., Pober, J. S., Mendrick, D. L., Cotran, R. S., and Gimbrone, M. A., Jr.**, Identification of an inducible endothelial-leukocyte adhesion molecule, *Proc. Natl. Acad. Sci. U.S.A.*, 84, 9238, 1987.

370. **Yednock, T. A. and Rosen, S. D.**, Lymphocyte homing, *Adv. Immunol.*, 44, 313, 1989.
371. **Stoolman, L. M.**, Adhesion molecules controlling lymphocyte migration, *Cell*, 56, 907, 1989.
372. **Lasky, L. A., Singer, M. S., Yednock, T. A., Dowbenko, D., Fennie, C., Rodriguez, H., Nguyen, T., Stachel, S., and Rosen, S. D.**, Cloning of a lymphocyte homing receptor reveals a lectin domain, *Cell*, 56, 1045, 1989.
373. **Siegelman, M. H., Van De Rijn, M., and Weissman, I. L.**, Mouse lymph node homing receptor cDNA clone encodes a glycoprotein revealing tandem interaction domains, *Science*, 243, 1165, 1989.
374. **Kishimoto, T. K., Jutila, M. A., Berg, E. L., and Butcher, E. C.**, Neutrophil Mac-1 and MEL-14 adhesion proteins inversely regulated by chemotactic factors, *Science*, 245, 1238, 1989.
375. **Smith, C. W., Kishimoto, T. K., Abbassi, O., McIntire, L. V., and Anderson, D. C.**, Human MEL-14 antigen contributes to the CD11/CD18-independent adhesion of neutrophils to endothelial cells, *Fed. Am. Soc. Exp. Biol. J.*, 4, A2840, 1990.
376. **Geng, J.-G., Bevilacqua, M. P., Moore, K. L., McIntyre, T. M., Prescott, S. M., Kim, J. M., Bliss, G. A., Zimmerman, G. A., and McEver, R. P.**, Rapid neutrophil adhesion to activated endothelium mediated by GMP-140, *Nature (London)*, 343, 757, 1990.
377. **Patel, K. D., Zimmerman, G. A., Prescott, S. M., McEver, R. P., and McIntyre, T. M.**, H_2O_2 and t-butylhydroperoxide induced endothelial cell (EC)-dependent neutrophil (PMN) adhesion: role of GMP140, *Fed. Am. Soc. Exp. Biol. J.*, 4, A558, 1990.
377a. **Patel, K. D. and McIntyre, T. M.**, personal communication.
378. **Suzuki, M., Grisham, M. B., and Granger, D. N.**, Hydrogen peroxide promotes neutrophil adherence in cat mesenteric venules, *Gastroenterology*, 98, 476A, 1990.
378a. **Granger, D. N.**, personal communication.
379. **Stahl, G. L., Terashita, Z., and Lefer, A. M.**, Role of platelet activating factor in propagation of cardiac damage during myocardial ischemia, *J. Pharmacol. Exp. Ther.*, 244, 898, 1988.
380. **Gillespie, M. N., Kojima, S., Kunitomo, M., and Jay, M.**, Coronary and myocardial effects of activated neutrophils in perfused rabbit hearts, *J. Pharmacol. Exp. Ther.*, 239, 836, 1986.
381. **Prasad, K., Kalra, J., Chan, W. P., and Chaudhary, A. K.**, Effect of oxygen free radicals on cardiovascular function at organ and cellular levels, *Am. Heart J.*, 117, 1196, 1989.
382. **Burton, K. P., McCord, J. M., and Ghai, G.**, Myocardial alterations due to free-radical generation, *Am. J. Physiol.*, 246, H776, 1984.
383. **Ytrehus, K., Myklebust, R., and Mjos, O. D.**, Influence of oxygen radicals generated by xanthine oxidase in the isolated perfused rat heart, *Cardiovascular Res.*, 20, 597, 1986.
384. **Jackson, C. V., Mickelson, J. K., Pope, T. K., Rao, P. S., and Lucchesi, B. R.**, O_2 free radical-mediated myocardial and vascular dysfunction, *Am. J. Physiol.*, 251, H1225, 1986.
385. **Trump, B. F. and Berezesky, I. K.**, Role of sodium and calcium regulation in toxic cell injury, in *Drug Metabolism and Drug Toxicity*, Mitchell, J. R. and Horning, M. G., Eds., Raven Press, New York, 1984, 261.
386. **Schanne, F. A. X.**, Calcium dependence of toxic cell death: a final common pathway, *Science*, 206, 700, 1979.
387. **Fliss, H.**, Oxidation of proteins in rat heart and lungs by polymorphonuclear leukocyte oxidants, *Mol. Cell. Biochem.*, 84, 177, 1988.
388. **Kim, M.-S. and Akera, T.**, O_2 free radicals: cause of ischemia-reperfusion injury to cardiac Na^+-K^+-ATPase, *Am. J. Physiol.*, 252, H252, 1987.

389. **Kramer, J. H., Mak, I. T., and Weglicki, W. B.**, Differential sensitivity of canine cardiac sarcolemmal and microsomal enzymes to inhibition by free radical-induced lipid peroxidation, *Circ. Res.*, 55, 120, 1984.

390. **Scherer, N. M. and Deamer, D. W.**, Oxidative stress impairs the function of sarcoplasmic reticulum by oxidation of sulfhydryl groups in the Ca^{2+}-ATPase, *Arch. Biochem. Biophys.*, 246, 589, 1986.

391. **Trimm, J. L., Salama, G., and Abramson, J. J.**, Sulfhydryl oxidation induces rapid calcium release from sarcoplasmic reticulum vesicles, *J. Biol. Chem.*, 261, 16092, 1986.

392. **Rowe, G. T., Manson, N. H., Caplan, M., and Hess, M. L.**, Hydrogen peroxide and hydroxyl radical mediation of activated leukocyte depression of cardiac sarcoplasmic reticulum, *Circ. Res.*, 53, 584, 1983.

393. **Kaminishi, T. and Kako, K. J.**, Sensitivity to oxidants of mitochondrial and sarcoplasmic reticular calcium uptake in saponin-treated cardiac myocytes, *Basic Res. Cardiol.*, 84, 282, 1989.

394. **Kato, M. and Kako, K. J.**, Na^+/Ca^{2+} exchange of isolated sarcolemmal membrane: effects of insulin, oxidants and insulin deficiency, *Mol. Cell. Biochem.*, 83, 15, 1988.

395. **Matsuoka, T., Kaminishi, T., Kato, M., and Kako, K. J.**, Prevention of free radical-induced dysfunction of membrane-bound (Na^+,K^+)ATPase and heart cell damage, *J. Mol. Cell. Cardiol.*, 19 (Suppl. IV), S78, 1987.

396. **Lenz, M., Michael, L. H., Smith, C. V., Hughes, H., Shappell, S. B., Taylor, A. A., Entman, M. L., and Mitchell, J. R.**, Glutathione disulfide formation and lipid peroxidation during cardiac ischemia and reflow in the dog *in vivo*, *Biochem. Biophys. Res. Commun.*, 164, 722, 1989.

397. **Noronha-Dutra, A. A. and Steen, E. M.**, Lipid peroxidation as a mechanism of injury in cardiac myocytes, *Lab. Invest.*, 47, 346, 1982.

398. **Aida, K., Onodera, T., Oguro, T., Ashraf, M., and Smith, R.**, Effects of H_2O_2 on cultured myocytes from adult rat heart, *Fed. Am. Soc. Exp. Biol. J.*, 4, A1142, 1990.

399. **Shepherd, M., Bruening, M., Auld, A. M., and Barritt, G. J.**, Effects of energy deprivation and hydrogen peroxide on contraction and myoplasmic free calcium concentrations in isolated myocardial muscle cells, *Biochem. Med. Metab. Biol.*, 38, 195, 1987.

400. **Mak, I. T., Kramer, J. H., and Weglicki, W. B.**, Potentiation of free radical-induced lipid peroxidative injury to sarcolemmal membranes by lipid amphiphiles, *J. Biol. Chem.*, 261, 1153, 1986.

401. **Whitmer, J. T., Idell-Wenger, J. A., Rovetto, M. J., and Neely, J. R.**, Control of fatty acid metabolism in ischemic and hypoxic hearts, *J. Biol. Chem.*, 253, 4305, 1978.

402. **Idell-Wenger, J. A., Grotyohann, L. W., and Neely, J. R.**, Coenzyme A and carnitine distribution in normal and ischemic hearts, *J. Biol. Chem.*, 253, 4310, 1978.

403. **Shaikh, N. A. and Downar, E.**, Time course of changes in porcine myocardial phospholipid levels during ischemia, *Circ. Res.*, 49, 316, 1981.

404. **Corr, P. B., Snyder, D. W., Lee, B. I., Gross, R. W., Keim, C. R., and Sobel, B. E.**, Pathophysiological concentrations of lysophosphatides and the slow response, *Am. J. Physiol.*, 243, H187, 1982.

405. **Chien, K. R., Han, A., Sen, A., Buja, L. M., and Willerson, J. T.**, Accumulation of unesterified arachidonic acid in ischemic canine myocardium: relationship to a phosphatidylcholine deacylation-reacylation cycle and the depletion of membrane phospholipids, *Circ. Res.*, 54, 313, 1984.

406. **Sen, A., Miller, J. C., Reynolds, R., Willerson, J. T., Buja, L. M., and Chien, K. R.**, Inhibition of the release of arachidonic acid prevents the development of sarcolemmal membrane defects in cultured rat myocardial cells during adenosine triphosphate depletion, *J. Clin. Invest.*, 82, 1333, 1988.

407. **McDonough, K. H., Henry, J. J., and Spitzer, J. J.**, Effects of oxygen radicals on substrate oxidation by cardiac myocytes, *Biochim. Biophys. Acta*, 926, 127, 1987.

408. **McDonough, K. H. and Spitzer, J. J.**, Effects of hypoxia and reoxygenation on adult rat heart cell metabolism, *Proc. Soc. Exp. Biol. Med.*, 173, 519, 1983.

409. **Goldhaber, J. I., Ji, S., Lamp, S. T., and Weiss, J. N.,** Effects of exogenous free radicals on electromechanical function and metabolism in isolated rabbit and guinea pig ventricle. Implications for ischemia and reperfusion injury, *J. Clin. Invest.,* 83, 1800, 1989.
410. **Dreyer, W. J., Michael, L. H., West, M. S., Smith, C. W., Rossen, R. D., Anderson, D. C., and Entman, M. L.,** Neutrophil localization in ischemic canine myocardium: time course and transmural distribution during reperfusion, *Circulation,* 82 (Suppl.), III, 275, 1990 (Abstr.).
411. **Youker, K., Entman, M. L., Taylor, A. A., Shappell, S. B., and Smith, C. W.,** H_2O_2 induced adherence of neutrophils to canine cardiac myocytes is PAF-dependent and involves stimulation of CD11/CD18 adherence to ICAM-1, *Circulation,* 82 (Suppl. 4), A1093, 1990 (Abstr.).
411a. **Entman, M. L.,** personal communication.
412. **Gibson, R. S.,** Current status of calcium channel-blocking drugs after Q wave and non-Q wave myocardial infarction, *Circulation,* 80 (Suppl.), IV107, 1989.
412a. **Smith, C. W., Entman, M. L., and Anderson, D. C.,** personal communication.
413. **Dorn, G. W., II, Gertler, A. S., Gordon, L., Usher, B. W., and Hendrix, G. H.,** Left ventricular dysfunction in symptomatic mitral valve prolapse, *Chest,* 95, 370, 1989.
414. **Youker, K., Smith, C. W., Shappell, S. B., Michael, L. H., Rossen, R. D., Anderson, D. C., and Entman, M. L.,** Interleukin-6 contained in post-ischemic cardiac lymph induces CD18-dependent binding of activated neutrophils to isolated cardiac myocytes, *Clin. Res.,* 38, 250A, 1990.
415. **Kreisle, R. A. and Parker, C. W.,** Specific binding of leukotriene B4 to a receptor on human polymorphonuclear leukocytes, *J. Exp. Med.,* 157, 628, 1983.
416. **Clancy, R. M., Dahinden, C. A., and Hugli, T. E.,** Arachidonate metabolism by human polymorphonuclear leukocytes stimulated by *N*-formyl-Met-Leu-Phe or complement component C5a is independent of phospholipase activation, *Proc. Natl. Acad. Sci. U.S.A.,* 80, 7200, 1983.
417. **Haines, K. A., Giedd, K. N., Rich, A. M., Korchak, H. M., and Weissmann, G.,** The leukotriene B_4 paradox: neutrophils can, but will not, respond to ligand-receptor interactions by forming leukotriene B_4 or its omega-metabolites, *Biochem. J.,* 241, 55, 1987.
418. **Osaki, M., Sumimoto, H., Takeshige, K., Cragoe, E. J., Jr., Hori, Y., and Minakami, S.,** Na^+/H^+ exchange modulates the production of leukotriene B_4 by human neutrophils, *Biochem. J.,* 257, 751, 1989.
419. **Haurand, M. and Flohé, L.,** Leukotriene formation by human polymorphonuclear leukocytes from endogenous arachidonate. Physiological triggers and modulation by prostanoids, *Biochem. Pharmacol.,* 38, 2129, 1989.
419a. **Smith, E. F., III,** personal communication.
420. **Jackson, W. T., Boyd, R. J., Froelich, L. L., Goodson, T., Bollinger, N. G., Herron, D. K., Mallet, B. E., and Gapinski, D. M.,** Inhibition of LTB$_4$ binding and aggregation of neutrophils by LY255283 and LY223982, *Fed. Am. Soc. Exp. Biol. J.,* 2, A1110, 1988 (Abstr.).
421. **Jackson, W. T., Froelich, L. L., Goodson, T., Herron, D. K., Mallett, B. E., and Gapinski, D. M.,** Inhibition of LTB$_4$ induced leukopenia by LY255283 and LY223982, *Pharmacologist,* 30, A206, 1988 (Abstr.).
422. **Jackson, R. H., Morrissey, M. M., Tivade, J. S., Hurt, S., Jarvis, M. F., and Sills, M. A.,** [³H]CGS 23131, a novel antagonist radioligand for the LTB$_4$ receptor on human polymorphonuclear neutrophils, *Fed. Am. Soc. Exp. Biol. J.,* 4(3), 4980A, 1990.
423. **Hahn, R. A., MacDonald, B. R., Simpson, P. J., Potts, B. D., and Parli, C. J.,** Antagonism of leukotriene B$_4$ receptors does not limit canine myocardial infarct size, *J. Pharmacol. Exp. Ther.,* 253, 58, 1990.
424. **Taylor, A., Gasic, A., Kitt, T., Shappell, S., Rui, J., Lenz, M., Smith, C. W., and Mitchell, J.,** A specific leukotriene B$_4$ antagonist protects against myocardial ischemia-reflow injury, *Clin. Res.,* 37, 528A, 1989 (Abstr.).

425. Gillard, J., Ford-Hutchinson, A. W., Chan, C., Charleson, S., Denis, D., Foster, A., Fortin, R., Leger, S., McFarlane, C. S., Morton, H., Piechuta, H., Riendeau, D., Rouzer, C. A., Rokach, J., Young, R., MacIntyre, D. E., Peterson, L., Bach, T., Eiermann, G., Hopple, S., Humes, J., Hupe, L., Luell, S., and Metzger, J., L-663,536 (MK-886) (3-[1-(4-chlorobenzyl)-3-t-butyl-thio-5-isopropylindol-2-yl]-2,2-dimethylpropanoic acid), a novel, orally active leukotriene biosynthesis inhibitor, *Can. J. Physiol. Pharmacol.*, 67, 456, 1989.

426. Rouzer, C. A., Ford-Hutchinson, A. W., Morton, H. E., and Gillard, J. W., MK886, a potent and specific leukotriene biosynthesis inhibitor blocks and reverses the membrane association of 5-lipoxygenase in ionophore-challenged leukocytes, *J. Biol. Chem.*, 265, 1436, 1990.

427. Guhlmann, A., Keppler, A., Kästner, S., Krieter, H., Brückner, U. B., Messmer, K., and Keppler, D., Prevention of endogenous leukotriene production during anaphylaxis in the guinea pig by an inhibitor of leukotriene biosynthesis (MK-886) but not by dexamethasone, *J. Exp. Med.*, 170, 1905, 1989.

428. Miller, D. K., Gillard, J. W., Vickers, P. J., Sadowski, S., Læeveillé, C., Mancini, J. A., Charleson, P., Dixon, R. A. F., Ford-Hutchinson, A. W., Fortin, R., Gauthier, J. Y., Rodkey, J., Rosen, R., Rouzer, C., Sigal, I. S., Strader, C. D., and Evans, J. F., Identification and isolation of a membrane protein necessary for leukotriene production, *Nature (London)*, 343, 278, 1990.

429. Dixon, R. A. F., Diehl, R. E., Opas, E., Rands, E., Vickers, P. J., Evans, J. F., Gillard, J. W., and Miller, D. K., Requirement of a 5-lipoxygenase-activating protein for leukotriene synthesis, *Nature (London)*, 343, 282, 1990.

430. Bassand, J.-P., Machecourt, J., Cassagnes, J., Lusson, J. R., Borel, E., and Schiele, F., Limitation of myocardial infarct size and preservation of left ventricular function by early administration of APSAC in myocardial infarction, *Am. J. Cardiol.*, 64, 18A, 1989.

431. Tanaka, M., Brooks, S. E., FitzHarris, G. P., Stoler, R. C., Jennings, R. B., and Reimer, K. A., Effect of the IB4 anti-CD18 antibody on myocardial PMN accumulation and infarct size in dogs, *Fed. Am. Soc. Exp. Biol. J.*, 4, A4378, 1990 (Abstr.).

431a. Dreyer, W. J., personal communication.

431b. Entman, M. L., Youker, K., Taylor, A. A. and Shappell, S. B., unpublished observations.

432. Kihlström, M., Protection effect of endurance training against reoxygenation-induced injuries in rat heart, *J. Appl. Physiol.*, 68, 1672, 1990.

433. Cupo, J. F., Allen, R. A., Jesaitis, A. J., and Bokoch, G. M., Reconstitution and characterization of the human neutrophil N-formyl peptide receptor and GTP binding proteins in phospholipid vesicles, *Biochim. Biophys. Acta*, 982, 31, 1989.

434. Wright, S. D., Reddy, P. A., Jong, M. T. C., and Erickson, B. W., C3bi receptor (complement receptor type 3) recognizes a region of complement protein C3 containing the sequence Arg-Gly-Asp, *Proc. Natl. Acad. Sci. U.S.A.*, 84, 1965, 1987.

435. Goldman, D. W., Regulation of the receptor system for leukotriene B4 on human neutrophils, *Ann. N.Y. Acad. Sci.*, 524, 187, 1988.

436. Baggiolini, M., Walz, A., and Kunkel, S. L., Neutrophil-activating peptide-1/interleukin 8, a novel cytokine that activates neutrophils, *J. Clin. Invest.*, 84, 1045, 1989.

437. Besemer, J., Hujber, A., and Kuhn, B., Specific binding, internalization, and degradation of human neutrophil activating factor by human polymorphonuclear leukocytes, *J. Biol. Chem.*, 264, 17409, 1989.

438. Hwang, S.-B., Specific receptors of platelet-activating factor, receptor heterogeneity, and signal transduction mechanisms, *J. Lipid Mediat.*, 2, 123, 1990.

430. Dent, G., Ukena, D., Chanez, P., Sybrecht, G., and Barnes, P., Characterization of PAF receptors on human neutrophils using the specific antagonist, WEB 2086: correlation between receptor binding and function, *FEBS Lett.*, 244, 365, 1989.

440. Entman, M. L., Youker, K., Shappell, S. B., Siegel, C., Rothlein, R., Dreyer, W. J., Schmalstieg, F. C., and Smith, C. W., Neutrophil adherence to isolated adult canine myocytes. Evidence for a CD18-dependent mechanism, *J. Clin. Invest.*, 85, 1497, 1990.

Chapter 6

BIOACTIVATION OF XENOBIOTICS BY THE RESPIRATORY BURST OF HUMAN GRANULOCYTES

Michael D. Corbett and Bernadette R. Corbett

TABLE OF CONTENTS

I. INTRODUCTION

Certain phagocytic cells of the hemopoietic system are probably more responsible for the production of reactive forms of oxygen than any other cell type in humans. The role that such active oxygen species play in the maintenance of health has been a frequent subject of review.[1-3] Of the mechanisms by which hemopoietic cells produce reactive oxygen species (ROS), the "respiratory burst" is likely the most important, certainly on a quantitative basis.[1-3] The respiratory burst phenomenon is best known for its occurrence in phagocytic cells, especially the neutrophils and macrophages (Figure 1). Cell types with lesser roles, such as phagocytes, but which possess a potent ability to display the respiratory burst, are the eosinophils and monocytes. All of these cell types are closely related, and play a similar function in homeostasis by killing foreign cells. The respiratory burst has an indispensable role in effecting such killing. There are some differences in the processes by which these cells form active oxygen species. A good example of a difference in available metabolic chemistry is the loss of the microbicidal enzyme myeloperoxidase from the monocytes as they take up permanent locations and mature into the various tissue macrophages. Following a minimal quantum of stimulation, both monocytes and macrophages can display the respiratory burst, but, since macrophages lack myeloperoxidase, the nature and amounts of the ultimate ROS produced by the two cell types should be significantly different. Such differences in reactive oxygen chemistry are likely to be an important determinant of the physiological function of these related cell types.

Extensive evidence also implicates the adverse role that such hemopoietic cells have in certain pathological conditions. The release of ROS and other radicals is thought to play an important role in the etiology of important inflammatory diseases, including such conditions as emphysema, ulcerative colitis, and adult respiratory distress (ARD) syndrome.[1,4] Thus, the evolution of such specialized cells with the ability to produce toxic species derived from molecular oxygen has provided higher organisms with an effective mechanism to protect against foreign organisms, but also, under some conditions, can present significant hazards to the parent organism.

II. EFFECT OF THE RESPIRATORY BURST ON XENOBIOTICS

Reactive species produced by the respiratory burst can react chemically with a wide assortment of biological and nonbiological compounds (or xenobiotics). We are particularly interested in the reactions of respiratory burst species with xenobiotics. In previous studies, we demonstrated that both peroxidases from a variety of sources and phagocytic cells undergoing the respiratory burst were capable of converting nontoxic xenobiotic chemicals into highly reactive intermediates.[5-7] The production of reactive intermediates

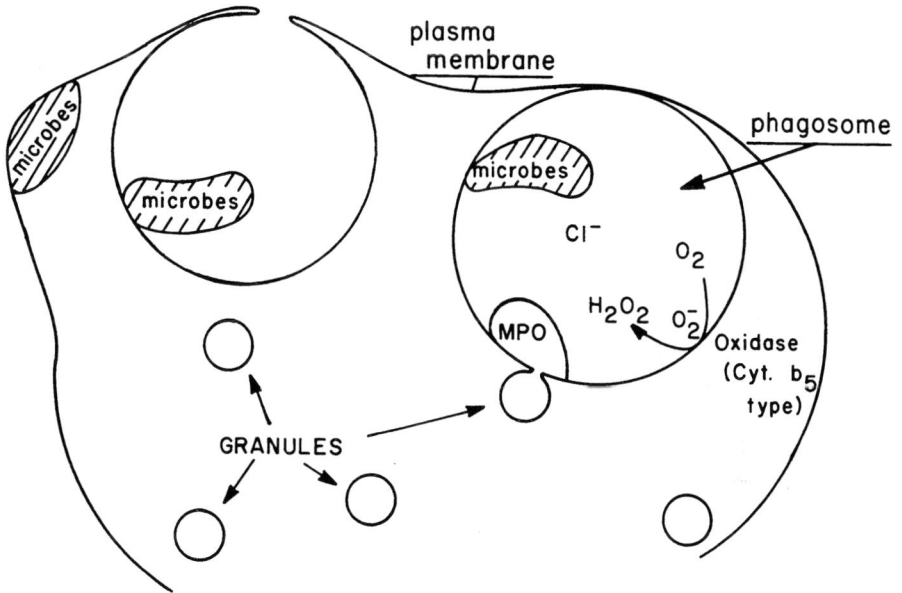

FIGURE 1. The phagocytic function of granulocytes.

from certain xenobiotics by phagocytic cells undergoing the respiratory burst, appears to be more complex, probably because of the greater complexity of the respiratory burst relative to individual reactions catalyzed by peroxidases. Our *in vitro* studies involved comparisons of the fate of the xenobiotics incubated with peroxidases or added to phagocytic cell cultures that had been induced to give the respiratory burst.[6,7] We found that the respiratory burst of granulocytes is more active in effecting the conversion of arylamines into other products than are most peroxidases, and that some of the xenobiotic metabolites have a strong ability to bind to cellular macromolecules. We have focused on the binding of arylamines to nucleic acids, but extensive binding to protein has also been noted. We have not attempted to investigate covalent binding of arylamines to sugars or lipids.

III. RESPIRATORY BURST-INDUCED BINDING OF ARYLAMINES TO NUCLEIC ACIDS

After our initial observations of the ability of granulocytes to effect the binding of several selected arylamines as the result of the respiratory burst,[6,7] we extended our studies in order to determine if this was a general phenomenon. As shown in Table 1, all compounds that possess an amino group attached to the aromatic ring were found to give varying amounts of binding to both DNA and RNA. In all cases, the amount of binding (expressed as

TABLE 1
Binding of Arylamines and Related Xenobiotics to
Granulocyte Nucleic Acids

Substrate	nmol Arylamine/mg nucleic acid			
	DNA	**$(n)^a$**	**RNA**	**$(n)^a$**
2-Aminofluorene	0.61 ± 0.40	(25)	2.97 ± 0.9	(21)
Aniline	0.45 ± 0.07	(4)	19.10 ± 1.5	(4)
N,N-Dimethylaniline	0.35 ± 0.10	(4)	4.70 ± 1.0	(4)
4-Methylaniline	0.26 ± 0.23	(12)	4.73 ± 1.8	(11)
Phenetidine	0.22 ± 0.06	(5)	13.80 ± 4.8	(3)
Aminophenol	0.16 ± 0.03	(4)	60.80 ± 14.8	(4)
Acetaminophen (^{14}C–Ar)	0.14 ± 0.10	(22)	3.90 ± 2.0	(21)
Acetaminophen (^{14}C=O)	0.13 ± 0.06	(8)	2.40 ± 1.2	(6)
4-Chloroaniline	0.11 ± 0.09	(36)	7.54 ± 3.4	(33)
3,4-Dichloroaniline	0.05 ± 0.05	(6)	4.55 ± 2.4	(6)
4-Nitroaniline	<0.01	(13)	5.15 ± 4.9	(13)
Phenacetin	<0.01	(3)	0.30 ± 0.2	(3)
Acetanilide	<0.01	(3)	<0.01	(3)
4-Chloroacetanilide	<0.01	(3)	<0.01	(3)
4-Nitroacetanilide	<0.01	(3)	<0.01	(3)
2-Acetylaminofluorene	<0.01	(3)	<0.01	(3)

Note: Granulocytes (2×10^6 cells/ml) were activated by addition of 0.1 μg/mL
phorbol myristate acetate (PMA) and then incubated with 10 μM radioactive
substrate for 30 min. The nucleic acids were isolated and the amount of
substrate incorporated was quantitated as previously described.[6]

[a]   Numbers in parentheses indicate the number *(n)* of observations.

specific activity) was higher for RNA than for DNA. Figure 2 illustrates
possible fates of an arylamine xenobiotic following its activation by some
product of the respiratory burst. The reactive metabolite may have to follow
a "tortuous path" before reaching and reacting with DNA. On the other hand,
RNA is freely available in the cytoplasm and is in much closer proximity to
where the reactive metabolites of the arylamines are produced.

Because of the inherent problems of reproducibility encountered from
using several human subjects as granulocyte donors over a 2-year-period, we
attempted to minimize these differences by employing 4-chloroaniline as a
substrate in most sets of experiments. The data were then normalized relative
to the amount of binding for this compound. Table 2 shows the amount of
binding of the various arylamine or arylamide substrates relative to 4-chlo-
roaniline. An important trend is that aromatic compounds with the strongest
electron donating substituents gave the highest amount of binding to DNA.

Note that amide derivatives of arylamines were not activated by the
respiratory burst of granulocytes, and that no nucleic acid binding occurred
in these cases (Table 1). This was true in all cases studied in which the
substrate lacked either a free aromatic amino group or a phenolic functional

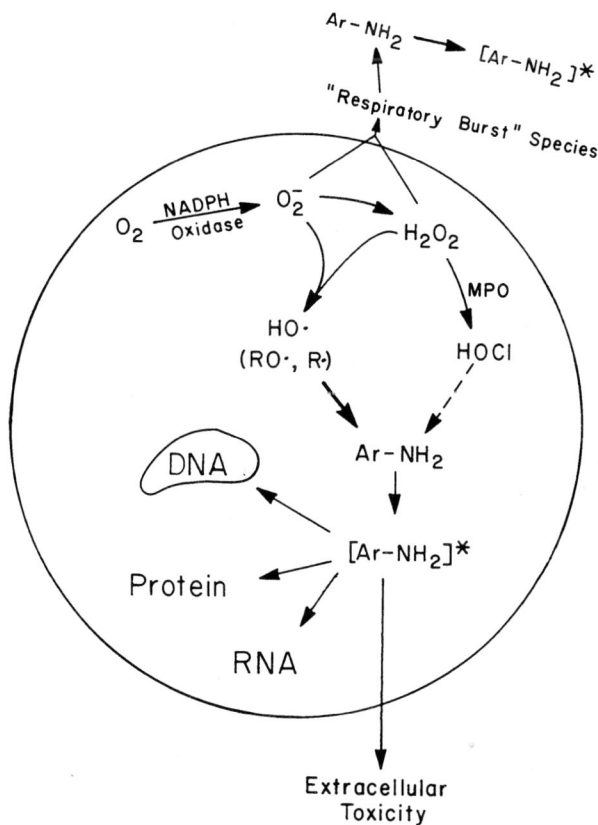

FIGURE 2.   Bioactivation of xenobiotic arylamines by the respiratory burst of phagocytic cells.

group. Even the polynuclear amide, 2-acetylaminofluorene, was recalcitrant to bioactivation by the respiratory burst of human granulocytes. Such a finding not only confirms the recalcitrance of amides to such bioactivation, but also demonstrates clearly that a relatively electron-rich polynuclear aromatic system is also inert to such an activation system. On the other hand, it does appear that higher polynuclear aromatic hydrocarbons are activated by the respiratory burst.[8] From an a priori basis, we expect that certain polynuclear aromatic hydrocarbons, even those that lack the amine or phenolic functional group, would probably form a 1 e^- oxidation product under the conditions of the respiratory burst. Whether such an oxidized hydrocarbon might interact and bind to DNA cannot be predicted at this time.

A notable exception to our observation that arylamides are not bioactivated by the respiratory burst is the case of 4-hydroxyacetanilide; this compound is also known as acetaminophen or paracetamol and is one of the most widely employed nonprescription analgesics available for human use. We have clearly

TABLE 2
Relative Binding of Arylamine Substrates to
Cellular Nucleic Acids

Substrate (n)[a]	nmol Arylamine/mg nucleic acid	
	DNA	RNA
2-Aminofluorene (26)	6.6 ± 3.4	0.46 ± 0.1
Phenetidine (4)	2.7 ± 0.8	3.10 ± 1.5
4-Aminophenol (4)	2.2 ± 1.0	10.40 ± 5
4-Aminoacetanilide (2)	2.2 ± 1.0	0.42 ± 0.2
Acetaminophen [^{14}C=O] (8)	1.7 ± 0.7	0.20 ± 0.1
Acetaminophen [^{14}C–Ar] (21)	1.5 ± 0.5	0.46 ± 0.2
4-Methylaniline (10)	1.4 ± 0.7	1.40 ± 0.8
4-Chloroaniline (37)	1.0	1.0
3,4-Dichloroaniline (3)	0.4 ± 0.1	0.59 ± 0.05
Phenacetin (3)	0	0.03 ± 0.01
4-Nitroaniline (7)	0	0.50 ± 0.2
Acetanilide (3)	0	0
4-Chloroacetanilide (3)	0	0
4-Nitroacetanilide (3)	0	0

Note: Granulocytes (2×10^6) cells/ml were activated by addition
of 0.1 μg/ml of phorbol myristate acetate (PMA) and then
incubated with 10 μ*M* of radioactive substrate for 30 min.
The experiments were paired with 4-chloroaniline and, after
quantitation,[6] the values of binding to DNA and RNA were
normalized to the amount of binding by 4-chloroaniline.

[a]   Numbers in parentheses indicate the number *(n)* of observations.

demonstrated that this chemical is bound to nucleic acids as the result of the
respiratory burst.[7] The aromatic ring and the acetyl group bind equally. This
contrasts markedly with other studies that have employed bioactivation systems including cytochrome P-450, common peroxidases, and prostaglandin
synthase.[9,10] Those studies generally concluded that acetaminophen was not
bioactivated to a metabolite that would bind to nucleic acids in an intact
manner (i.e., without prior loss of the acetyl group). Of course, acetaminophen
is a special case for amides in that it possesses a phenolic functional group,
which may be as susceptible to metabolic activation by the respiratory burst
as is the free aromatic amine functional group.[7]

   Of considerable interest was the observation that *N,N*-dimethylaniline
gave rather extensive covalent binding to granulocyte nucleic acids following
the respiratory burst (Table 1). This discovery raises a question concerning
whether or not arylamines that bind to nucleic acids must possess at least one
free N–H bond. If an N–H bond is not necessary in order for the respiratory
burst to effect nucleic acid binding, then a feasible mechanism for arylamine
activation is by simple extraction of an electron from an electron-rich aromatic

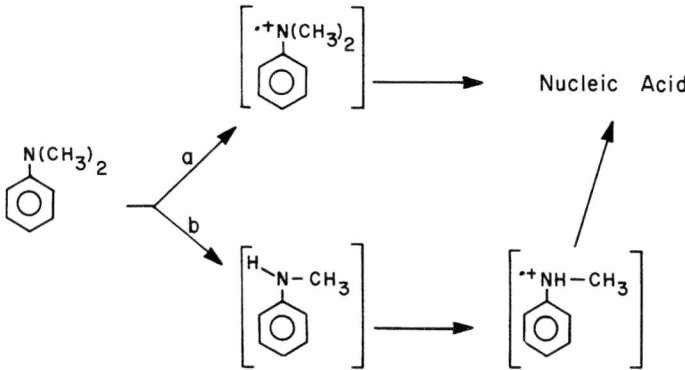

SCHEME 1. Possible mechanisms for the covalent binding of dimethylaniline to nucleic acids.

system (a characteristic common to most arylamines). Such a mechanism (Scheme 1, path a) is consistent with our observation that acylated amines (arylamides) are not activated by the respiratory burst. The deactivating effect of the carbonyl group could result from the delocalization of the electron pair on the nitrogen atom into the carbonyl oxygen atom.

On the other hand, our data from amides does not preclude the alternative mechanism for arylamine activation which involves H atom abstraction (Scheme 1, path b). Even though [ring ^{14}C]-N,N-dimethylaniline binds quite well to nucleic acids, we cannot rule out the possibility that bioactivation via H atom abstraction follows oxidative demethylation by some process. Studies with labeled methyl groups in N,N-dimethylaniline have yet to be conducted to answer this question. However, we have been unable, using HPLC methods, to detect the intermediary production of N-methylaniline or aniline during the respiratory burst action on N,N-dimethylaniline, which results in nearly the same amount of nucleic acid binding. At this time, our available data tends to suggest path a in Scheme 1 for the binding of N,N-dimethylaniline to nucleic acids.

IV. SIGNIFICANCE OF RESPIRATORY BURST-INDUCED BIOACTIVATION OF XENOBIOTICS

Since the 1960 reports of the Millers proposing mechanisms by which cytochrome P-450, in concert with certain additional enzymes, causes the metabolic activation of arylamines and related compounds,[11] extensive research has been conducted in this area. A unifying concept of these mechanisms is the production of either the hydroxylamine or hydroxamic acid analogs of the amine. Such active metabolites, upon further enzyme-catalyzed conjugation, lead to highly reactive electrophilic species, possibly even a nitrenium ion species which serves as the ultimate toxic species by virtue of

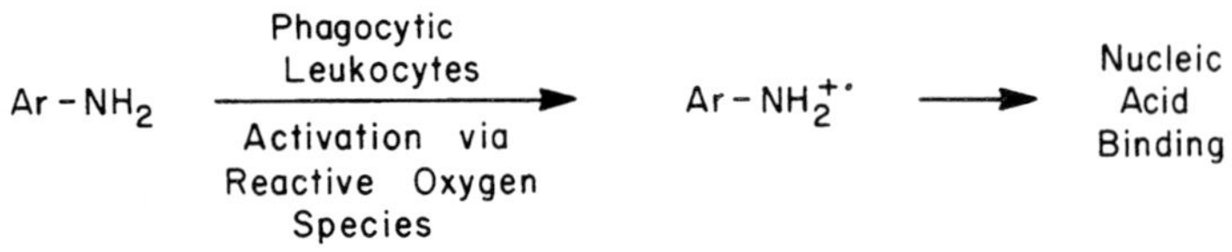

$$Ar-NH_2 \xrightarrow[\substack{\text{Activation via} \\ \text{Reactive Oxygen} \\ \text{Species}}]{\substack{\text{Phagocytic} \\ \text{Leukocytes}}} Ar-NH_2^{+\cdot} \longrightarrow \substack{\text{Nucleic} \\ \text{Acid} \\ \text{Binding}}$$

SCHEME 2. One-electron oxidation of arylamines to reactive electrophilic radicals.

a strong ability to form covalent bonds to critical cellular macromolecules. Our studies with phagocytic-type cells have indicated that such a pathway is not the mechanism by which the respiratory burst causes bioactivation of arylamines. We have been unable to detect the production of hydroxylamines from arylamines, and amides are not even activated by this process. This has led us to propose a totally different mechanism by which arylamines are activated by the respiratory burst of granulocytes. We believe that a process, such as a 1 e^- abstraction by an electron-deficient oxygen metabolite (e.g., hydroxyl radical) operates during the respiratory burst (Scheme 2). Such a process would produce an electrophilic radical species, which could react with various available biological nucleophiles, including nucleic acids. More important than just providing another pathway for macromolecule binding is the potential ability of such a radical-induced reaction to activate arylamines that are not bioactivated by the 2 e^- cytochrome P-450 pathway of arylamine activation. Many of the compounds in Table 1, especially aniline and anilines with electron withdrawing groups on the aromatic ring, will not give nucleic acid binding via the classical cytochrome P-450 pathway, but do give significant levels of nucleic acid binding as the result of the respiratory burst, which the data clearly shows.

The significance of the potential existence of a totally different mechanism for arylamine bioactivation is great. Not only does this new pathway allow for the potential genotoxicity of arylamines originally thought to lack such a property, but it is also relevant to such pharmacological factors as the site of activation and subsequent organ specificity. It is possible that many arylamines could be activated at sites where the respiratory burst occurs, such as sites of infection or chronic inflammation. A critical question to answer is whether specific arylamines mitigate or exacerbate the effect of ROS upon the cells with which they come into contact. Research on this question is a major activity in our laboratory.

ACKNOWLEDGMENT

This research was supported by Grant ES03631 from the National Institute of Environmental Health Sciences, DHHS.

REFERENCES

1. **Klebanoff, S. J.**, Oxygen metabolism and the toxic properties of phagocytes, *Ann. Intern. Med.*, 105, 480, 1980.
2. **Babior, B. M. and Crowley, C. A.**, Chronic granulomatous disease and other disorders of oxidative killing by phagocytes, in *Metabolic Basis of Inherited Disease*, Stanbury, J. B., Wyngaarden, J. B., Frederickson, D. S., Goldstein, J. L., and Brown, M. S., Eds., McGraw-Hill, New York, 1983, 1956.
3. **Sbarra, A. J. and Strauss, R. R., Eds.**, *The Respiratory Burst and Its Physiological Significance*, Plenum Press, New York, 1988.
4. **Brain, J. D.**, Toxicological aspects of alteration of pulmonary macrophage function, *Annu. Rev. Pharmacol. Toxicol.*, 26, 547, 1986.
5. **Corbett, M. D., Corbett, B. R., and Batchelor, A. O.**, Action of chloroperoxidase on 4-chloroaniline: *N*-oxidation and halogenation, *Biochem. J.*, 187, 893, 1980.
6. **Corbett, M. D. and Corbett, B. R.**, Nucleic acid binding of arylamines during the respiratory burst of human granulocytes, *Chem. Res. Toxicol.*, 1, 356, 1988.
7. **Corbett, M. D., Corbett, B. R., Hannothiaux, M.-H., and Quintana, S. J.**, Metabolic activation and nucleic acid binding of acetaminophen and related arylamine substrates by the respiratory burst of human granulocytes, *Chem. Res. Toxicol.*, 2, 260, 1989.
8. **Trush, M. A., Seed, J. L., and Kensler, T. W.**, Oxidant-dependent metabolic activation of polycyclic aromatic hydrocarbons by phorbol ester-stimulated human polymorphonuclear leukocytes: possible link between inflammation and cancer, *Proc. Natl. Acad. Sci. U.S.A.*, 82, 5194, 1985.
9. **Hinson, J. A.**, Biochemical toxicology of acetaminophen, in *Reviews in Biochemical Toxicology*, Hodgson, E., Bend, J. R., and Philpot, R. M., Eds., Elsevier, New York, 1980, 103.
10. **Larsson, R., Ross, D., Berlin, T., Olsson, L. I., and Moldeus, P.**, Prostaglandin synthase catalyzed metabolic activation of *p*-phenetidine and acetaminophen by microsomes isolated from rabbit and human kidney, *J. Pharmacol. Exp. Ther.*, 235, 475, 1985.
11. **Miller, E. C. and Miller, J. A.**, Some historical perspectives on the metabolism of xenobiotic chemicals to reactive electrophiles, in *Bioactivation of Foreign Compounds*, Anders, M. W., Ed., Academic Press, San Diego, 1985, 3.

Chapter 7

ROLE OF IRON AND REACTIVE OXYGEN SPECIES IN ASBESTOS-INDUCED LUNG INJURY

Susan Shull, Muniraj Manohar, Joanne P. Marsh, Yvonne M. W. Janssen, and Brooke T. Mossman

TABLE OF CONTENTS

I. ASBESTOS AND LUNG INJURY

Asbestos is a family of minerals of diverse chemical and physical composition. The common feature of these hydrated silicates is their fibrous geometry. There are two major classes of asbestos fibers, amphibole and serpentine. Crocidolite, a characteristic amphibole, is needlelike in appearance. Chrysotile is curved and threadlike, as its serpentine designation suggests. In addition to their distinct chemical composition, asbestos fibers are typically associated with other mineral ores and frequently contaminated with inorganic and organic solvents from commercial processing (for a recent review see Reference 1).

Inhalation of asbestos causes both malignant and nonmalignant diseases of the respiratory tract. Bronchogenic carcinoma can occur in the respiratory epithelium after a latency period of 20 to 30 years. The occurrence of this type of lung cancer in asbestos workers is closely linked to other factors, especially smoking. Mesotheliomas occur in the pleura and in the peritoneum. This disease also has a protracted latency period of 35 to 40 years. The occurrence of malignant mesotheliomas in humans is evidently not enhanced by smoking. Nonmalignant alveolar fibrosis, known as asbestosis, can also occur in asbestos workers (for review see Reference 2).

In experimental models of disease, a common sequence of events ensues in the respiratory tract in response to asbestos, regardless of the route of administration (inhalation, intratracheal instillation, or injection into the lung). Asbestos causes an inflammatory response with characteristic infiltration of alveolar macrophages (AM), polymorphonuclear leukocytes (PMN), and lymphocytes. These recruited cells and resident phagocytes can interact directly with asbestos fibers by phagocytosis, although this process is incomplete with longer asbestos fibers.[3] Accompanying lysosomal enzyme activity results in attempted dissolution of the fibers, cell death, and release of lysosomal constituents into the extracellular milieu. The resident cell population of the lower respiratory tract may interact directly with the asbestos fibers or react to the various substances evolved by the recruited inflammatory and phagocytic cells — an "innocent bystander" effect.

II. INVOLVEMENT OF REACTIVE OXYGEN SPECIES IN ASBESTOS-MEDIATED LUNG CELL INJURY

A variety of lung cells exhibit decreased viability in response to asbestos exposure *in vitro*. These cytotoxic responses are dependent upon the fibrous geometry of asbestos, as well as the fiber length. Long asbestos fibers (>10 μm) cause a more extensive cytotoxic response in hamster tracheal epithelial cells (HTE) than do short fibers (<2 μm).[4,5] Furthermore, freshly isolated hamster and rat AM release superoxide ($O_2^- \cdot$) in a dose-dependent manner when exposed to crocidolite asbestos, but not the chemically identical, nonfibrous analog, riebeckite.[6]

Several studies suggest that asbestos-induced lung cell injury involves reactive oxygen species (ROS). For example, endogenous superoxide dismutase (SOD) activity increases in response to asbestos in HTE cells, rat lung fibroblasts, and AM isolated from the lungs of hamsters and rats after inhalation of chrysotile or crocidolite.[4-6] In contrast to its inability to prevent cell damage caused by glass fibers (Code 100 fibrous glass), exogenously added SOD protects tracheal epithelial cells from injury due to asbestos exposure. The protective effect of SOD is dependent on the active form of the enzyme[5] and cannot be mimicked by nonspecific protein additions, such as bovine serum albumin (BSA).[7] These data indicate a role for $O_2^-\cdot$ or other ROS in asbestos-mediated lung cell injury.

Involvement of ROS in asbestos-mediated cell damage occurs at a variety of metabolic levels. Lipid peroxidation was observed by Gabor and Anca[8] in red cell membranes after exposure to asbestos. Asbestos-induced lipid peroxidation has also been demonstrated in rat lung microsomes,[9] phospholipid preparations,[10] in C3H10T1/2 cell cultures,[11] and in rat lung cells.[12] Most recently, we have demonstrated by HPLC, enhanced lipid peroxidation in inflammatory cells in bronchoalveolar lavage fluid of rats exposed by inhalation to asbestos.[13] The peroxidation of membrane lipids is inhibitable by treatment of asbestos with iron chelators, such as desferroxamine[10,14,15] or EDTA,[16] and, therefore, is probably mediated by the hydroxyl radical (HO·).

Asbestos-mediated DNA damage also appears to involve ROS. Kasai and Nishimura[17] measured hydroxylation at C-8 of the guanine residues in DNA incubated in a neutral, aqueous medium containing asbestos and hydrogen peroxide (H_2O_2). DNA strand breaks were observed in this system and in the DNA of C3H10T1/2 cells cultured in the presence of asbestos.[14] Treatment with desferroxamine inhibited the asbestos-mediated DNA strand breaks, again suggesting an oxidant-mediated process.

The involvement of ROS in asbestos-induced cell damage also has been elucidated by direct measurement of ROS. Leanderson et al.[18] have demonstrated, in the absence of added H_2O_2, the asbestos-induced hydroxylation of 2-deoxyguanosine to 8-hydroxydeoxyguanosine, a process mediated by hydroxyl radical. Using a cell-free system, Elstner et al.[19] examined several enzyme systems that catalyze oxygen reduction via HO· and found crocidolite caused an increase in ROS. Using electron paramagnetic resonance spectroscopy (EPR) and spin-trap reagents, Zalma et al.[20] demonstrated HO· production by asbestos in a cell-free system. In a human mesothelial cell system, however, Gabrielson et al.[21] did not observe either DNA strand breaks or ROS as detected by EPR when cells were exposed to amosite asbestos.

Within an aqueous, cell-free medium, asbestos is capable of generating both HO· and $O_2^-\cdot$ radicals from H_2O_2.[22-24] These studies, in addition to those of Costa et al.,[25] indicate that the reducing activity present on the surface of asbestos depends on the availability of Fe^{2+} and H_2O_2. H_2O_2 is readily available from phagocytic cells through the purposeful, partial reduction of oxygen. The generation of oxygen radicals by asbestos through the decomposition of H_2O_2 might occur by Fenton-Haber-Weiss chemistry:[26,27]

$$H_2O_2 + (\text{crocidolite}) - Fe^{2+} \longrightarrow HO\cdot + HO^- + (\text{crocidolite}) - Fe^{3+}$$

and

$$(\text{crocidolite}) - Fe^{3+} + H_2O_2 \longrightarrow (\text{crocidolite}) - Fe^{2+} + 2H^+ + O_2^-\cdot$$

or

$$HO\cdot + H_2O_2 \longrightarrow H_2O + H^+ + O_2^-\cdot$$

Well-established biological routes of ROS generation exist, such as the "respiratory burst" of phagocytic cells where $O_2^-\cdot$ is produced catalytically by NADPH oxidase present in cell membranes and the xanthine oxidase reaction in endothelial cells.[28-32] Xanthine oxidase derives from the NAD^+-dependent xanthine dehydrogenase, which produces no radical intermediate, through an irreversible proteolytic conversion. The resultant inflammatory response to inhalation of asbestos provides the necessary extracellular mediators capable of ROS generation, a process enhanced by participation of recruited and resident lung cells.[3]

Both resident cells within the lung, as well as infiltrating phagocytic cells, are capable of responding to asbestos by generation of ROS. Asbestos stimulates PMN metabolic activity and generation of ROS when exposed *in vitro*.[33,34] However, human peripheral blood monocytes do not release ROS and demonstrate suppressed metabolic activity after exposure to asbestos.[35] Mouse and hamster AM also respond to asbestos exposure *in vitro* by increased generation of ROS.[36,37]

We have shown that rodent AM (hamster and rat) generate $O_2^-\cdot$ when exposed to crocidolite or chrysotile asbestos *in vitro*.[6,38] Long chrysotile fibers were more potent than short ones (<2 μm) in generating $O_2^-\cdot$.[38]

The addition of mannitol or dimethylthiourea (DMTU), scavengers of ROS, decreases cytotoxicity in HTE cells due to crocidolite and chrysotile asbestos. Catalase and 1,4-diazobicyclo[2.2.2]octane (DABCO), however, are not effective in protecting tracheal epithelial cells from cell injury by asbestos *in vitro*.[4,5] The cytotoxicity of tracheal epithelial cells caused by glass fibers (Code 100 fibrous glass) could not be prevented by addition of exogenous SOD, catalase, DABCO, DMTU or mannitol.

Goodglick and Kane[39] also found that crocidolite asbestos elicited ROS from both mouse peritoneal macrophages and a macrophagelike cell line. Furthermore, asbestos-induced cell death could be prevented by addition of antioxidant enzymes. Both SOD and catalase inhibited asbestos-induced cytotoxicity in mouse macrophages.

III. PROTECTION OF ASBESTOS-MEDIATED LUNG CELL INJURY BY IRON CHELATORS

Weitzman and Graceffa[22] demonstrated that asbestos-catalyzed oxygen

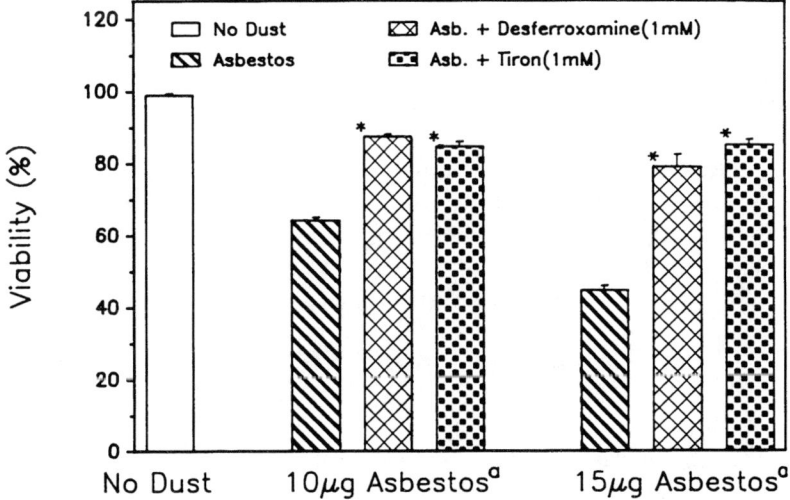

FIGURE 1.   Viability of rat lung fibroblasts after addition of asbestos or asbestos pretreated
with iron chelators. The diploid rat lung fibroblast cell line (RL-82) derived from Fischer 344
rats, was obtained from Dr. Marlene Absher, University of Vermont College of Medicine.[7] Cells
were grown in Ham's F 12 medium (GIBCO, Grand Island, NY) supplemented with 2 mM L-
glutamine, 10% fetal bovine serum, and 100 μg/mL chlortetracycline and incubated at 37°C in
an atmosphere of 5% CO_2/95% air. U.I.C.C. crocidolite asbestos fibers were incubated for 24
h in serumless Ham's F 12 medium containing either 1 mM desferroxamine (Desferal mesylate,
CIBA Pharmaceutical Co., Summit, NJ) or 1 mM Tiron (Sigma, St. Louis, MO). Fibers were
washed three times in medium prior to addition to RL-82 cells at 10 or 15 μg crocidolite/cm^2
dish. Viable cells were assayed from individual dishes (n = 3 per group) using the trypan blue
exclusion technique at 24 h after addition of fibers. a, U.I.C.C. crocidolite asbestos
[$Na_2(Fe^{3+})_2(Fe^{2+})_3Si_8O_{22}(OH)_2$]; *, significantly different ($p<0.001$) compared to unchelated
fibers as determined by two-way analysis of variance.

radical generation in a cell-free system could be inhibited by pretreatment of
asbestos with the iron chelator, desferroxamine. If iron available on the surface
of asbestos fibers participates in the observed cytotoxicity of lung cells *in
vitro*, then chelation of iron on asbestos fibers should attenuate this response.
In our studies, rat lung fibroblasts were exposed to either 10 or 15 μg of
crocidolite asbestos or identical concentrations of crocidolite pretreated with
either of the Fe^{3+} chelators, desferroxamine or Tiron (Figure 1). Exposure
to unchelated crocidolite caused a dose-dependent decrease in the viability of
rat lung fibroblasts. Pretreatment of asbestos fibers with iron chelators, however, resulted in a significant inhibition of this cytotoxic reaction, at both
concentrations of crocidolite.[40]

The recurring theme emerging from current research is that asbestos
induces alterations in cellular functions at multiple levels. Evidence cited
above indicates that many changes in response to asbestos are mediated by
ROS. Both direct measurement of ROS in lung cells exposed to asbestos
fibers and protection by exogenous antioxidant enzymes and scavengers of

ROS support this concept.[40] Another emerging theme is that asbestos-mediated cellular damage can be inhibited or attenuated by rendering iron groups unavailable on the surface of asbestos fibers. As described above, treatment of asbestos with iron chelators prevents or attenuates lipid peroxidation and damage to DNA, manifested by hydroxylation or strand breaks.

Understanding the damage to lung cells and tissue caused by exposure to asbestos yields insight into mechanisms of prevention and treatment. Do lung cells possess the necessary machinery to prevent or recover from asbestos induced damage? Moreover, do lung cells contain adequate antioxidant enzyme levels essential for removal of ROS and are these enzymes inducible?

IV. REGULATION OF ANTIOXIDANT ENZYMES BY REACTIVE OXYGEN SPECIES

Cells have available a cadre of antioxidant enzymes: catalase, glutathione peroxidase (GPX), and the superoxide dismutases (both manganese [MnSOD] and copper-zinc [CuZnSOD] containing), for protection from ROS. We have demonstrated that cells of the respiratory tract respond to asbestos by increased production of ROS, a process that is inhibitable by addition of exogenous antioxidant enzymes. Furthermore, activities of antioxidant enzymes in cells of the respiratory tract increase in response to asbestos[5] and ROS[41] *in vitro* and after inhalation of asbestos.[42] Do lung cells respond to injury by asbestos or ROS by regulation of antioxidant enzymes through a common mechanism? To answer this question, we examined the response at the gene level (gene expression) of antioxidant enzymes in HTE cells exposed to xanthine and xanthine oxidase — a reaction generating ROS.

HTE cells, isolated and characterized as described by Mossman et al.[43] were exposed to 50 μM xanthine and 0.05 U xanthine oxidase (nontoxic concentrations). After 18 h the cells were collected, and total cellular RNA was extracted using the procedure of Chomczynski and Sacchi.[44] RNA was electrophoretically separated in an agarose-formaldehyde gel and then transferred to nitrocellulose (Northern blotting). This blot was then probed and reprobed with ^{32}P-labeled cDNA of the antioxidant enzymes, MnSOD and CuZnSOD (pSP65-RMS and pUC13-RCS were obtained from Dr. Ye-Shih Ho, Duke University Medical Center, NC) and GPX [pmGPX-9A(3)x and pmGPX-5A(2) were obtained from Dr. Guy Mullenbach, Chiron Corp., CA]. As seen in Figure 2, the messenger RNA for MnSOD showed an increase after exposure of HTE cells to the xanthine + xanthine oxidase generating system. This response was observed regardless of the growth phase of the cells, although cells growing exponentially demonstrated a greater response. HTE cells responded to ROS by increasing mRNA levels of MnSOD by greater than 14-fold over control (cells exposed only to xanthine) in the log phase or nearly 8-fold greater than control cells when exposed at confluence.[44a] The level of mRNA of GPX in HTE cells doubled after exposure to the generating system. In contrast to MnSOD, the mRNA levels for CuZnSOD in HTE cells remained unchanged, or slightly decreased in response to ROS.

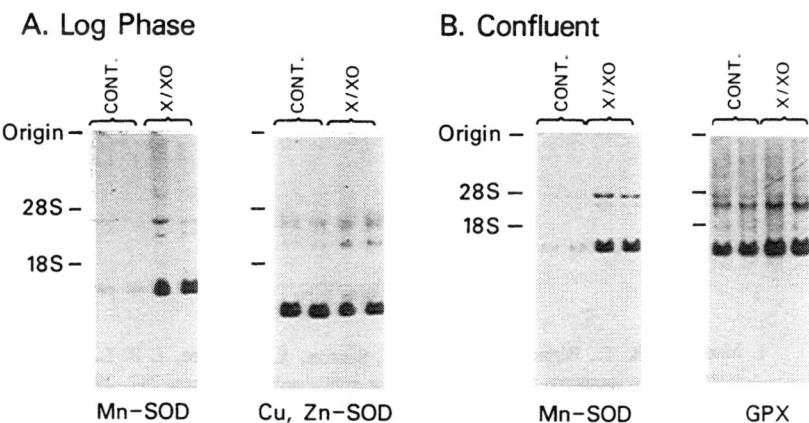

A. Log Phase   B. Confluent

Mn–SOD   Cu, Zn–SOD   Mn–SOD   GPX

FIGURE 2. Steady-state levels of MnSOD mRNA in HTE cells exposed either during log phase (A) or at confluence (B) to the ROS generating system, xanthine and xanthine oxidase. HTE[43] cells were grown in Ham's F 12 medium containing 10% fetal bovine serum (Sigma, St. Louis, MO) and 10 U/ml penicillin and 10 μg/ml streptomycin. Xanthine oxidase (Calbiochem. Corp., San Diego, CA) was added to appropriate dishes at a final concentration of 0.05 U/ml in medium containing 50 μM xanthine. Control cells were exposed to medium containing xanthine alone. Each lane represents a separate cell culture. Northern blots were probed with the labeled cDNA of the antioxidant enzyme listed beneath. (A) HTE cells were growing exponentially during exposure. (B) HTE cells reached confluence prior to exposure. Cont., control cell cultures; X/XO, cells exposed to xanthine and xanthine oxidase; Mn-SOD, manganese-containing superoxide dismutase; Cu, Zn-SOD = copper-zinc-containing superoxide dismutase; GPX, glutathione peroxidase.

The xanthine + xanthine oxidase system continuously generates $O_2^-\cdot$ as well as more reduced species (HO·, H_2O_2). In the neutral, aqueous environment of the extracellular space in tissue culture, $O_2^-\cdot$ can spontaneously dismutate to H_2O_2. Of the possible species formed from $O_2^-\cdot$, H_2O_2 is both the more mobile, due to its uncharged state, and possesses the longer half-life.

These data clearly demonstrate the independent regulation of each of the antioxidant enzymes in lung cells in response to oxidant stress. Our observations are consistent with previously observed modulation of mRNA levels of MnSOD by interleukin-1 (IL-1) and tumor necrosis factor (TNF)[45,46] and lipopolysaccharide.[47] We are currently exploring whether or not HTE cells can elaborate or respond to these cytokines. The induction of MnSOD mRNA by oxidant stress might occur via cytokines or through the same signaling apparatus as the cytokines. Examination of the elements in the rat MnSOD gene that are specifically responsive to oxidant stresses is ongoing in our laboratory. In addition, we are investigating the effect of asbestos on antioxidant enzyme gene induction in both lung cell culture systems and in a rat inhalation model of asbestosis.[3]

ACKNOWLEDGMENTS

Funded with support from The National Heart, Lung and Blood Institute
(RO1 HL-39469 and SCOR P50 HL-14212).

REFERENCES

1. **Mossman, B. T., Bignon, J., Corn, M., Seaton, A., and Gee, J. B. L.,** Asbestos:
    scientific developments and implications for public policy, *Science,* 247, 294, 1990.
2. **Mossman, B. T. and Gee, J. B. L.,** Asbestos-related diseases, *New Engl. J. Med.,*
    320(26), 1721, 1989.
3. **Mossman, B. T., Marsh, J. P., Sesko, A., Hill, S., Shatos, M. A., Doherty, J.,
    Petruska, J., Adler, K. B., Hemenway, D., Mickey, R., Vacek, P., and Kagan, E.,**
    Inhibition of lung injury, inflammation, and interstitial pulmonary fibrosis by polyethylene
    glycol-conjugated catalase in a rapid inhalation model of asbestosis, *Am. Rev. Respir.
    Dis.,* 141, 1266, 1990.
4. **Mossman, B. T. and Landesman, J. M.,** Importance of oxygen free radicals in asbestos-
    induced injury to airway epithelial cells, *Chest,* 83S, 50S, 1983.
5. **Mossman, B. T., Marsh, J. P., and Shatos, M. A.,** Alteration of superoxide dismutase
    activity in tracheal epithelial cells by asbestos and inhibition of cytotoxicity by antioxi-
    dants, *Lab Invest.,* 54, 204, 1986.
6. **Hansen, K. and Mossman, B. T.,** Generation of superoxide from alveolar macrophages
    exposed to asbestiform and nonfibrous particles, *Cancer Res.,* 47, 1681, 1987.
7. **Shatos, M. A., Doherty, J. M., Marsh, J. P., and Mossman, B. T.,** Prevention of
    asbestos-induced cell death in rat lung fibroblasts and alveolar macrophages by scavengers
    of active oxygen species, *Environ. Res.,* 44, 103, 1987.
8. **Gabor, S. and Anca, Z.,** Effect of asbestos on lipid peroxidation in the red cells, *Br.
    J. Ind. Med.,* 32, 39, 1975.
9. **Gulumian, M., Sardianos, F., Kilroe-Smith, T., and Ockerse, G.,** Lipid peroxidation
    in microsomes induced by crocidolite fibres, *Chem. Biol. Interact.,* 44, 111, 1983.
10. **Weitzman, S. A. and Weitberg, A. B.,** Asbestos-catalysed lipid peroxidation and its
    inhibition by desferroxamine, *Biochem. J.,* 225, 259, 1985.
11. **Brown, R. C., Poole, A., and Turver, C. J.,** *In vitro* correlates of mineral dust toxicity,
    Fd. Chem. Toxicol., 24, 535, 1986.
12. **Jajte, J., Lao, I., and Wisniewska-Knypl, J. M.,** Enhanced lipid peroxidation and
    lysosomal enzyme activity in the lungs of rats with prolonged pulmonary deposition of
    crocidolite asbestos, *Br. J. Ind. Med.,* 44, 180, 1987.
13. **Petruska, J. and Mossman, B. T.,** Detection of malondialdehyde (MDA) in bron-
    choalveolar lavage (BAL) of rats exposed to crocidolite asbestos, *Am. Rev. Respir. Dis.,*
    139, A212, 1989.
14. **Turver, C. J. and Brown, R. C.,** The role of catalytic iron in asbestos induced lipid
    peroxidation and DNA-strand breakage in C3H10T1/2 cells, *Br. J. Cancer,* 56, 133,
    1987.
15. **Fontecave, M., Mansuy, D., Jaouen, M., and Pezerat, H.,** The stimulatory effects
    of asbestos on NADPH-dependent lipid peroxidation in rat liver microsomes, *Biochem.
    J.,* 241, 561, 1987.
16. **Gulumian, M. and Kilroe-Smith, T. A.,** Crocidolite-induced lipid peroxidation in rat
    lung microsomes. I. Role of different ions, *Environ. Res.,* 43, 267, 1987.

17. **Kasai, H. and Nishimura, S.,** DNA damage induced by asbestos in the presence of hydrogen peroxide, *Gann,* 75, 841, 1984.
18. **Leanderson, P., Soderkvist, P., Tagesson, C., and Axelson, O.,** Formation of 8-hydroxydeoxyguanosine by asbestos and man made mineral fibers, *Br. J. Ind. Med.,* 45, 309, 1988.
19. **Elstner, E. F., Schutz, W., and Vogl, G.,** Cooperative stimulation by sulfite and crocidolite asbestos fibres of enzyme catalyzed production of reactive oxygen species, *Arch. Toxicol.,* 62, 424, 1988.
20. **Zalma, R., Bonneau, L., Guignard, J., Pezerat, H., and Jaurand, M. C.,** Formation of oxy radicals by oxygen reduction arising from the surface activity of asbestos, *Can. J. Chem.,* 65, 2338, 1987.
21. **Gabrielson, E. W., Rosen, G. M., Grafstrom, R. C., Strauss, K. E., and Harris, C. C.,** Studies on the role of oxygen radicals in asbestos-induced cytopathology of cultured human lung mesothelial cells, *Carcinogenesis,* 7, 1161, 1986.
22. **Weitzman, S. A. and Graceffa, P.,** Asbestos catalyzes hydroxyl and superoxide radical generation from hydrogen peroxide, *Arch. Biochem. Biophys.,* 228, 373, 1984.
23. **Eberhardt, M. K., Roman-Franco, A. A., and Quiles, M. R.,** Asbestos-induced decomposition of hydrogen peroxide, *Environ. Res.,* 37, 287, 1985.
24. **Pezerat, H., Zalma, R., Guignard, J., and Jaurand, M. C.,** Production of oxygen radicals by the reduction of oxygen arising from the surface activity of mineral fibers, *IARC Sci. Publ.,* 90, 100, 1989.
25. **Costa, D., Guignard, J., Zalma, R., and Pezerat, H.,** Production of free radicals arising from the surface activity of minerals and oxygen. Part I. Iron mine ores, *Toxicol. Ind. Health,* 5(6), 1061, 1989.
26. **Fridovich, I.,** Oxygen radicals, hydrogen peroxide, and oxygen toxicity, in *Free Radicals in Biology,* Pryor, W. A., Ed., Academic Press, San Diego, 1976, 239.
27. **Halliwell, B. and Gutteridge, J. M. C.,** *Free Radicals in Biology and Medicine,* Clarendon Press, Oxford, 1989, 1.
28. **Fantone, J. C. and Ward, P. A.,** Role of oxygen-derived free radicals and metabolites in leukocyte-dependent inflammatory reactions, *Am. J. Pathol.,* 107, 397, 1982.
29. **Weiss, S. J. and LoBuglio, A. F.,** Phagocyte-generated oxygen metabolites and cellular injury, *Lab. Invest.,* 47, 5, 1982.
30. **Freeman, B. A. and Crapo, J. D.,** Free radicals and tissue injury, *Lab. Invest.,* 47, 412, 1982.
31. **Baggiolini, M. and Wymann, M. P.,** Turning on the respiratory burst, *TIBS,* 15, 69, 1990.
32. **Sibille, Y. and Reynolds, H. Y.,** Macrophages and polymorphonuclear neutrophils in lung defense and injury, *Am. Rev. Respir. Dis.,* 141, 471, 1990.
33. **Doll, N. J., Stankus, R. P., Goldbach, S., and Salvaggio, J. E.,** In vitro effect of asbestos fibers on polymorphonuclear leukocyte function, *Int. Arch. Allergy Appl. Immunol.,* 68, 17, 1982.
34. **Hedenborg, M. and Klockars, M.,** Production of reactive oxygen metabolites induced by asbestos fibers in human polymorphonuclear leucocytes, *J. Clin. Pathol.,* 40, 1189, 1987.
35. **Doll, N. J., Bozelka, B. E., Goldbach, S., Anoi ve-Lopez, E., and Salvaggio, J. E.,** Asbestos-induced alteration of human peripheral blood monocyte activity, *Int. Arch. Allergy Appl. Immunol.,* 69, 302, 1982.
36. **Donaldson, K. and Cullen, R. T.,** Chemiluminescence of asbestos-activated macrophages, *Br. J. Exp. Pathol.,* 65, 81, 1984.
37. **Case, B. W., Ip, M. P. C., Padilla, M., and Kleinerman, J.,** Asbestos effects on superoxide production. An in vitro study of hamster alveolar macrophages, *Environ. Res.,* 39, 299, 1986.

38. **Mossman, B. T., Hansen, K., Marsh, J. P., Brew, M. E., Hill, S., Bergeron, M., and Petruska, J.,** Mechanisms of fibre-induced superoxide release from alveolar macrophages and induction of superoxide dismutase in the lungs of rats inhaling crocidolite, *IARC Sci. Publ.,* 90, 81, 1989.
39. **Goodglick, L. A. and Kane, A. B.,** Role of reactive oxygen metabolites in crocidolite asbestos toxicity to mouse macrophages, *Cancer Res.,* 46, 5558, 1986.
40. **Mossman, B. T., Marsh, J. P., Shatos, M. A., Doherty, J., Gilbert, R., and Hill, S.,** Implication of active oxygen species as second messengers of asbestos toxicity, *Drug Chem. Toxicol.,* 10, 157, 1987.
41. **Mossman, B. T. and Marsh, J. P.,** Mechanisms of toxic injury by asbestos fibers: role of oxygen-free radicals, in *In Vitro Effects of Mineral Dusts,* NATO ASI Series, Vol G3, Springer-Verlag, Berlin, 1985, 66.
42. **Janssen, Y. W. M., Marsh, J. P., Absher, M., Borm, P. J. A., and Mossman, B. T.,** Increases in endogenous antioxidant enzymes during asbestos inhalation in rats, *Free Radical Res. Commun.,* 11, 53, 1990.
43. **Mossman, B. T., Ezerman, E. B., Adler, K. B., and Craighead, J. E.,** Isolation and spontaneous transformation of cloned lines of hamster tracheal epithelial cells, *Cancer Res.,* 40, 4403, 1980.
44. **Chomczynski, P. and Sacchi, N.,** Single-step method of RNA isolation by acid guanidinium thiocyanate-phenol-chloroform extraction, *Anal. Biochem.,* 162, 156, 1987.
44a. **Shull, S. Heintz, N. H., Periasamy, M., Manohar, M., Janssen, Y. M. W., Marsh, J. P., and Mossman, B. T.,** Differential regulation of antioxidant enzymes in response to oxidants, *J. Biol. Chem.,* in press.
45. **Masuda, A., Longo, D. L., Kobayashi, Y., Appella, E., Oppenheim, J. J., and Matsushima, K.,** Induction of mitochondrial manganese superoxide dismutase by interleukin 1, *Fed. Am. Soc. Exp. Biol. J.,* 2, 3087, 1988.
46. **Wong, G. H. W. and Goeddel, D. V.,** Induction of manganous superoxide dismutase by tumor necrosis factor: possible protective mechanism, *Science,* 242, 941, 1988.
47. **Visner, G. A., Dougall, W. C., Wilson, J. M., Burr, I. A., and Nick, H. S.,** Regulation of manganese superoxide dismutase by lipopolysaccharide, interleukin-1, and tumor necrosis factor, *J. Biol. Chem.,* 265, 2856, 1990.

Chapter 8

THE VASCULAR ENDOTHELIUM IN OXIDANT-INDUCED LUNG INJURY

Stephen J. Elliott and William P. Schilling

TABLE OF CONTENTS

In this chapter the morphologic features of pulmonary oxygen toxicity and its significance in human disease will be briefly examined. Second, the deleterious effects of oxidant stress on the vascular endothelium will be described with particular emphasis on the role of endothelial dysfunction in the genesis of observed pathophysiology. Finally, we will review a model developed in our laboratory for the study of endothelial cell dysfunction induced by oxidant stress.

I. PULMONARY OXIDANT INJURY

A. THE AIR-BLOOD BARRIER

Under normal conditions the blood-gas barrier, which consists of (1) the alveolar epithelial cell, (2) a thin basement membrane, and (3) the capillary endothelial cell, separates inspired oxygen within the alveoli from the vascular space.[1] This anatomic configuration facilitates the passive diffusion of inspired oxygen into circulating red blood cells.[2] In a variety of pulmonary diseases, fluid, proteinaceous deposits, and cells may increase the width of the air-blood barrier, thereby limiting the diffusion of O_2 and risking insufficient O_2 delivery to the tissues. As one of the few options available to the clinician, an increase in the fraction of inspired oxygen concentration will serve to increase the diffusion gradient from the gas-filled alveoli to the pulmonary capillaries. Such a maneuver may come at a cost, in the form of pulmonary oxygen toxicity.[3] For many patients, the therapeutic strategies employed may contribute to the development of chronic lung disease.[4]

Administration of hyperoxic gas mixtures is a significant part of the therapeutic strategy in various respiratory diseases, and represents one of the most common oxidant stresses encountered clinically. The effect of hyperoxia has been examined in various murine, rodent, canine, lupine, and primate models (for review, see Ref. 5), and has been shown to comprise two phases: exudative and proliferative. Oxygen toxicity is initially characterized by morphologic endothelial cell damage, noted histologically by increased intercellular gaps, and functionally by the nonhydrostatic movement of fluid from the vasculature into the perivascular space. Extravasated fluid and protein together form hyaline membranes,[6-8] which act to increase the air-blood barrier and further impair the diffusion of O_2 from the alveoli to the blood. In the proliferative phase, hyperplasia of alveolar type II cells occurs.

B. THE NEONATAL INFANT

Critically ill infants are subjected to multiple physiological stresses, only one of which is hyperoxia. In particular, the exposure of the lungs to positive pressure ventilation has been associated with the development of chronic lung disease. Despite the multifactorial nature of this disease process, it appears that oxygen toxicity may play a significant role in its genesis.[8]

In the case of the premature infant, alveolar and acinar vascular development is not completed by the time of birth and, compared to the adult, interstitial tissue is relatively prominent. These features result in a relative decrease in the surface area available for oxygen diffusion and increased air-blood barrier width. In view of these anatomic considerations, it is not surprising that the preterm infant may require supplemental oxygen in order to maintain an arterial partial pressure of oxygen similar to that found in healthy, term infants. However, in preterm infants the inspiration of normoxic (21%) gas may itself be a biologically significant oxidant stress. Evidence for this hypothesis is provided by our recent studies in preterm infants breathing room air (21% O_2) in which plasma glutathione disulfide concentrations were increased compared to those measured in term infants.[9] Indeed, compared to term infants, preterm babies have a higher incidence of chronic lung disease, possibly related in part to the concentration of oxygen in inspired gas mixtures.[10-13] Previous work by Frank et al.[14] has suggested that antioxidant enzyme activities in the lungs of prematurely born infants may not have reached adult levels by the time of birth. For this reason, the preterm infant may be particularly susceptible to hyperoxic lung injury.

Thus, acute lung injury and chronic lung disease, including those initiated by chemical mechanisms, represent significant causes of morbidity and mortality, especially in the pediatric age group. Understanding of the chemical and molecular mechanisms of oxidant-induced cellular dysfunction may lead to significant improvements in the prevention and management of these disease processes.

C. VASCULAR REACTIVITY

Altered vascular tone and reactivity are key elements in the pathophysiology of oxidant-induced vascular injury. Perfusion of isolated rabbit lungs with medium containing purine/xanthine oxidase stimulates vasoconstriction.[15] Gurtner et al.[16] similarly found that perfusion of isolated lungs with tert-butylhydroperoxide (*t*-bu-OOH) induces a vasopressor response, the magnitude of which correlates with the ratio of thromboxane B_2 to prostaglandin I_2 release. The effect of oxidant stress on vascular tone may vary dependent upon (1) the etiologic agent, (2) the vascular bed under study, and (3) the experimental conditions. For example, exposure of rabbits *in vivo* to 100% oxygen for 48 to 60 h markedly *reduces* the pressor response when the isolated lungs are subsequently perfused with the chemical oxidant, *t*-bu-OOH.[17]

The effects of oxidant stress on vascular tone and reactivity may be endothelium-dependent or may occur via a direct effect on the vascular smooth muscle. Hydrogen peroxide relaxes precontracted bovine pulmonary arterial rings by a mechanism independent of either the endothelium or prostanoid mediators.[18] In contrast, Coflesky and Evans[19] found that hyperoxia inhibits the relaxant effect of acetylcholine on precontracted rat pulmonary arteries, yet has no effect on the direct-acting muscle relaxants, nitroprusside and papaverine.

The central role of the endothelial cell in oxidant-induced vascular injury is supported by the changes observed in vascular permeability, which reflect compromise of endothelial barrier function. Indicative of increased vascular permeability, weights of isolated rabbit lungs increase during perfusion with the oxygen radical-generating system, purine/xanthine oxidase.[15] Furthermore, there is substantial evidence that albumin flux across cultured endothelial monolayers increases in the presence of various oxidants, including molecular oxygen and hydrogen peroxide.[20,21]

It has long been recognized that infiltration of leukocytes, cells that normally circulate within the vascular space, occurs early in the process of lung injury, especially that induced by hyperoxia.[22] The neutrophilic leukocyte may injure endothelial cells via the secretion of reactive species,[23] thus acting to increase the "oxidant burden" placed on the endothelium. However, the etiologic role of this cell type in oxygen toxicity is controversial. Regardless of the oxidative actions of neutrophils, their mere presence increases the air-blood barrier width and may further limit oxygen diffusion and increase oxygen requirements.

II. OXIDANT STRESS AND THE VASCULAR ENDOTHELIAL CELL

The endothelium forms the intimal lining of the blood vessel wall. In the past, vascular endothelial cells were thought to simply provide a barrier function, limiting the leakage of fluid and proteins from the vascular space into the surrounding tissues. It is now recognized, however, that these cells actively contribute to the regulation of vascular tone and reactivity. In response to a number of hematogenous substances, endothelial cells secrete vasoactive compounds such as prostaglandin I_2,[24] nitrogen monoxide,[25] and endothelin,[26] which alter the contractile function of subjacent smooth muscle cells. In addition, endothelial cells actively regulate capillary permeability by the adjustment of intercellular pore size.

Because of its unique location, the pulmonary vascular endothelial cell (VEC) may be subjected to various forms of oxidative stress. The luminal aspect of the cell may be exposed to blood-borne reactive metabolites, oxidant species secreted by polymorphonuclear leukocytes, and dissolved molecular oxygen. When gas mixtures with elevated F_IO_2 are inspired, the abluminal aspect of the endothelial cell is confronted with hyperoxia, and presumably increased concentrations of oxygen-derived radical species. Various structural and metabolic parameters have been investigated in the study of oxidant stress in vascular endothelial cells. Hyperoxia interrupts protein synthesis in endothelial cells by inhibition of polypeptide chain elongation[27] and by DNA strand breakage.[28] Oxidant stress induces rearrangement of cytoskeletal elements, including disruption of peripheral bands,[28a] and reduces fluidity of the hydrophobic core of the VEC plasma membrane, as determined by polarization of the fluorescent probe, diphenylhexatriene.[29]

Oxidant stress has been demonstrated to alter the secretion of several paracrine substances by VECs. Hydrogen peroxide stimulates the synthesis of platelet-activating factor (PAF) by cultured endothelial cells, measured via [^{3}H]acetate incorporation into PAF.[30] In addition, the H_2O_2-generating system, glucose-glucose oxidase, stimulates release of the prostacyclin metabolite, 6-keto-PGF$_{1\alpha}$ from cultured VECs.[31] Finally, cell injury may be manifested by the oxidation of sulfhydryl groups,[32] some of which may be critically located on functionally and topographically important proteins, such as surface membrane Ca^{2+}-ATPase or ion conductance channels.[33,34]

III. CALCIUM SIGNALING IN VASCULAR ENDOTHELIAL CELLS

Changes in cytosolic Ca^{2+} concentration ($[Ca^{2+}]_i$) play a significant role in the response of arterial endothelial cells to hematogenous, vasoactive compounds. Bradykinin, an agonist which stimulates endothelial cell surface receptors, and A23187, a Ca^{2+} ionophore, are both potent stimuli for the secretion of endothelial-derived relaxation factor.[35] Activation of the bradykinin receptor results in a biphasic change in $[Ca^{2+}]_i$, consisting of an initial peak, followed by a prolonged elevation (plateau phase) during which the value of $[Ca^{2+}]_i$ slowly declines toward baseline.[36,37] These changes in $[Ca^{2+}]_i$ activate various intracellular enzymes, and putatively provide the link between receptor stimulation and secretion of vasoactive substances. In this way, cytosolic Ca^{2+} acts as a "second messenger" within the cell.

To study changes in $[Ca^{2+}]_i$, cells may be incubated with the Ca^{2+}-sensitive fluorescent probe, fura-2/AM. After diffusion into the endothelial cell, intracellular esterases cleave the acetoxymethyl groups from the parent fura-2 molecule. The probe is then trapped inside the cell, and has fluorescent characteristics which depend upon its binding to Ca^{2+}. Under conditions of saturating concentrations of Ca^{2+}, fura-2 exhibits maximal 510 nm emission intensity when excited at 340 nm. In the absence of Ca^{2+}, maximal emission intensity occurs at an excitation wavelength of 380 nm. Cells loaded with fura-2 can be excited at rapidly alternating excitation wavelengths to provide a ratio value which is dependent upon $[Ca^{2+}]_i$.[38]

Studies which have utilized fura-2 have shown that bradykinin induces the release of Ca^{2+} from internal stores within the cell, and stimulates the influx of Ca^{2+} from the extracellular space. The initial peak $[Ca^{2+}]_i$ observed in fura-2-loaded cells stimulated with bradykinin reflects, in part, release of Ca^{2+} from internal stores. The more prolonged plateau phase of elevated $[Ca^{2+}]_i$ reflects agonist-stimulated influx of Ca^{2+}. These two phases may be temporally separated by stimulation of fura-2-loaded endothelial cells in the absence of extracellular Ca^{2+}. Under this condition, the bradykinin-stimulated increase in $[Ca^{2+}]_i$ reflects solely the release of Ca^{2+} from internal stores. The subsequent addition of Ca^{2+} to the extracellular medium results in a large increase in $[Ca^{2+}]_i$, and is indicative of influx of Ca^{2+} from the extracellular space.

TABLE 1
**Effect of t-bu-OOH on Bradykinin-
Stimulated Cytosolic Free Calcium ($[Ca^{2+}]_i$)**

Condition	Basal $[Ca^{2+}]_i$ (nM)	Peak $[Ca^{2+}]_i$ (nM)	Plateau $[Ca^{2+}]_i$ (nM)
Control	82 ± 9[a]	293 ± 17	174 ± 6
t-bu-OOH			
1 h	88 ± 6	222 ± 15	127 ± 9
2 h	114 ± 7	183 ± 19	135 ± 12
3 h	184 ± 11	244 ± 9	214 ± 15

[a]	Values are mean $\pm$ SE; n, 4 to 9.

IV. OXIDANT STRESS AND CALCIUM SIGNALING

In our laboratory, we have been characterizing the effect of oxidant stress on receptor-mediated Ca^{2+} signaling in cultured pulmonary arterial endothelial cells.[39,40] We hypothesized that the effect of oxidant stress on vascular tone may occur via an effect on receptor-mediated Ca^{2+} signaling within the endothelial cell. We selected the short-chained t-bu-OOH as the model oxidant since this agent readily diffuses through the cell membrane into the cytosol. Its reduction to t-butanol is linked to the oxidation of reduced glutathione, a reaction catalyzed by glutathione peroxidase. Reduced glutathione is regenerated via glutathione reductase at the expense of reducing equivalents provided by NADPH.

To examine the effect of oxidant stress on Ca^{2+} signaling, fura-2-loaded cells were incubated with t-bu-OOH (0.4 mM) before stimulation with bradykinin. The response of cultured calf pulmonary artery endothelial cells to stimulation with bradykinin was determined after the cells were incubated with the oxidant for various lengths of time. Peak $[Ca^{2+}]_i$ response to bradykinin stimulation decreases from a control value of 293 ± 17 nM to 222 ± 15 nM in cells incubated with t-bu-OOH for 1 h ($p<0.05$) (Table 1). In addition, $[Ca^{2+}]_i$ in t-bu-OOH-treated cells declines more rapidly toward baseline after the peak. Since the sustained phase of the $[Ca^{2+}]_i$ response reflects Ca^{2+} influx from the extracellular space, the more rapid return toward basal levels in the oxidant-treated cells suggests that bradykinin-stimulated influx of Ca^{2+} is decreased under these conditions. These effects of the oxidant during the first 1 h of incubation occur without significant change in the basal $[Ca^{2+}]_i$ level.

After more prolonged incubation, basal $[Ca^{2+}]_i$ increases to 114 ± 7.0 nM and 184 ± 11 nM in cells treated with t-bu-OOH for 2 and 3 h, respectively. The fold increase in $[Ca^{2+}]_i$ produced by bradykinin progressively decreases through the same time frame. Bradykinin stimulates a 1.3-fold increase in cells treated for 3 h, compared to a 3.5-fold increase in controls. Thus, the cells become progressively unresponsive to bradykinin when incubated for increasing periods in t-bu-OOH.

To characterize the effect of t-bu-OOH on each component of the bradykinin-stimulated response (i.e., the release of Ca^{2+} from internal stores and the influx of Ca^{2+} across the cell membrane), cells were incubated with the peroxide for 1, 2, or 3 h prior to suspension in buffer containing 0.2 mM EGTA and no added Ca^{2+}.

The bradykinin-stimulated release of Ca^{2+} from internal stores, determined by the fold increase in $[Ca^{2+}]_i$ in cells in Ca^{2+}-free/EGTA buffer, progressively declines with increasing incubation durations with t-bu-OOH. Thus, in comparison to a fold increase of 2.28 ± 0.21 observed in control cells, bradykinin stimulates a fold increase of only 1.44 ± 0.08 cells incubated with t-bu-OOH for 2 h, and 1.08 ± 0.03 in cells incubated for 3 h.

The effect of t-bu-OOH on agonist-stimulated Ca^{2+} influx was measured by the fold increase in $[Ca^{2+}]_i$ upon readdition of Ca^{2+} to the cell suspension after addition of bradykinin. In contrast to control cells, in which addition of Ca^{2+} results in a 3.6-fold increase in $[Ca^{2+}]_i$, cells treated with t-bu-OOH for 2 h respond with a 2.5 ± 0.3-fold change. Notably, after an incubation period of 3 h, addition of Ca^{2+} after bradykinin produces a marked increase in $[Ca^{2+}]_i$ which continues to increase slowly with time.

It is to be noted that elevated basal $[Ca^{2+}]_i$ observed after prolonged (>1 h) incubation with t-bu-OOH is also observed when measurements are performed in Ca^{2+}-free/EGTA buffer. Compared to a control level of 51 ± 6 nM, basal $[Ca^{2+}]_i$ of cells treated with the oxidant for 2 and 3 h increases to 66 ± 7 nM and 85 ± 4 nM, respectively.

Thus, these results in fura-2-loaded VECs suggest that the effects of oxidant stress on agonist-stimulated Ca^{2+} signaling occur in three phases. The first phase is characterized by inhibition of agonist-stimulated Ca^{2+} influx, and occurs at a time when there is no change in basal $[Ca^{2+}]_i$. The second phase is characterized by decreased peak response to bradykinin and increased basal $[Ca^{2+}]_i$, suggesting that agonist-stimulated release of Ca^{2+} from internal stores becomes inhibited. Finally, the cells become essentially unresponsive to bradykinin and Ca^{2+} progressively accumulates in the cytosolic compartment.

To further examine this sequence of events, the effect of t-bu-OOH on bradykinin-stimulated uptake and efflux of $^{45}Ca^{2+}$ was examined after cells were incubated with the oxidant for various lengths of time. Cells were treated with t-bu-OOH for 1, 2, or 3 h prior to measurement of $^{45}Ca^{2+}$ uptake under basal and stimulated conditions. Association of $^{45}Ca^{2+}$ into cells at 37°C was determined in cells in monolayer culture. Uptake was initiated by addition of $^{45}Ca^{2+}$ with or without the agonist. Values are relative to basal uptake at 5 min in control cells, which is normalized to 100%. Thus, in comparison to control basal uptake (100%), basal uptake at 5 min increased slightly to 126% in cells incubated with the oxidant for 1 h, and 135% in cells incubated for 2 h.

Under stimulated conditions, control uptake at 5 min was more than twice that of basal uptake. In cells incubated with *t*-bu-OOH for 1 h, bradykinin-stimulated uptake was significantly decreased, and further decreased by incubation with the oxidant for 2 h.

Bradykinin-stimulated $^{45}Ca^{2+}$ efflux from VECs reflects release of Ca^{2+} from internal stores.[41] Incubation of cells with *t*-bu-OOH for 2 h decreased bradykinin-stimulated $^{45}Ca^{2+}$ efflux, such that the initial rate constant after addition of bradykinin was decreased in treated cells to approximately 50% that of control cells. Incubation of cells with the oxidant for 3 h further reduced the agonist-stimulated efflux rate to approximately 30% of controls.

The results of the radioisotopic experiments described are in agreement with those obtained in the fura-2 studies and support the conclusion that the effects of *t*-bu-OOH in VEC signal transduction are seen in three phases. These three phases comprise (1) inhibition of agonist-stimulated Ca^{2+} influx pathway, (2) inhibition of Ca^{2+} release from internal stores, and (3) progressive accumulation of Ca^{2+} in the cytosolic compartment. Inhibition of the agonist-stimulated influx pathway appears to occur as an initial event before there is any effect on basal $[Ca^{2+}]_i$ or release of Ca^{2+} from internal stores. The molecular mechanism for this inhibition remains to be determined, but it is possible that the oxidant either directly alters the putative Ca^{2+} channel, or prevents channel opening via an effect on a functionally important membrane-associated messenger. It is clear, however, that the inhibitory effect of the oxidant on the influx pathway does not depend upon an elevation in $[Ca^{2+}]_i$.

Inhibition of release of Ca^{2+} from internal stores, which appears to be the characteristic effect of the second phase of oxidant stress in endothelial cells, may be related to an effect of *t*-bu-OOH on inositol polyphosphate metabolism. Bellomo et al.[42] found that oxidant stress inhibits hormonally and nonhormonally stimulated increases in inositol polyphosphate production in cultured hepatocytes, but found no effect on basal levels. Whether inositol polyphosphate-mediated signaling is inhibited by oxidant stress in vascular endothelial cells is unclear. It is possible that *t*-bu-OOH directly alters the internal store membrane integrity via lipid peroxidation. Such an effect may not, therefore, be directly related to cytosolic messengers such as inositol trisphosphate which stimulates the release of Ca^{2+} from the endoplasmic reticulum.

The final phase of the effect of *t*-bu-OOH on Ca^{2+} homeostasis in endothelial cells is characterized by a progressive rise in basal $[Ca^{2+}]_i$. At this point, the cells clearly have little response to stimulation with bradykinin, but whether they have sustained irreversible cell injury is unknown. Regardless, altered Ca^{2+} signaling occurs prior to sustained elevation in basal $[Ca^{2+}]_i$.

V. INHIBITION OF GLUTATHIONE REDUCTASE

Bis(2-chloroethyl)-1-nitrosourea (BCNU) selectively inhibits glutathione reductase,[43-47] thereby interrupting glutathione redox cycle activity. Under

TABLE 2
Effect of *t*-bu-OOH on Bradykinin-Stimulated Cytosolic Free Calcium [Ca²⁺]ᵢ in BCNU-Treated Endothelial Cells

Condition	Basal [Ca²⁺]ᵢ (nM)	Peak [Ca²⁺]ᵢ (nM)	Plateau [Ca²⁺]ᵢ (nM)
Control	103 ± 6[a]	367 ± 22	174 ± 6
BCNU	129 ± 10	381 ± 17	189 ± 9
BCNU + *t*-bu-OOH	148 ± 7	244 ± 7	181 ± 22

[a] Values are mean $\pm$ SE, n, 9 to 20.

several different experimental conditions,[48-52] BCNU enhances the effects of oxidative stress. To determine the role of glutathione redox cycling in oxidant-induced inhibition of Ca^{2+} signaling, and to determine whether the time-dependent effects of *t*-bu-OOH could be accelerated, cells were incubated with BCNU prior to treatment with *t*-bu-OOH. Agonist-stimulated Ca^{2+} signaling was determined in cells loaded with fura-2, via the probe's acetoxymethyl form, fura-2/AM.

Incubation of cells with BCNU (75 μM) for 20 min slightly increases basal [Ca²⁺]ᵢ from 103 ± 6 nM to 129 ± 10 nM, but has no effect on the response of the cells to stimulation with bradykinin.[40] Treatment of cells with *t*-bu-OOH (0.4 mM) for 30 min results in Ca^{2+} signaling responses not significantly different from control. In cells incubated sequentially with BCNU and *t*-bu-OOH, basal [Ca²⁺]ᵢ is significantly elevated to 148 ± 7 nM, and peak [Ca²⁺]ᵢ in response to bradykinin is significantly decreased (Table 2). The response of cells incubated with BCNU prior to treatment with *t*-bu-OOH for 30 min is similar to that observed in cells treated with *t*-bu-OOH alone for 2 h.

The increased basal [Ca²⁺]ᵢ in cells treated with BCNU suggests that it alters the mechanism which determines the steady state between Ca^{2+} influx and efflux. The marked decrease in bradykinin-stimulated peak [Ca²⁺]ᵢ in oxidant-stressed cells pretreated with BCNU indicates that oxidant stress has an enhanced effect on Ca^{2+} signaling in endothelial cells when glutathione reductase activity is decreased.

Irreversible cell injury in endothelial cells may be related to elevation of resting [Ca²⁺]ᵢ, although an etiological relationship between elevated [Ca²⁺]ᵢ and cell death is difficult to prove.[53,54] Schanne et al.[53] noted that extracellular Ca^{2+} was necessary for A23187 and nine toxins to decrease viability (determined by exclusion of trypan blue) in isolated rat hepatocytes, and suggested that derangements in Ca^{2+} homeostasis may represent the "final common pathway" in toxic cell death. In support of this, we have found increased basal [Ca²⁺]ᵢ only at the time points which follow prolonged incubations with the oxidant.

The primary and central importance of cellular calcium in the process of cell injury has been emphasized by Trump et al.[54] who suggested that cell injury reduces ATP levels within the cell, leading to inhibition of Ca^{2+} sequestration mechanisms by mitochondria and endoplasmic reticulum. In this way, cytosolic Ca^{2+} concentration increases, and results in alteration of cytoskeletal components, while activation of Ca^{2+}-dependent phospholipases increases surface membrane permeability to Ca^{2+}. However, Lemasters et al.[55] employed digitized, low-light video microscopy to show that after hepatocytes were exposed to cyanide and iodoacetate, $[Ca^{2+}]_i$ does not change during surface membrane bleb formation or before loss of cell viability, suggesting that, at least in hepatocytes exposed to these two toxins, $[Ca^{2+}]_i$ may not be the key factor in irreversible cell injury. On the other hand, increased protein degradation directly results from elevated $[Ca^{2+}]_i$ in skeletal muscle cells of mdx mice.[56] Thus, the role of Ca^{2+} in cell injury and death remains unclear. The effect of t-bu-OOH on bradykinin-stimulated changes in $[Ca^{2+}]_i$ occurs before sustained increases in $[Ca^{2+}]_i$. Thus, measurement of recovery of endothelial cells from oxidant stress, i.e., recovery of responsiveness to bradykinin, may provide information concerning the mechanisms involved in irreversible cell damage.

VI. CONCLUSION

Oxidant-induced inhibition of Ca^{2+} signaling in endothelial cells may explain, in part, the effect of oxidant stress on vascular tone, reactivity, and permeability. Identification of the chemical and molecular events responsible for the findings described above will be required to further our understanding of the basic mechanisms of oxidative injury. Thus, advances in the prevention and treatment of oxidant-induced injury in the clinical setting depend, in part, on further and adequate characterization of the intracellular molecular mechanisms responsible for damage. In short, acute lung injury and chronic pulmonary disease, each linked to various forms of oxidant injury, await advances at the subcellular level.

ACKNOWLEDGMENT

This work was supported by Grant 89G-190 from the American Heart Association (TX Affiliate), Grant 871317 from the American Heart Association (National), and Grants HL-37044 and HL-02595 from the National Institutes of Health.

REFERENCES

1. **Murray, J. F.**, in *The Normal Lung*, W.B. Saunders, Philadelphia, 1986.
2. **Staub, N. C.**, Alveolar-arterial oxygen tension gradient due to diffusion, *J. Appl. Physiol.*, 18, 673, 1963.
3. **Kafer, E. R.**, Pulmonary oxygen toxicity: a review of the evidence for acute and chronic oxygen toxicity in man, *Br. J. Anaesth.*, 43, 687, 1971.
4. **Avery, M. E., Tooley, W. H., Keller, J. B., Hurd, S. S., Bryan, M. H., Cotton, R. B., Epstein, M. F., Fitzhardinge, P. M., Hansen, C. B., Hansen, T. N., Hodson, W. A., James, L. S., Kitterman, J. A., Nielsen, H. C., Poirier, T. A., Truog, W. E., and Wung, J. T.**, Is chronic lung disease in low birth weight infants preventable? A survey of eight centers, *Pediatrics*, 79, 26, 1987.
5. **Clark, J. M. and Lambertsen, C. J.**, Pulmonary oxygen toxicity: a review, *Pharmacol. Rev.*, 23, 37, 1971.
6. **Kapanci, Y., Weibel, E. R., and Kaplan, H. P.**, Pathogenesis and reversibility of the pulmonary lesions of oxygen toxicity in monkeys. II. Ultrastructural and morphometric studies, *Lab. Invest.*, 20, 101, 1969.
7. **Kistler, G. S., Caldwell, P. R., and Weibel, E. R.**, Development of fine structural damage to alveolar and capillary lining cells in oxygen-poisoned rat lungs, *J. Cell Biol.*, 33, 605, 1967.
8. **Barter, R. A., Finlay-Jones, L. R., and Walters, M. N. I.**, Pulmonary hyaline membrane: sites of formation in adult lungs after assisted respiration and inhalation of oxygen, *J. Pathol. Bacteriol.*, 95, 481, 1968.
9. **Smith, C. V., Martin, N. E., Smith, H. W., Elliott, S. J., and Hansen, T. N.**, Oxidant stress responses in premature infants during exposure to hyperoxia, *Fed. Am. Soc. Exp. Biol. J.*, 4, A350, 1990.
10. **Shanklin, D. R. and Wolfson, S. L.**, Therapeutic oxygen as a possible cause of pulmonary hemorrhage in premature infants, *N. Engl. J. Med.*, 277, 833, 1967.
11. **Westgate, H. D., Fisch, R. O., and Langer, L. O., et al.**, Pulmonary and respiratory function changes in survivors of hyaline membrane disease, *Dis. Chest.*, 55, 465, 1969.
12. **Edwards, D. K., Dyer, W. M., and Northway, W. H.**, Twelve years experience with bronchopulmonary dysplasia, *Pediatrics*, 59, 839, 1977.
13. **Rhodes, P. G., Hall, R. T., and Leonidas, J. C.**, Chronic pulmonary disease in neonates with assisted ventilation, *Pediatrics*, 55, 788, 1975.
14. **Frank, L. and Groseclose, E. E.**, Preparation for birth into an O_2-rich environment: the antioxidant enzymes in the developing rabbit lung, *Pediatr. Res.*, 18, 240, 1984.
15. **Tate, R. M., Vanbenthuysen, K. M., Shasby, D. M., McMurty, I. F., and Repine, J. E.**, Oxygen-radical-mediated permeability edema and vasoconstriction in isolated perfused rabbit lungs, *Am. Rev. Respir. Dis.*, 126, 802, 1982.
16. **Gurtner, G. H., Knoblauch, A., Smith, P. L., Sies, H., and Adkinson, N. F.**, Oxidant and lipid-induced pulmonary vasoconstriction mediated by arachidonic acid metabolites, *J. Appl. Physiol.*, 55, 949, 1983.
17. **Gurtner, G. H., Michael, J. R., Farrukh, I. S., Sciuto, A. M., and Adkinson, N. F.**, Mechanism of hyperoxia-induced pulmonary vascular paralysis: effect of antioxidant pretreatment, *J. Appl. Physiol.*, 59, 953, 1985.
18. **Burke, T. M. and Wolin, M. S.**, Hydrogen peroxide elicits pulmonary arterial relaxation and guanylate cyclase activation, *Am. J. Physiol.*, 252, H721, 1987.
19. **Coflesky, J. T. and Evans, J. N.**, Pharmacologic properties of isolated proximal pulmonary arteries after seven-day exposure to *in vivo* hyperoxia, *Am. Rev. Respir. Dis.*, 138, 945, 1988.
20. **Phillips, P. G. and Tsan, M.-F.**, Hyperoxia causes increased albumin permeability of cultured endothelial monolayers, *J. Appl. Physiol.*, 64, 1196, 1988.
21. **Shasby, D. M., Lind, S. E., Shasby, S. S., Goldsmith, J. C., and Hunninghake, G. W.**, Reversible oxidant-induced increases in albumin transfer across cultured endothelium: alterations in cell shape and calcium homeostasis, *Blood*, 65, 605, 1985.

22. **Hammerschmidt, D. E.**, Leukocytes in lung injury, *Chest*, 83, 16S, 1983.
23. **Harlan, J. M., Levine, J. D., Callahan, K. S., and Schwartz, B. R.**, Glutathione redox cycle protects cultured endothelial cells against lysis by extracellularly generated hydrogen peroxide, *J. Clin. Invest.*, 73, 706, 1984.
24. **Needleman, P., Turk, J., Jakschik, B. A., Morrison, A. R., and Lefkowith, J. B.**, Arachidonic acid metabolism, *Annu. Rev. Biochem.*, 55, 69, 1986.
25. **Palmer, R. M., Ferridge, A. G., and Moncada, S.**, Nitric oxide release accounts for the biological activity of endothelium-derived relaxing factor, *Nature (London)*, 327, 524, 1987.
26. **Yanigasawa, M., Kurihara, H., Kimura, S., Tomobe, Y., Kobayashi, M., Mitsui, Y., Yazaki, Y., Goto, K., and Masaki, T.**, A novel potent vasoconstrictor peptide produced by vascular endothelial cells, *Nature (London)*, 332, 411, 1988.
27. **Jornot, L., Mirault, M. E., and Junod, A. F.**, Protein synthesis in hyperoxic endothelial cells: evidence for translational defect, *J. Appl. Physiol.*, 63, 457, 1987.
28. **Junod, A. F., Jornot, L., and Petersen, H.**, Differential effects of hyperoxia and hydrogen peroxide on DNA damage, polyadenosine diphosphate-ribose polymerase activity, and nicotinamide adenine dinucleotide and adenosine triphosphate contents in cultured endothelial cells and fibroblasts, *J. Cell. Physiol.*, 140, 177, 1989.
28a. **Phillips, P. G., Higgins, P. J., Malik, A. B. and Tsan, M.-F.**, Effect of hyperoxia on the cytoarchitecture of cultured endothelial cells, *Am. J. Pathol.*, 132, 59, 1988.
29. **Block, E. R., Patel, J. M., Angelides, K. J., Sheridan, N. P., and Garg, L. C.**, Hyperoxia reduces plasma membrane fluidity: a mechanism for endothelial cell dysfunction, *J. Appl. Physiol.*, 60, 826, 1986.
30. **Lewis, M. S., Whatley, R. E., Cain, P., McIntyre, T. M., Prescott, S. M., and Zimmerman, G. A.**, Hydrogen peroxide stimulates the synthesis of platelet-activating factor by endothelium and induces endothelial cell-dependent neutrophil adhesion, *J. Clin. Invest.*, 82, 2045, 1988.
31. **Harlan, J. M. and Callahan, K. S.**, Role of hydrogen peroxide in the neutrophil-mediated release of prostacyclin from cultured endothelial cells, *J. Clin. Invest.*, 74, 442, 1984.
32. **Meister, A.**, Selective modification of glutathione metabolism, *Science*, 220, 472, 1983.
33. **Bellomo, G., Mirabelli, F., Richelmi, P., and Orrenius, S.**, Critical role of sulfhydryl groups in ATP-dependent Ca^{2+} sequestration by the plasma membrane fraction from rat liver, *FEBS Lett.*, 163, 136, 1983.
34. **Srivastava, S. K., Ansari, N. H., Lui, S., Izban, A., Das, B., Szabo, G., and Bhatnagar, A.**, The effects of oxidants on biomembranes and cellular metabolism, *Mol. Cell. Biochem.*, 91, 149, 1989.
35. **Gordon, J. L. and Martin, W.**, Endothelium-dependent relaxation of the pig aorta: relationship to stimulation of Rb efflux from isolated endothelial cells, *Br. J. Pharmacol.*, 79, 531, 1983.
36. **Colden-Stanfield, M., Schilling, W. P., Ritchie, A. K., Eskin, S. G., Navarro, L. T., and Kunze, D. L.**, Bradykinin-induced increases in cytosolic calcium and ionic currents in cultured bovine aortic endothelial cells, *Circ. Res.*, 61, 632, 1987.
37. **Schilling, W. P., Ritchie, A. K., Navarro, L. T., and Eskin, S. G.**, Bradykinin-stimulated calcium influx in cultured bovine aortic endothelial cells, *Am. J. Physiol.*, 255, H219, 1988.
38. **Grynkiewicz, G., Poenie, M., and Tsien, R. Y.**, A new generation of Ca^{2+} indicators with greatly improved fluorescence properties, *J. Biol. Chem.*, 260, 3440, 1985.
39. **Elliott, S. J., Eskin, S. G., and Schilling, W. P.**, Effect of t-butyl-hydroperoxide on bradykinin-stimulated changes in cytosolic calcium in vascular endothelial cells, *J. Biol. Chem.*, 264, 3806, 1989.
40. **Elliott, S. J. and Schilling, W. P.**, Carmustine augments the effects of tert-butyl hydroperoxide on calcium signaling in cultured pulmonary artery endothelial cells, *J. Biol. Chem.*, 265, 103, 1990.

41. **Freay, A., Johns, A., Adams, D. J., Ryan, U. S., and Van Breeman, C.,** Bradykinin and inositol 1,4,5-trisphosphate-stimulated calcium release from intracellular stores in cultured bovine endothelial cells, *Pfluegers Arch.,* 414, 377, 1989.

42. **Bellomo, G., Thor, H., and Orrenius, S.,** Alterations in inositol phosphate production during oxidative stress in isolated hepatocytes, *J. Biol. Chem.,* 262, 1530, 1987.

43. **Frischer, H. and Ahmad, T.,** Severe generalized glutathione reductase deficiency after antitumor chemotherapy with BCNU, *J. Lab. Clin. Med.,* 89, 1080, 1977.

44. **Babson, J. R. and Reed, D. J.,** Inactivation of glutathione reductase by 2-chloroethyl nitrosurea-derived isocyanates, *Biochem. Biophys. Res. Commun.,* 83, 754, 1978.

45. **Shinohara, K. and Tanaka, K. R.,** Mechanism of inhibition of red blood cell glutathione reductase activity by BCNU, *Clin. Chim. Acta,* 92, 147, 1979.

46. **Kehrer, J. P.,** The effect of BCNU on tissue glutathione reductase activity, *Toxicol. Lett.,* 17, 63, 1983.

47. **Reed, D. J.,** in *Oxidative Stress,* Sies, H., Ed., Academic Press, San Diego, 1985, 115.

48. **Miccadei, S., Kyle, M. E., Gilfor, D., and Farber, J. L.,** Toxic consequence of the abrupt depletion of glutathione in cultured rat hepatocytes, *Arch. Biochem. Biophys.,* 265, 311, 1988.

49. **Paraidathathu, T., Combs, A. B., and Kehrer, J. P.,** In vivo effects of 1,3-bis(2-chloroethyl)-1-nitrosourea and doxorubicin on the cardiac and hepatic glutathione systems, *Toxicology,* 35, 113, 1985.

50. **Riley, M. V.,** A role for glutathione and glutathione reductase in control of corneal hydration, *Exp. Eye Res.,* 39, 751, 1984.

51. **Smith, C. V.,** Effect of BCNU pretreatment on diquat-induced oxidant stress and hepatotoxicity, *Biochem. Biophys. Res. Commun.,* 144, 415, 1987.

52. **Tsokos-Kuhn, J. O.,** Lethal injury by diquat redox cycling in an isolated hepatocyte model, *Arch. Biochem. Biophys.,* 265, 415, 1988.

53. **Schanne, F. A., Kane, A. B., Young, E. E., and Farber, J. L.,** Calcium dependence of toxic cell death: a final common pathway, *Science,* 206, 700, 1979.

54. **Trump, B. F., Berezesky, I. K., Laiho, K. U., Osornio, A. R., Mergner, W. J., and Smith, M. W.,** The role of calcium in cell injury: a review, *Scanning Electron Microsc.,* II, 437, 1980.

55. **Lemasters, J. L., DiGuiseppi, J., Nieminen, A.-L., and Herman, B.,** Blebbing, free Ca^{2+} and mitochondrial membrane potential preceding cell death in hepatocytes, *Nature (London),* 325, 78, 1987.

56. **Turner, P. R., Westwood, T., Regen, C. M., and Steinhardt, R. A.,** Increased protein degradation results from elevated free calcium levels found in muscle from mdx mice, *Nature (London),* 335, 735, 1988.

Chapter 9

OXYGEN-INDUCED RETINOPATHY IN THE RAT: A PROPOSED ROLE FOR PEROXIDATION REACTIONS IN THE PATHOGENESIS

John S. Penn

TABLE OF CONTENTS

I. THE RETINA AND PEROXIDATION

Vitamin E has been known for many years to be an effective antioxidant in biological systems, yet vitamin E deficiency in human beings was, until recently, relatively unknown, and descriptions of deficiency symptoms were, with few exceptions,[1] limited to experimental animals. Over the past few years, however, reports on the efficacy of vitamin E in treating certain types of human diseases have increased, especially concerning diseases in which lipid peroxidation may be involved in the pathogenesis. This trend is evidenced by a recently inaugurated large-scale study of vitamin E supplementation of patients with early Parkinson's disease,[2] and a recent evaluation of vitamin E as a suppressant of free radical damage during cardiopulmonary bypass operations.[3] Research on potential medical applications for vitamin E has now become one of the most active areas of clinical investigation.

Under normal conditions, the retina is susceptible to damaging oxidative reactions because of its very high levels of polyunsaturated fatty acids, its constant bombardment by light, and a relatively high oxygen flux across its membranes. Very little is known of the biochemical mechanisms of retinal oxygen damage, although there has been much recent interest in the biochemistry of oxygen toxicity in general. There has been speculation that cell damage is due to the action of oxygen-derived free radicals — molecules that contain an odd number of electrons, rendering them chemically reactive.[4] An abundance of excellent substrates for oxidative reactions is present in the retina. Particularly susceptible are polyunsaturated fatty acids, which are found in higher concentrations in retinal tissue than in any other.[5,6] Polyunsaturated fatty acids can react with an oxygen radical *initiator* to produce a lipid-free radical in the membrane bilayer. This alkyl radical reacts rapidly with molecular oxygen to form a lipid peroxyl radical, which can then react with another polyunsaturated fatty acid to generate yet another alkyl radical. This process is called chain propagation (see Figure 1). As many as 100 lipid molecules may be peroxidized in this manner from a single oxygen radical initiator. Lipid peroxidation can be terminated by several means. The products of lipid peroxidation include aldehydes, among them, malondialdehyde (MDA), which are toxic to the cell. The uncontrolled production of these toxic compounds leads to cell death.

Interest in the antioxidant vitamin E as a treatment for retinopathy of prematurity (ROP) stems from the premise that free radical damage to cells is the causal factor and that, by acting as a free radical scavenger, vitamin E may prevent cellular damage.[7] Since vitamin E is an integral part of the lipid bilayer, it can react with lipid radicals there by transferring a hydrogen atom to the radical. Vitamin E then becomes a free radical, but, because of its relative stability, is unable to continue the chain propagation of lipid peroxidation. The vitamin E radical may be terminated by reacting with another radical of vitamin E, a lipid radical, or by transfer of the unpaired electron to some other electron acceptor.

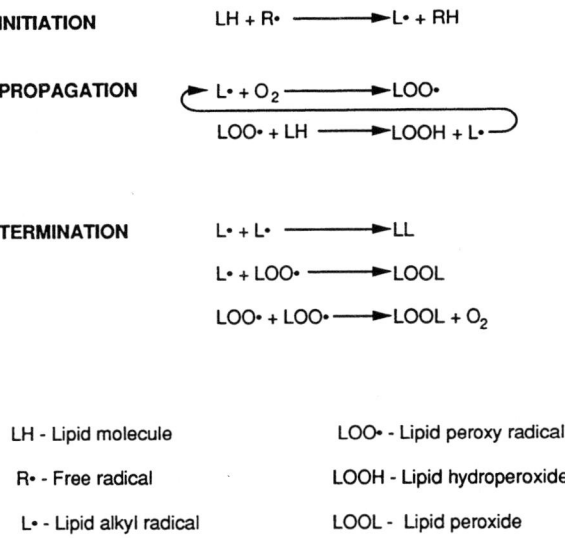

INITIATION $\quad$ LH + R· $\longrightarrow$ L· + RH

PROPAGATION $\quad$ L· + O_2 $\longrightarrow$ LOO·

$\qquad$ LOO· + LH $\longrightarrow$ LOOH + L·

TERMINATION $\quad$ L· + L· $\longrightarrow$ LL

$\qquad$ L· + LOO· $\longrightarrow$ LOOL

$\qquad$ LOO· + LOO· $\longrightarrow$ LOOL + O_2

LH - Lipid molecule $\qquad$ LOO· - Lipid peroxy radical

R· - Free radical $\qquad$ LOOH - Lipid hydroperoxide

L· - Lipid alkyl radical $\qquad$ LOOL - Lipid peroxide

FIGURE 1. Schematic of initiation, propagation, and termination of lipid peroxidation reactions. (From Penn, J. S. and Thum, L. A., *Oxygen Radicals in Biology and Medicine*, Plenum Press, New York, 1989, 1025. With permission.)

The defense mechanisms against free radicals also include various enzyme systems, namely superoxide dismutases, catalases, and peroxidases.[8] Superoxide dismutase (SOD) converts superoxide radicals to hydrogen peroxide and oxygen. The hydrogen peroxide produced is metabolized by catalase and glutathione peroxidase. Vitamin E,[9,10] SOD,[11,12] and glutathione peroxidase[13] are all present in the retina. However, this is not the limit of the retina's antioxidant system. Tappel[14] has proposed a scheme, which was later considered in detail by others[15,16] whereby a coupled exchange of unpaired electrons brings vitamin C and the glutathione system into the battery of antioxidants, as shown in Figure 2. Although the scheme is not completely defined for *in vivo* systems, some of its components have been characterized *in vitro* by Winkler.[17]

Direct evidence from animal models that free radicals are involved in retinal oxygen toxicity has come from several studies. Using kittens aged 1 to 11 d and exposed to 72 h of 80% oxygen, Bougle and co-workers demonstrated a decrease in retinal SOD activity, inferring a reduced ability of the kitten retina to defend itself against free radicals of oxygen. Pretreatment with vitamin E protected against this effect.[18] Taki[19] has reported that the retinas of kittens exposed to hyperoxia show evidence of lipid peroxidation, suggesting a free radical reaction with the lipids of cell membranes, where polyunsaturated fatty acids are found. Further evidence from Taki[20] takes the form of increased levels of lipid peroxides in the blood of kittens exposed to 48 h of 70% oxygen from day three of life. The effects were reduced by a

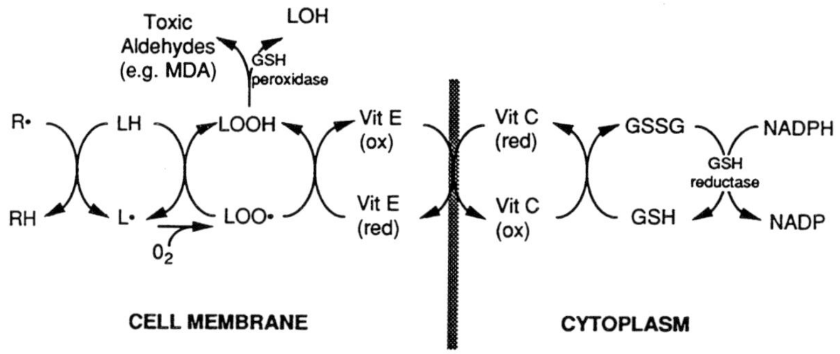

FIGURE 2.   Schematic of proposed retinal antioxidant strategy. (From Anderson, R. E. and Penn, J. S., *Vitamin (Jpn.)*, 61, 237, 1987. With permission.)

subcutaneous administration of tocopherol acetate during and after exposure. Yagi and co-workers have reported elevated lipoperoxide levels in both the blood and the retina of chick embryos after exposure to 95% oxygen for 12 or 24 h.[21] Finally, Hiramatsu and colleagues have also measured increases in lipoperoxides in rabbit retinas following 12 h at 90 to 95% oxygen.[22]

II. RETINOPATHY OF PREMATURITY (ROP)

Through a decade of research, peroxidation has been implicated in human retinal degenerations.[23] Ocular siderosis, or iron in the eye, has been known to result in retinal degeneration. Certain retinotoxic drugs, such as adriamycin, are presumed to act through photooxidative processes. Senile macular degeneration is yet another candidate for peroxidation-induced degeneration. Finally, retinopathy of prematurity (ROP) occurs only in premature infants housed in elevated levels of atmospheric oxygen. Although the agent in each of these insults may be different, cell destruction and death apparently proceed through pathways common to lipid or protein oxidation.

Retinopathy of prematurity (ROP) is a condition evidenced by a degeneration of retinal blood vessels. It occurs in premature infants who, due to pulmonary complications, are housed in hyperoxic conditions. The development of the retinal vasculature is not complete in premature infants. The result of hyperoxia is the obliteration of a portion of the existing vessels and the abnormal development of new vessels, which eventually become fibrotic, contract, and cause retinal detachment.

The history of ROP in the U.S. shows two distinct "epidemics." The first occurred in the 1940s and 1950s, after oxygen treatment was suggested for premature infants, but before it was confirmed as a causal factor in the retinopathy. Once the oxygen risk was determined, the epidemic was eliminated by more careful dose administration and monitoring. A second epidemic is now underway and has been brought about, ironically, by our improved ability to care for premature infants. As survival, particularly of very low birth-weight infants, has increased, so has the incidence of ROP. It should be emphasized that, although it has been nearly 40 years since oxygen was recognized as a causal factor, the precise mechanism of cytotoxicity of oxygen on retinal cells still remains a mystery.

III. PATHOGENESIS OF ROP

The effects of supplemental oxygen on incompletely vascularized retinas can be divided into two phases: an initial vasoconstrictive response that occurs during exposure, and a subsequent phase of vasoproliferation that takes place after removal to room air. There is evidence that the vasoproliferation may depend on local retinal hypoxia produced by previous vasobliteration.[24] It has been suggested that, under hyperoxic conditions, the choroid may furnish excessive concentrations of oxygen to the inner retina, producing retinal vasobliteration.[25] The inner retina may be particularly susceptible to this condition at the developmental stage where photoreceptors are not yet fully formed and are unable to create an adequate oxygen sink.[26] Upon return to room air, the choroidal circulation is inadequate to overcome the relative hypoxia.[27]

Others have speculated that retinal vasoconstriction may be a physiological protective response to hyperoxia.[28,29] When aspirin, which produces an inhibition of prostaglandin-induced retinal vasodilation, was administered to beagle puppies prior to exposure to 95 to 100% oxygen, a persistent vasodilation was observed, with subsequent severe retinopathy.[28] It seems likely that retinal vascular damage results primarily from cytotoxic effects of oxygen on endothelial cells, rather than secondarily from the vasoconstriction and diminished blood flow as was originally postulated.[30] If retinal vascular channels are dilated while increased oxygen levels are present, then endothelial vessel walls are in greater contact with toxic oxygen radicals, which are known to damage cell membranes. Indeed, the first event that can be detected by electron microscopy is a selective injury to the endothelial cells of the immature vessels, with no obvious changes in the neuronal elements of the retina.[30] These results support a direct injury to the retinal endothelium. Studies have shown that cells gradually exposed to hyperoxia will build up resistance by increased intracellular production of antioxidants.[31] Therefore, it is tempting to postulate that immature vessels are at higher risk because they have not sufficiently matured to produce protective antioxidants or have not had the necessary gradual exposure to oxygen to induce an adequate complement of these molecules.

Clinically, a controversy has been generated over the supplementation of vitamin E to premature infants as therapy for ROP. Within the past decade there has been a nearly equal split between those clinical trials that showed protective effects of vitamin E[32-34] and those that refuted its efficacy.[35-37]

The work of Kretzer and associates[38] includes an elaborate proposal for the cellular mechanism of ROP and the protective role of vitamin E. This proposal is based on the finding that oxygen-treated infants have a higher incidence of spindle cell gap junctions than healthy normoxic infants. These authors postulated that gap junctions alter the normal vasoformative process. Premature infants are born with relatively low levels of retinal vitamin E, particularly in the avascular region.[39] If administered at the appropriate time, vitamin E supplementation results in a rapid increase in retinal levels. Infants supplemented in this manner display a reduction in spindle cell gap junctions and, according to these authors, a reduction in the severity of the retinopathy.[38]

IV. OXYGEN-INDUCED RETINOPATHY IN THE RAT

Past research has shown hyperoxia-induced abnormalities in the retinal blood vessels of several species. The rat,[40,41] mouse,[42] cat,[43] dog,[28] and rabbit,[44] all have demonstrated ROP-like alterations of retinal vasculature when placed in high oxygen environments during development. No model has proved completely reliable, but the newborn rat offers several attractive features that have drawn the attention of investigators for 35 years. These features include: (1) the ontogeny of the rat retina is similar to that of the human; of particular importance is the fact that rats have a clearly demonstratable spindle cell complement; (2) the retinal vasculature of the rat develops postnatally, facilitating its study; (3) the newborn rat is vitamin E deficient, like the premature infant; and (4) the rat is an inexpensive and easily maintained tissue source for which pertinent techniques have already been perfected.

The development of the rat retinal vasculature has been described many times.[45-48] The retinal vessels are first seen as undifferentiated spindle cells derived from the outer wall of the hyaloid vessel at about 14 d gestation.[46] The cells migrate into luminized vessels in the nerve fiber layer of the retina.[47] By postnatal day 15, the vessels have reached the periphery of the retina, and the capillary sprouting is complete.[48] Vessel growth from advancing mesenchymal cells is also seen in the human, but the contribution of these cells is equivocal in the kitten and rabbit, where endothelial budding is implicated.[46] The distinction between vessel proliferation from undifferentiated spindle cells and from preexisting capillaries is emphasized because the mechanism for the development and plasticity of the vessels may be different.[25,49]

V. RECENT STUDIES OF ROP IN RATS

In 1985, Yabe[50] used the rat to examine the efficacy of vitamin E in oxygen-induced retinopathy. He found that the retinal vessels of vitamin E-deficient newborn rats showed an increased oxygen sensitivity, which he attributed to the observation that retinal vascular development is delayed in vitamin E-deficient rats. Yabe concluded that vascular immaturity is the important variable in predisposing a rat to oxygen-induced retinopathy. We are also developing a rat model for ROP, in order to conduct a careful and systematic analysis of the biochemical factors that affect both normal retinal vascular development and susceptibility of this process to damage by oxygen. Our aim is to identify factors which are responsible for protecting retinas from oxygen damage and to seek ways to compensate for the absence of these factors or their ineffectiveness in newborn hyperoxic animals.

Immediately after birth, rat pups and their mothers were placed in an atmosphere of 60% oxygen for 14 d. At that time, some animals were used for analysis of retinal tissue and the remainder were allowed to survive for longer periods of time in room air. At prescribed times, pups from this group were anesthetized and retinal function determined by electroretinography. For those rats sacrificed immediately after oxygen exposure, eyes were removed for morphological and/or biochemical examinations. Some rats were perfused with India ink to allow for determination of the effects of treatment on retinal vasculature. This report will be limited to a brief description of the morphological examination, and a more detailed account of the biochemical findings.

Figure 3a shows a flat-mounted retina of a rat maintained from birth for 14 d in room air and perfused with India ink. Note the high degree of vascularization, with capillary networks well-formed and patent. Figure 3b shows the retina of a rat pup maintained in 60% oxygen for 14 d. There is a noticeable loss of retinal capillaries due to 2 weeks presence in the high levels of oxygen. This loss is primarily the result of: (1) retardation of the progression of vasoformation to the retinal periphery; and (2) vasobliteration of the central and periarterial capillaries.

Included in our battery of tests is a determination of the endogenous levels of several important retinal antioxidants. Upon conducting these assays, we found a significant reduction in the levels of retinal ascorbic acid and retinal α-tocopherol in oxygen-exposed rats (Table 1). Control retinas contained 20 and 35% higher levels, respectively. No differences were seen in the levels of glutathione peroxidase, glutathione *S*-transferase, or glutathione reductase. We are currently examining retinal superoxide dismutase and catalase in order to determine if either is compromised.

The reduced levels of retinal vitamin E in oxygen-reared rats encouraged us to follow Yabe's investigation with our own attempt at α-tocopherol manipulation. A notable difference was that we attempted to *supplement, as well as deprive,* the rat pups through the diet of the mothers. Dietary supplementation or deprivation began when breeding pairs were combined and lasted through pregnancy and the treatment period (a total of 35 to 39 d). The two diets were manufactured by Dyets, Inc. (Bethlehem, PA), and consisted of 1.0 g α-tocopherol acetate per kilogram of food or no α-tocopherol acetate. Using this method, we were able to cause three- to fourfold differences in the retinal vitamin E levels of the newborn animals, regardless of oxygen level. Supplemented rats had retinal α-tocopherol levels of 1.5 to 2.0 mmol/ mol phospholipid, while levels in deprived rats were 0.4 to 0.6 mmol/mol phospholipid.

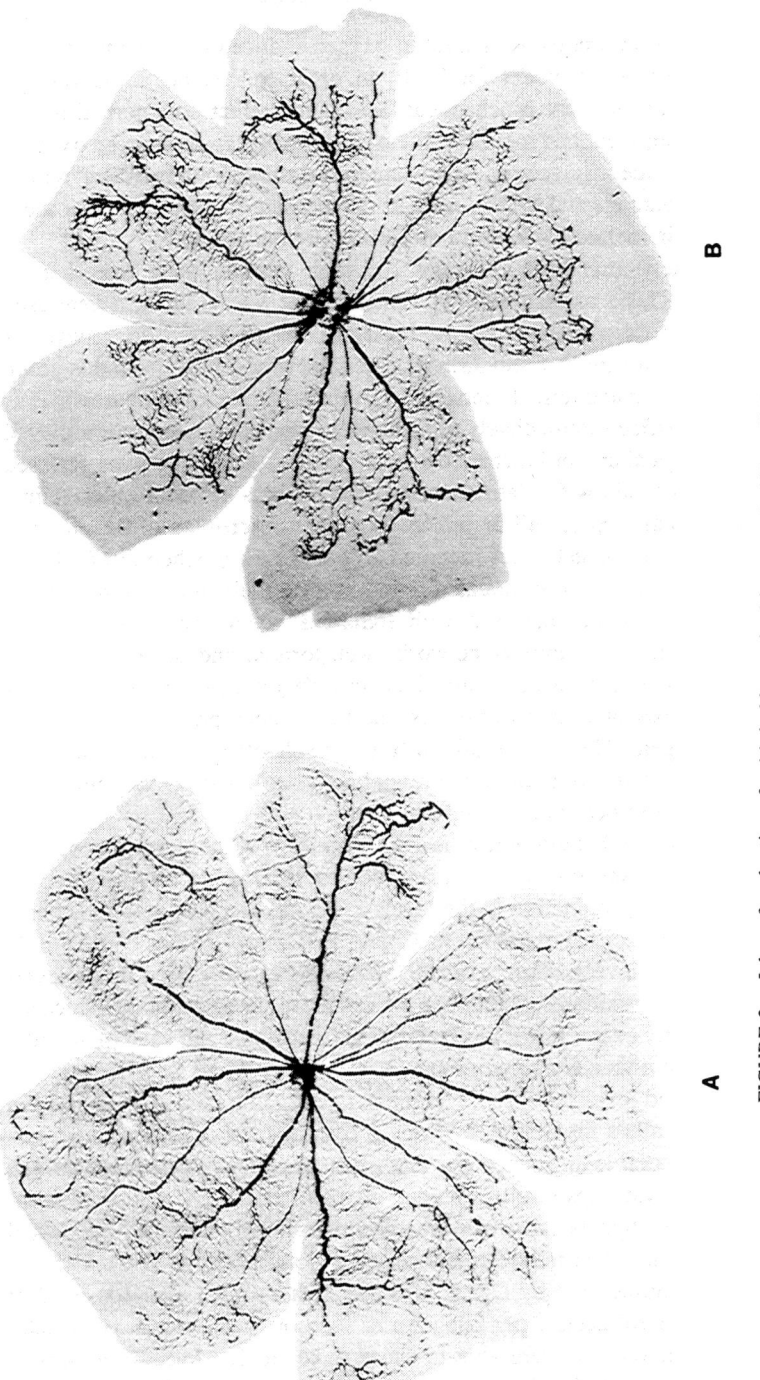

B

A

FIGURE 3.   Ink-perfused retina of a 14-d-old rat raised in room air (A) or 60% oxygen (B).

These dietary α-tocopherol treatments, in combination with oxygen treatment, affected the levels of other antioxidants as well. Of note was a significant increase in the retinal level of glutathione peroxidase in both the vitamin E-supplemented and -deprived rats of the 60% oxygen-exposed group. More important, however, was that vitamin E supplementation lessened the severity of oxygen-induced vasobliteration. Interestingly, both vitamin E-supplemented and -deprived rats raised in 60% oxygen sustained less vasobliteration than oxygen-exposed rats with no vitamin E manipulation. We plan to examine the cause of this, beginning with the interrelationship of vitamin E and glutathione peroxidase.

The actual contribution made by glutathione peroxidase in defending the immature retina from oxygen-induced retinopathy is unknown for humans or animals. But its drastic alteration by the above treatment serves as a reminder that one must consider *all* retinal antioxidants when manipulating any single one. The interactions of these molecules, as the scheme in Figure 2 indicates, are extremely complex, and an alteration in the level of any one is likely to effect the relative contributions of others, as well.

We have also attempted to supplement retinal vitamin C levels in newborn rats. This endeavor was carried out by daily subcutaneous injections of mothers with 0.5 g/kg sodium ascorbate on one occasion, and by daily intraperitoneal injections of pups on another. Neither of these efforts proved successful. It was possible to raise the serum level of vitamin C in the mothers and in the pups, but not the retinal levels. It should be noted that ascorbic acid is not a true vitamin to rats, since they produce it, and therefore rats cannot be made scorbutic — a factor that may have prevented a more positive outcome.

Diminution of endogenous antioxidant levels by deficient diets or augmentation by exogenous antioxidant addition constitute one strategy for establishing the participation of peroxidation in oxygen-induced retinopathy. A second strategy is the measurement of a selective loss of appropriate substrate molecules for peroxidation reactions. Still a third strategy is to measure the buildup of the typical products of peroxidation within the retina. Any one of these three strategies alone yields some indication of the participation of peroxidation in oxygen-induced retinopathy. In concert, they constitute a powerful argument.

We have measured the specific loss of docosahexaenoic acid in the retinas of 60% oxygen-exposed newborn rats. This molecule is the predominant fatty acid in the retina, making up more than one third of the total that is bound by phospholipid. Since it has six double bonds, it is much more susceptible to free radical attack than is a saturated fatty acid. The specific loss of docosahexaenoic acid in oxygen-treated rat pups was significant, reaching 30% by 14 d of exposure (Table 1).

An increase in one form of peroxidation product was also apparent in the retinas of 60% oxygen-exposed rats (Table 1). The method by which this was

TABLE 1
Indicators of Retinal Peroxidation in Oxygen-Reared Rats

Indicator	Room Air	60% Oxygen	p Value
Malondialdehyde (nmol/mg protein)	11.1 ± 3.5	23.2 ± 1.3	<0.001
Docosahexaenoic acid (mole%)	15.5 ± 0.8	11.0 ± 1.1	<0.02
Vitamin E (mmol/mol phospholipid)	0.96 ± 0.19	0.71 ± 0.15	<0.005
Vitamin C (nmol/mg cytoplasmic protein)	69.3 ± 7.9	58.6 ± 11.0	<0.005

determined is the thiobarbituric acid conjugation method. Although this test is the most frequently used analysis for this purpose, it is quite nonspecific. However, only a comparison of *relative* values was required, and this comparison concerned two groups of rats for which the only variable was inspired oxygen. We were not interested in developing absolute levels of thiobarbituric acid-reactive species for either treatment alone. Our determinations revealed levels twice as high in retinas from oxygen-treated rats than in those from room air counterparts (23.2 vs 11.1 nmol/mg retinal protein, respectively).

An understanding of the pathogenesis of ROP is still in its primitive stages. However, the identification of substances that protect the retina against oxygen damage, and the determination of the relationship between their retinal concentrations and the effect of high atmospheric oxygen, is an important step. The true challenge will be translating the knowledge about biochemical mechanisms of ROP gained from animal models into rational therapeutic approaches that are applicable to human infants.

ACKNOWLEDGMENTS

The author wishes to thank L. Thum, A. Riley, L. Larrazabal, and B. Tolman for their technical assistance. This work was supported by grants from the National Eye Institute, Knights Templar Eye Foundation, the Arkansas Science and Technology Authority, and Research to Prevent Blindness, Inc.

REFERENCES

1. **Farrell, P. M., Bieri, J. G., and Wood, R. E.,** The occurrence and effects of human vitamin E deficiency, *J. Clin. Invest.,* 60, 233, 1977.
2. **Lewin, R.,** Drug trial for Parkinson's, *Science,* 236, 1420, 1987.
3. **Cavarocchi, N. C., England, M. D., and O'Brien, J. F., et al.,** Superoxide generation during cardiopulmonary bypass: is there a role for vitamin E? *J. Surg. Res.,* 40, 519, 1986.
4. **Handelman, G. J. and Dratz, E. A.,** The role of antioxidants the retinal pigment epithelium and the nature of prooxidant-induced damage, *Adv. Free Radical Biol. Med.,* 2, 1, 1986.

5. **Anderson, R. E. and Andrews, L. M.**, Biochemistry of retinal photoreceptor membranes in vertebrates and invertebrates, in *Visual Cells in Evolution,* Westfall J., Ed., Raven Press, New York, 1981, 1.

6. **Fliesler, S. J. and Anderson, R. E.**, Chemistry and metabolism of lipids in the vertebrate retina, *Prog. Lipid Res.,* 22, 79, 1983.

7. **Dormandy, T. L.**, Free radical oxidation and antioxidants, *Lancet,* 1, 647, 1978.

8. **Fridovich, I.**, Oxygen: aspect of its toxicity and elements of its defense, *Curr. Eye Res.,* 3, 1, 1984.

9. **Dilley, R. and McConnell, D.**, Alpha-tocopherol in the retinal outer segment of bovine eyes, *J. Membr. Biol.,* 2, 317, 1970.

10. **Stone, W. L., Katz, M. L., and Lurie, M., et al.**, Effects of dietary vitamin E and selenium on light damage to the rat retina, *Photochem. Photobiol.,* 29, 725, 1979.

11. **Hall, M. and Hall, D.**, Superoxide dismutase of bovine and frog outer segments, *Biochem. Biophys. Res. Commun.,* 67, 1199, 1975.

12. **Bensinger, R. E., Crabb, J. W., and Johnson, C. M.**, Purification and properties of superoxide dismutase from bovine retina, *Exp. Eye Res.,* 34, 623, 1982.

13. **Penn, J. S., Naash, M. I., and Anderson, R. E.**, Effect of light history on retinal antioxidants and light damage susceptibility in the rat, *Exp. Eye Res.,* 44, 779, 1987.

14. **Tappel, A. L.**, Will antioxidant nutrients slow aging processes? *Geriatrics,* 23, 97, 1968.

15. **Packer, J. E., Slater, T. F., and Willson, R. L.**, Direct observation on a free radical interaction between vitamin E and vitamin C, *Nature (London),* 278, 737, 1979.

16. **Anderson, R. E. and Penn, J. S.**, The role of vitamin E in retinal degenerative diseases, *Vitamin (Jpn.),* 61, 237, 1987.

17. **Winkler, B. S.**, *In vitro* oxidation of ascorbic acid and its prevention by GSH, *Biochim. Biophys. Acta,* 925, 258, 1987.

18. **Bougle, D., Vert, P., Reichart, E., Hartemann, D., and Heng, E. L.**, Retinal superoxide dismutase activity in newborn kittens exposed to normobaric hyperoxia: effect of vitamin E, *Pediatr. Res.,* 16, 400, 1982.

19. **Taki, M.**, Causal relationship of lipid peroxide with experimental retinopathy of prematurity. 1. Morphological distribution of lipid peroxide in kitten retina after oxygen administration, *Nagoya Med. J.,* 28, 115, 1983.

20. **Taki, M.**, Causal relationship of lipid peroxide with experimental retinopathy of prematurity. 2. Lipid peroxide and platelet aggregation in kitten blood after oxygen administration, *Nagoya Med. J.,* 28, 121, 1983.

21. **Yagi, K., Matsuoka, S., Ohkawa, H., Ohishi, N., Takevchi, Y., and Kakai, H.**, Lipoperoxide level of the retina of chick embryo exposed to high concentration of oxygen, *Clin. Chim. Acta,* 80, 355, 1977.

22. **Hiramatsu, T., Harata, K., Nishigaki, I., and Yaki, K.**, The formation of lipoperoxide in the retina of rabbits exposed to high concentration of oxygen, *Experientia,* 32, 622, 1976.

23. **Wiegand, R. D., Jose, J. G., Rapp, L. M., and Anderson, R. E.**, Free radicals and damage to ocular tissues, in *Free Radicals in Biology and Aging,* Armstrong, D., et al., Eds., Raven Press, New York, 1984, 317.

24. **Johnson, L.**, Retrolental fibroplasia: a new look at an unsolved problem, *Hosp. Pract.,* 16, 109, 1981.

25. **Patz, A.**, Current concepts on the effects of oxygen on the developing retina, *Curr. Eye Res.,* 3, 159, 1984.

26. **Kretzer, F. L. and Hittner, H. M.**, Initiating events in the development of retinopathy of prematurity, in *Retinopathy of Prematurity,* Silverman, W. A. and Flynn, J. T., Eds., Blackwell Scientific, Boston, 1985, 127.

27. **Ernest, J. T. and Goldstick, T. K.**, Retinal oxygen tension and oxygen reactivity in retinopathy of prematurity in kittens, *Invest. Ophthalmol. Vis. Sci.,* 25, 1129, 1984.

28. **Flower, R. W. and Blake, D. A.**, Retrolental fibroplasia: role of the prostaglandin cascade in the pathogenesis of oxygen induced retinopathy in the newborn beagle, *Pediatr. Res.,* 15, 1293, 1981.

29. **Dollery, C. T., Hill, D. W., Mailer, C. M., and Ramalho, P. S.,** High oxygen pressure and the retinal blood vessels, *Lancet,* 2, 291, 1964.
30. **Ashton, N. and Pedler, C.,** Studies on developing retinal vessels. IX. Reaction of endothelial cells to oxygen, *Br. J. Ophthalmol.,* 46, 257, 1962.
31. **Crapo, J. D. and McCord, J. M.,** Oxygen-induced changes in pulmonary superoxide dismutase assayed by antibody titrations, *Am. J. Physiol.,* 231, 1196, 1976.
32. **Hittner, H. M., Godio, L. B., and Rudolph, A. J., et al.,** Retrolental fibroplasia: efficacy of vitamin E in a double-blind clinical study of preterm infants, *N. Engl. J. Med.,* 305, 1365, 1981.
33. **Johnson, L., Schaffer, D., and Quinn, G., et al.,** Vitamin E supplementation in retinopathy of prematurity, *Ann. N.Y. Acad. Sci.,* 393, 473, 1982.
34. **Finer, N. N., Schindler, R. F., and Peters, K. L., et al.,** Vitamin E and retrolental fibroplasia: improved visual outcome with early vitamin E, *Ophthalmology,* 90, 428, 1983.
35. **Puklin, J. E., Simon, R. M., and Ehrenkranz, R. A.,** Influence of retrolental fibroplasia of intramuscular vitamin E administration during respiratory distress syndrome, *Ophthalmology,* 89, 96, 1982.
36. **Watts, J. O., Milner, R. A., and McCormick, A. Q.,** Failure of vitamin E to prevent RLF, *Clin. Invest. Med.,* 8, A176, 1985.
37. **Phelps, D. L., Rosenbaum, A. L., Isenberg, S. J., Leake, R. D., and Dorey, F. J.,** Tocopherol efficacy and safety for preventing retinopathy of prematurity: a randomized controlled, double-masked trial, *Pediatrics,* 79, 489, 1987.
38. **Kretzer, F. L., Hittner, H. M., and Johnson, A. T.,** Vitamin E and retrolental fibroplasia: ultrastructural support of clinical efficacy, *Ann. N.Y. Acad. Sci.,* 393, 145, 1982.
39. **Nielsen, J. C., Naash, M. I., and Anderson, R. E.,** The regional distribution of vitamins E and C in human adult and preterm infant retinas, *Invest. Ophthalmol. Vis. Sci.,* 29, 22, 1988.
40. **Patz, A.,** Oxygen studies in retrolental fibroplasia. IV. Clinical and experimental observations, *Am. J. Ophthalmol.,* 38, 291, 1954.
41. **Ricci, B.,** Effects of hyperbaric, normobaric and hypobaris supplementation on retinal vessels in newborn rats: a preliminary study, *Exp. Eye Res.,* 44, 459, 1987.
42. **Michaelson, I. C., Herz, H., Lewkowitz, E., and Kertesz, D.,** Effect of increased oxygen on the development of the retinal vessels, *Br. J. Ophthalmol.,* 38, 577, 1954.
43. **Phelps, D. L. and Rosenbaum, A. L.,** The role of tocopherol in oxygen-reduced retinopathy: kitten model, *Pediatrics,* 59, 988, 1977.
44. **Ashton, N., Tripathi, B., and Knight, G.,** Effect of oxygen on the developing retinal vessels of the rabbit. I. Anatomy and development of the retinal vessels of the rabbit, *Exp. Eye Res.,* 14, 214, 1972.
45. **Henkind, P. and De Oliveira, L. F.,** Development of retinal vessels in the rat, *Invest Ophthalmol.,* 6, 520, 1967.
46. **Ashton, N.,** Oxygen and the growth and development of retinal vessels, *Am. J. Ophthalmol.,* 62, 412, 1966.
47. **Shakib, M., De Oliveira, L. F., and Henkind, P.,** Development of retinal vessels. II. Earliest stages of vessel formation, *Invest. Ophthalmol.,* 7, 689, 1968.
48. **Ashton, N. and Blach, R.,** Studies in developing retinal vessels. VIII. Effect of oxygen on the vessels of the ratling, *Br. J. Ophthalmol.,* 45, 321, 1961.
49. **Ashton, N.,** The pathogenesis of retrolental fibroplasia, symposium: RLF, *Ophthalmology,* 86, 1695, 1970.
50. **Yabe, H.,** Effects of vitamin E deficiency on oxygen induced retinopathy in rat, *Nippon Ganka Gakkai Zasshi,* 89, 624, 1985.

Chapter 10

FREE RADICALS IN MUSCULAR DYSTROPHY

James P. Kehrer and Michael E. Murphy

TABLE OF CONTENTS

I. OVERVIEW

The muscular dystrophies are a group of inherited diseases with both similarities and differences. Duchenne muscular dystrophy is the most common and severe form of these diseases in humans. A major advance in our understanding of Duchenne muscular dystrophy has been the recent identification of a 427-kDa membrane protein, termed dystrophin, as the product of the defective locus. Muscle levels of dystrophin inversely correlate with the severity of the disease. However, the function of this protein is unknown and thus the mechanism of muscle degeneration has not been established. Oxidative stress, a term describing the damage to tissues caused by reactive oxygen species (ROS), has long been hypothesized as the cause of many of the structural, functional, and biochemical changes characteristic of inherited muscular dystrophy in animals and humans. Support for this hypothesis is indirect but extensive, including measurements which show an increased content of by-products of oxidative damage, compensatory increases in cellular antioxidants, changes in the proportions and metabolism of cellular lipids, abnormal functions of cellular membranes, altered activities of membrane-bound enzymes, and disturbances in cellular protein turnover and energy production. In addition, increasing free radical fluxes in tissues by nutritional manipulations or other means can induce muscle injury which mimics the biochemistry and pathology of muscular dystrophy. It must be remembered, however, that any endogenous oxidative stress can either cause or result from dystrophic alterations in cellular physiology and biochemistry. Establishing the proper causal relationship between tissue pathology and free radical production is extremely difficult. In the case of muscular dystrophy, current evidence does not support the hypothesis that free radicals play the causal role in muscle damage, although it remains possible that this mechanism is involved in the secondary tissue injury associated with this disease.

II. DUCHENNE MUSCULAR DYSTROPHY (DMD)

Muscular dystrophy is a generic term referring to a group of inheritable human diseases (Table 1). The various conditions are alike in that they lead to muscular weakness, but differ from each other in the pattern of inheritance, age of onset, rate of progression of weakness, the muscles that are affected, and probably the mechanism by which the damage occurs. Duchenne muscular dystrophy (DMD) affects 1 in about 4000 male births. This disease confines patients to a wheelchair around the age of 8 to 10, and usually causes death before the age of 20. This makes it both the most common and most severe form of these diseases, and explains why it has received the most attention over the years. DMD, and associated animal models, are the focus of the current chapter.

TABLE 1
Forms of Muscular Dystrophy in Humans

Type	Age at onset	Cause
Duchenne (DMD)	1—5	X-linked recessive; lack of dystrophin
Becker (mild form of DMD)	5—25	X-linked recessive; lack of dystrophin
Limb-girdle	10—30	Autosomal recessive
Facioscapulohumeral	Any age	Autosomal dominant
Distal	40—60	Autosomal dominant
Ocular	Any age	Autosomal dominant
Oculopharyngeal	Any age	Autosomal dominant

TABLE 2
Animal Models of Muscular Dystrophy

Species	Strain	Cause
Dog	xmd	Lacks dystrophin; X-linked recessive mimics clinical symptoms of DMD
Mouse	mdx	Lacks dystrophin; X-linked recessive; mild clinical symptoms
	dy (various alleles)	Autosomal recessive; not a lack of dystrophin; increased proteolysis
	129 ReJ	Increased proteolysis?
Cat		Lacks dystrophin; X-linked recessive; mild clinical symptoms
Chicken	S11, lines 413,433	Unknown; not X-linked; not a lack of dystrophin
	Nutritional model	Lack of vitamin E and sulfur amino acids
Hamster	BIO 14.6	Unknown; cardiomyopathy more severe than skeletal muscle
Mink, turkey, sheep, duck, cow		Unknown or nutritional

III. ANIMAL MODELS

Animal models are widely used to investigate muscular dystrophy. A disadvantage of all animal models has been the uncertainty as to whether they are equivalent to any of the human disorders. Recently, a major advance in our understanding of Duchenne muscular dystrophy was made by identifying the specific protein which is absent or defecting in patients with this disease.[1] Now known as dystrophin, the muscle levels of this 427-kDa protein are low or absent in patients with Duchenne or Becker's muscular dystrophy.[2]

Among the many animal models used over the years, it appears that at least one mouse (mdx), one cat, and one dog (canine X-linked muscular dystrophy; CXMD) model of muscular dystrophy are caused by a deficiency of the same protein (Table 2).[1,3-5] However, in contrast to the human and dog diseases which are lethal, cats and mdx mice survive without obvious distress despite evidence of muscle degeneration. A major pathological difference between cats or mdx mice and either DMD or CXMD muscle is the absence of extensive fibrosis. Although this difference has been hypothesized to explain the improved clinical status of cats and mdx mice,[1] other factors such as maturation or differing regeneration capacity may be involved. For example, studies in the mdx mouse have shown extensive muscle fiber necrosis beginning at 3 weeks of age, followed by rapid regeneration such that by 5 weeks of age nearly complete recovery was evident.[5] An important task continues to be to distinguish between the trivial and important comparisons made among the different model systems used to study muscular dystrophy. In the meantime, animal models of the disease provide opportunities for research not possible with biopsy samples and provide the means to assess potential mechanisms of muscle degeneration.

IV. HYPOTHESES REGARDING THE CAUSE OF DMD

Although the protein which is deficient in DMD has been isolated, its function remains unknown. Thus, the mechanism of muscle damage in the various inherited muscular dystrophies has not been established. The evidence which is available has shown that dystrophin is a very large protein, similar to spectrin and α-actinin, which is localized in and may stabilize cellular membranes.[1,4,6] This location and the likelihood of a general structural role would explain the evidence that the plasma membrane is abnormal in DMD and that much of the pathology associated with this disease has a membrane origin. However, additional research will be required to determine how such a general effect can lead to the specific changes characteristic of muscular dystrophy, or how it could result in the evidence of oxidative stress in dystrophic muscle described below.

Many aspects of the physiology of normal muscle function have been characterized. In most instances, these normal functions have been found to be altered in dystrophic muscle. The current consensus regarding these changes is that they represent secondary responses of muscle tissue to the primary defect arising from a deficiency of dystrophin. However, this does not preclude these secondary alterations being important, or even critical, to the muscle damage which develops. Based on studies of these changes, numerous theories were advanced over the years to try to explain why muscle cells degenerate in the muscular dystrophies.[7] With the discovery of dystrophin, some are clearly no longer viable, while a reconsideration of others may help unravel the function of this protein.

A. NEUROGENIC THEORIES

Several different versions of the neurogenic theory have been proposed for DMD and muscular dystrophy in animals.[8-12] Transplantation and parabiotic experiments clearly disprove the original hypotheses in the case of the animal models. The placement of muscle cells obtained from young dystrophic chickens or mice into muscles of normal animals, and the reverse, showed the pathological symptoms were based on the genotype of the muscle cells, and not on some neurogenic factor.[13-15] Parabiotic exchanges of motor nerves in young mice supported the same conclusion.[15,16]

The experiments described above do not negate the possible involvement of improper muscle-nerve interaction and communication, as the result of the absence of dystrophin, as pathogenic mechanisms. A revised version of this hypothesis takes this idea into account by suggesting (indirectly) that dystrophin may play an essential role in muscle growth and development (perhaps at the time of embryogenesis) through its effects on innervation processes.[17] The neurological damage in the inherited muscular dystrophies may, therefore, be a secondary effect of abnormal muscle development. Alternatively, this type of damage may be the result of myofibril degeneration and inadequate regeneration, or may be a coincident but nonpathogenic expression of the underlying biochemical defect expressed in nonmuscle tissue. The former explanation is difficult to prove, but the latter may reflect the result of an underlying oxidative stress since a nutritionally induced oxidative stress (e.g., vitamin E deficiency) leads to neuropathies as well as myopathies in many animals.[18,19]

B. VASCULAR ISCHEMIA AND HORMONAL THEORIES

Several groups have proposed that the repeated occurrence of temporary arteriolar constriction leads to periods of ischemia in small regions of skeletal muscle in DMD patients which, over time, results in muscle destruction.[8,10,11,20] Some histological features of DMD resemble features of microvascular ischemia induced in animal models following the injection of small dextrin particles[8,20] but the discovery of dystrophin would appear to negate this idea. The same is true of the several theories developed on the role of growth hormone, thyroxin, serotonin, and insulin as factors in inherited muscular dystrophy.[10-14,21,22] Pharmacological intervention can alter the levels or activities of these hormones in dystrophic animals and humans but, although the treatments may temporarily alleviate symptoms in some models of the disease, none provides a cure.[23] The underlying basis of many hormonal changes in muscular dystrophy may be due to the overall changes in body chemistry and stress that accompany muscle destruction and wasting.

C. DEVELOPMENTAL BLOCK THEORY

Several investigators have suggested dystrophic muscle is subject to a developmental block which leads to muscle destruction by delaying the final steps in differentiation.[12,14] This concept has received additional support in recent years as the neurogenic theories have been modified to include the idea of abnormal maturation.[17] The fact that muscles from DMD patients, as well as dystrophic mice and chickens, only show substantial necrosis after they mature is evidence for the development-dependent expression of the lethal effects of the disease. The similarities between dystrophic muscle and immature fibers in the relative activities of metabolic pathways and in the specific isoenzyme patterns of structural protein support both the idea that mature fibers are susceptible to degeneration and that immature fibers degenerate without reaching the final stage of development.[12] So little is known about the normal biochemical events in the nucleus that control differentiation that few studies have addressed this development theory directly. It remains possible that the deficiency of dystrophin affects the process of muscle development, or affects a function which is unique, or particularly important, to mature muscle fibers.

D. IMMUNE SYSTEM FUNCTION

Alterations in leukocyte function have been reported in muscular dystrophy.[24,25] These alterations appear likely to be normal changes associated with the ongoing inflammatory events of muscle necrosis. However, even if functioning normally, the immune system may play a role in the pathology of muscular dystrophy. Detailed, comparative studies of muscular diseases have found that T cells, accompanied by macrophages, focally surrounded and invaded necrotic fibers in DMD. Only in rare instances were similar findings observed in nonnecrotic fibers.[26]

Normal functioning of the immune system involves, in part, the production of superoxide and hydrogen peroxide by activated leukocytes. This can probably account for some of the evidence of oxidative stress observed since these oxidants are capable of damaging nearly all muscle protein and lipid components. Although not studied in detail, leukocyte-mediated oxidative injury may account for a portion of the inhibition of metabolic processes, as well as other changes characteristic of muscular dystrophy.

E. PROTEOLYSIS

An enhanced activity of various proteases has been reported in dystrophic muscles from humans and animals.[27-29] Furthermore, successful treatment has been reported with several different protease inhibitors in the dy/dy strain of mice.[30,31] These findings have supported the hypothesis that the breakdown of dystrophic muscle fibers is caused by the excess activity of endogenous pathways and that, at least in dy/dy mice, proteolysis is a critical step in the pathogenesis of this disease. The basis for such an enhanced proteolysis has not been clearly established. One suggestion supported by studies in the 129 ReJ mouse strain is that dystrophic tissues contain lower than normal levels of protease inhibitors.[32] In contrast, in the mdx mouse strain, the deficiency of dystrophin has been correlated with a significant rise in intracellular free calcium which may, in turn, activate various proteases resulting in the degradation of muscle proteins.[33] Whether such pathways play a significant pathologic role in humans is unclear. The inability to achieve therapeutic success with protease inhibitors in humans,[23] conflicting data on the role of proteolytic enzymes in some animal models,[34] and the discovery of dystrophin suggests that while excess proteolysis is a critical defect in the dy/dy strain of mice, in other models the enhanced proteolysis is simply a reflection of ongoing muscle degeneration resulting from the primary defect.

TABLE 3
Similarities between Oxidative Stress and Muscular Dystrophy

	Species			
Evidence	Human	Mouse	Chicken	Hamster
1. Products of lipid peroxidation	↑	↑	↑	↑
2. Peroxidizable fatty acids	↔	↔	↔	↑
3. Activities of antioxidant enzymes	↑	↑	↑	↑
4. Activity of SR Ca^{2+}-ATPase	↓	↓	↓	Not done
5. Similarity to vitamin E deficiency	√	√	√	√
6. Content of tocopherols	Inconclusive	Inconclusive	↔	Not done
7. Content of tocopheryl quinones	Not done	None done	↑	Not done

F. FREE RADICAL THEORY

Free radical-induced oxidative injury to muscle cells is one hypothesis which can explain most, if not all, of the structural, functional, and biochemical changes characteristic of many neuromuscular disorders including muscular dystrophy.[29,35] Furthermore, oxidative damage consistent with free radical-induced injury has been identified in humans and animal models of muscular dystrophy (Table 3). The studies supporting this hypothesis have been recently reviewed in detail[35] and will not be repeated here. Since the basic defect leading to muscular dystrophy is apparently not the same in all model systems, it would appear that free radical-induced damage is secondary to the underlying pathology of muscular degeneration. Furthermore, antioxidant therapy has not proven to be of significant benefit to victims of the disease. Nevertheless, it is possible that free radical-induced damage plays an important role in the ultimate tissue injury.

In general, an increased level of oxidative injury could arise subsequent to an increase in the generation of radicals or a decrease in protective systems. In addition, it is possible that alterations in the susceptibility of tissue molecules to oxidation could arise secondary to changes in structure or orientation. The likelihood of a structural role for dystrophin would be most consistent with this latter idea.

Measurements in dystrophic muscle fibers have shown an increased content of by-products of oxidative damage, compensatory increases in cellular antioxidants, changes in the proportions and metabolism of cellular lipids, abnormal functions of cellular membranes, altered activities of membrane-bound enzymes such as Ca^{2+}-ATPase, disturbances in cellular protein turnover and energy production, and a variety of other changes (Figure 1). Although free radical-mediated reactions can explain all of these changes, the inhibition of Ca^{2+}-ATPase due to oxidative damage is of particular interest since this effect may be responsible for the increased intracellular free calcium reported in most models of this disease. Furthermore, this effect would provide a link between free radical-mediated processes and the increased degradative activity characteristic of the disease. However, even when considered to-

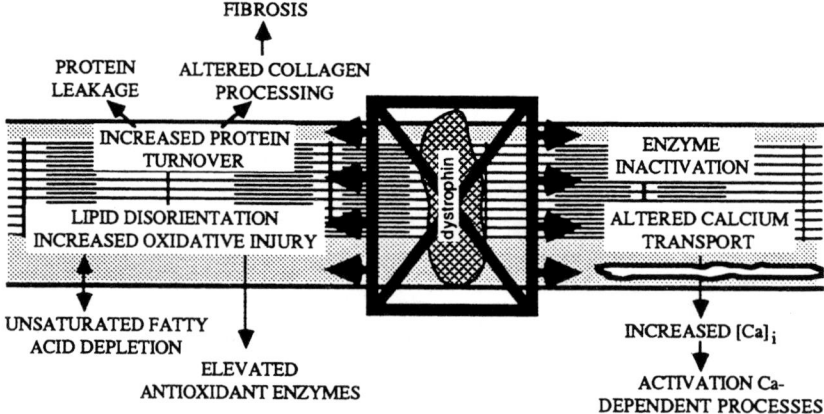

FIGURE 1.   Potential relationships between dystrophin and the damaging process identified in dystrophic muscle fibers. The absence of dystrophin may result in a disorientation of normal membrane structure. This, in turn, may make membrane lipids and proteins more susceptible to inactivation and oxidative injury. In conjuction with other processes, the resulting cascade could lead to muscle fiber degeneration and subsequent fibrosis.

gether, all of these oxidative changes still only represent indirect evidence of free radical-induced pathology.

1. Oxidative Stress in the Chicken Model of Muscular Dystrophy

Many of the tissue alterations which can be linked to free radicals were identified in patients or animals in advanced stages of the disease. This has made it difficult to differentiate changes which cause tissue injury from those which are the result of the numerous changes in cellular physiology and biochemistry during muscle degeneration. The chicken model of muscular dystrophy shares several characteristics with inherited muscular dystrophy in humans, although it is not caused by a deficiency of dystrophin.[4] Both are conditions caused by the defect of a single recessive gene, both lead to the degeneration of muscle tissues over time, and the underlying biochemical basis for cell death is unknown in both disorders.

The chicken model has a number of advantages over mammalian models where research is hampered by the heterogenous mixture of fiber types and the presence of cells at various stages of maturation. In particular, birds, but not mammals, exhibit an anatomical separation of fast (affected by the disease) and slow (unaffected) twitch muscle fiber types which enables the examination of virtually pure populations of the fibers. Furthermore, although the disease develops within several weeks of hatching, early stages before tissue pathology is evident can easily be studied. Last, the dramatically lower cost of the dystrophic chicken eggs as compared to genetically affected mice or dogs makes studying large numbers of samples possible.

A variety of methods have been developed to assess whether free radical processes are occurring in a tissue, although electron spin resonance remains

TABLE 4
Muscle Antioxidants in Muscular Dystrophy

	Human	Chicken	Mouse	Hamster
Vitamin E	↔	↔	—	—
Vitamin C	↔	—[a]	—[a]	—[a]
Glutathione	↑	↑	↑	—
Superoxide dismutase	↔, ↓	↑	—	↔
Catalase	↑	↑	—	↑
Glutathione peroxidase	↑	↑	↑	↔
Glutathione reductase	↑	↑	↑	↑
Citrate dehydrogenase	↑	—	—	—
Glucose 6-phosphate dehydrogenase	↑	↑	↑	↑
6-Phosphuglyconate dehydrogenase	—	↑	↑	—
Glutathione S-transferase	—	↑	—	—
Thioredoxin reductase	—	—	—	↑

[a] Not a vitamin in these species.

the only technique to directly measure radicals. Indirect methods include assessing the *in vitro* generation of radicals by subcellular fractions, the measurement of molecular changes consistent with oxidative processes (lipid peroxidation, protein oxidation, thiol oxidation), and measurement of a cell's enzymatic and nonenzymatic antioxidant status.

2. Tissue Antioxidants

Studies from several laboratories have shown that endogenous antioxidants and antioxidant enzymes are present at elevated levels in muscles from DMD patients and animal models (Table 4).[35,36] Such increases can reflect a normal compensatory response of a tissue to an oxidative stress. This may not be the basis for the response of dystrophic muscles, but the results show that, at the least, this tissue is not deficient in these factors.

Despite the apparent adequacy of tissue antioxidants, and the absence of an enhanced production of radicals by dystrophic tissues,[37,38] other data are suggestive that oxidative stress occurs in dystrophic muscles. Compared with normal muscle, significantly higher contents of glutathione disulfide, protein-glutathione mixed disulfides, and oxidized proteins were found in pectoralis major muscle (the muscle affected by this disease) of genetically dystrophic chickens at 4 weeks of age (before muscle degeneration is evident).[39] Other tissues did not show such disease-related differences. An examination of the overall magnitude of the oxidative changes in affected muscle revealed that they were relatively small and did not seem sufficient to explain the tissue injury which eventually appears. However, thiol residues of the protein pool in normal, but not dystrophic, pectoralis major muscle were relatively less oxidized in relation to the GSH pool as compared to all other tissues studied.[39] The mechanism and purpose of this unique maintenance of virtually no oxidized thiols in the proteins of normal pectoralis major muscle is not known, but its disappearance may be related to the disease process. It is, therefore, interesting that the development of nutritional myopathies in chickens is dependent not only on the deficiency of vitamin E, but also on a simultaneous lack of cysteine and other sulfur-containing amino acids.[40] The associated depletion of protein, but not soluble, thiol pools may be the underlying basis of this synergism.[40,41] One can speculate that the protein thiol loss common to both nutritional and inherited muscular dystrophy in chickens may affect the cysteine-rich domain of the dystrophin homolog in these birds. The function of this domain in mammalian dystrophin is not known, but is unlikely to be structural.

3. Vitamin E

As a result of studies on the nutritional myopathies, it has been known for many years that vitamin E, including both α- and γ-tocopherol (α-TH and γ-TH), is critical to normal muscle function.[42,43] Because of the absence of changes in tissue levels of vitamin E, and the failure of antioxidant therapy to ameliorate the genetic disorder, researchers have generally rejected the theory, first formulated 41 years ago,[44] that a defect in vitamin E metabolism in muscle tissue is the underlying basis of muscular dystrophy.

Early investigations into the levels and, more importantly, the antioxidant activity of vitamin E in muscle were limited by the relatively insensitive techniques used to measure the tocopherols and the tocopheryl quinones (α-TQ), which can be formed upon the oxidation of vitamin E. Using improved methodology,[45] changes in tissue levels of vitamin E and its oxidation product, α-TQ, consistent with an oxidative stress, were measured in chickens.[46] Although these changes were relatively small, the molar ratio of vitamin E to unsaturated fatty acids of 1:1200[47] suggests that each molecule of vitamin E must protect a large number of potential sites of oxidative attack. The efficiency of this protection, as well as the relatively slow turnover of vitamin E, implies some sort of recycling process is needed to avoid having to replace the entire vitamin E molecule every time it scavenges a free radical—a process likely to occur quite often.

Vitamin E cycling has been hypothesized to involve a "free radical reductase" and an activity consistent with this hypothesis has been identified in rat liver microsomes and cytosol.[48-52] A deficiency in vitamin E "activity" rather than "quantity" could provide a molecular explanation for the pathology of muscular dystrophy. Investigations into this hypothesis revealed that the activity of these antioxidant factors was low in chicken liver, and no difference could be measured between tissues from normal or dystrophic animals.[53] Furthermore, neither the chicken, mouse, nor rat had any significant activities of the factors in normal or dystrophic muscle.[53] Thus, it appears that this "free radical reductase" activity is specific for rat liver microsomes and perhaps nuclei.[53,54] These results, and our inability to demonstrate enhanced lipid peroxidation in dystrophic chicken muscle,[46] do not support a causal role for oxidative stress in this model of muscular dystrophy. It remains possible that the methodology to identify vitamin E cycling activity is not yet sufficiently developed since, intuitively, some sort of regeneration seems likely to occur in all membranes.

V. CONCLUSIONS

Exogenously generated free radicals or nutritional antioxidant deficiencies can clearly induce tissue injury which mimics the pathology of many disorders, including muscular dystrophy. The generalized toxicity of free radicals explains in part why, in addition to muscular dystrophy, several diseases of unknown origin are thought possibly to be caused by these reactive species. However, establishing a link, even secondarily, between free radicals and a specific disorder is difficult. With the possible exceptions of oxygen and carbon tetrachloride toxicity, free radicals have not been conclusively shown to cause any disease or pathology.

Based on the extensive studies published over the last 40 years, free radical-induced injury appears unlikely to play a primary role in the tissue injury associated with muscular dystrophy. Endogenous antioxidants are not deficient in dystrophic muscle and an enhanced production of radicals by dystrophic tissues could not be measured. Furthermore, therapeutic strategies involving antioxidants have not resulted in significant benefits to dystrophic animals or humans.[23]

The absence of dystrophin is clearly the critical defect in Duchenne and Becker's muscular dystrophy, as well as the mdx mouse, cat, and CXMD dog models of this disease. However, the function and pathologic role this protein plays in different forms of the disease remains unknown. The membrane location of dystrophin suggests that the absence of this protein may predispose these structures to oxidative injury. However, despite changes indicative of free radical injury in dystrophic muscle, current evidence does not support the hypothesis that free radicals play a causal role in muscle damage although it remains possible that this mechanism is involved in the secondary tissue injury associated with some forms of muscular dystrophy. The inconclusive role for oxidative stress in muscular dystrophy, despite extensive research, reemphasizes the problems encountered in establishing free radicals as a significant factor in human diseases.

REFERENCES

1. **Hoffman, E. P., Brown, R. H., and Kunkel, L. M.,** Dystrophin: the protein product of the Duchenne muscular dystrophy locus, *Cell,* 51, 919, 1987.
2. **Hoffman, E. P., Fischbeck, K. H., Brown, R. H., Johnson, M., Medori, R., Loike, J. D., Harris, J. B., Waterston, R., Brooke, M., Specht, L., Kupsky, W., Chamberlain, J., Caskey, T., Shapiro, F., and Kunkel, L. M.,** Characterization of dystrophin in muscle-biopsy specimens from patients with Duchenne's of Becker's muscular dystrophy, *N. Engl. J. Med.,* 318, 1363, 1988.
3. **Cooper, B. J., Winand, N. J., Stedman, H., Valentine, B. A., Hoffman, E. P., Kunkel, L. M., Scott, M.-O., Fischbeck, K. H., Kornegay, J. N., Avery, R. J., Williams, J. R., Schmickel, R. D., and Sylvester, J. E.,** The homologue of the Duchenne locus is defective in X-linked muscular dystrophy of dogs, *Nature (London),* 334, 154, 1988.
4. **Kunkel, L. M. and Hoffman, E. P.,** Duchenne/Becker Muscular dystrophy *Br. Med. Bull.,* 45, 630, 1989.
5. **Cooper, B. J.,** Animal models of Duchenne and Becker muscular dystrophy, *Br. Med. Bull.,* 45, 703, 1989.
6. **Zubrzycka-Gaarn, E. E., Bulamn, D. E., Karpati, G., Burghes, A. H. M., Belfall, B., Klamut, H. J., Talbot, J., Hodges, R. S., Ray, P. N., and Worton, R. G.,** The Duchenne muscular dystrophy gene product is localized in sarcolemma of human skeletal muscle, *Nature (London),* 333, 466, 1988.
7. **Murphy, M. E. and Kehrer, J. P.,** Free radicals: a potential pathogenic mechanism in inherited muscular dystrophy, *Life Sci.,* 39, 2271, 1986.
8. **Appel, S. H. and Roses, A. D.,** The muscular dystrophies, in *The Metabolic Basis of Inherited Disease,* Stanbury, J. B., Wyngaarden, J. B., Fredrickson, D. S., Goldstein, J. L., and Brown, M. S., Eds., McGraw-Hill, New York, 1983, 1970.
9. **Moser, H.,** Duchenne muscular dystrophy: pathogenic aspects and genetic prevention, *Hum. Genet.,* 66, 17, 1984.
10. **Harris, J. B. and Slater, C. R.,** Animal models: what is their relevance to the pathogenesis of human muscular dystrophy? *Br. Med. Bull.,* 36, 193, 1980.
11. **Elbrink, J. and Malhotra, S. K.,** The pathogenesis of Duchenne muscular dystrophy: significance of experimental observations, *Med. Hypoth.,* 17, 375, 1985.
12. **Vrbova, G.,** Duchenne dystrophy viewed as a disturbance of nerve-muscle interactions, *Muscle Nerve,* 6, 671, 1983.
13. **Cosmos, E., Butler, J., Mazliah, J., and Allard, E. P.,** Animal models of muscle diseases. part I: avial dystrophy, *Muscle Nerve,* 3, 252, 1980.
14. **Cosmos, E., Butler, J., Allard, P., and Mazliah, J.,** Factors that influence the phenotypic expression of genetically normal and dystrophic muscles, *Ann. N.Y. Acad. Sci.,* 317, 571, 1979.
15. **Cosmos, E., Butler, J., Mazliah, J., and Allard, E. P.,** Animal models of muscle diseases. part II: murine dystrophy, *Muscle Nerve,* 3, 350, 1980.
16. **Douglas, W. B. and Montgomery, A.,** Parabiosis: an appraisal of the technique and its role in the study of muscle diseases in animals, *Ann. N.Y. Acad. Sci.,* 317, 611, 1979.
17. **McComas, A. J., Preswick, G., and Garner, S.,** The sick motoneurone hypothesis of muscular dystrophy, *Prog. Neurobiol.,* 30, 309, 1988.
18. **Mason, K. E. and Korwitt, M. K.,** Effects of deficiency in animals, in *The Vitamins: Chemistry, Physiology, Pathology, Methods,* Vol. 5, Sebrell, W. H. and Harris, R. S., Eds., Academic Press, San Diego, 1972, 272.
19. **Nelson, J.,** Neuropathological studies of chronic vitamin E deficiency in mammals including humans in *Biology of Vitamin E,* Ciba Foundation Symposium #101, Pitman Books, London, 1983, 92.
20. **Engel, W. K.,** Duchenne muscular dystrophy: a histological based ischemia hypothesis and comparison with experimental ischemia myopathy, in *The Striated Muscle,* Pearson, C. M. and Mostofi, F. K., Eds., Williams and Wilkins, Baltimore, MD, 1973, 453.

21. **King, D. B., King, C. R., and Jacaruso, R. B.**, Avian muscular dystrophy: thyroidal influence on pectoralis muscle growth and glucose-6-phosphate dehydrogenase activity, *Life Sci.*, 28, 577, 1981.

22. **Ellis, D. A.**, Intermediary metabolism of muscle in Duchenne muscular dystrophy, *Br. Med. Bull.*, 36, 165, 1980.

23. **Kingston, W. J. and Moxley, R. T.**, Treatment of muscular dystrophies, *Gen. Pharmacol.*, 20, 263, 1989.

24. **Goldsmith, B. M. and Gruemer, H.**, Systemic membrane defect and the inhibition of lymphocyte capping in Duchenne muscular dystrophy, *Clin. Chim. Acta*, 164, 33, 1987.

25. **Karagol, U., Gerdner-Medwin, D., and Mastaglia, F. L.**, Neutrophil function in Duchenne muscular dystrophy, *J. Neurol. Sci.*, 73, 73, 1986.

26. **Engle, A. G. and Arahata, K.**, Mononuclear cells in myopathies: quantitation of functionally distinct subsets, recognition of antigen-specific cell-mediated cytotoxicity in some diseases, and implications for the pathogenesis of the different inflammatory myopathies, *Human Pathol.*, 17, 704, 1986.

27. **Nagy, B. and Samaha, F. J.**, Membrane defects in Duchenne dystrophy: protease affecting sarcoplasmic reticulum, *Ann. Neurol.*, 20, 50, 1986.

28. **Gopalan, P., Dufresne, M. J., and Warner, A. H.**, Evidence for a defective thiol protease inhibitor in skeletal muscle of mice with hereditary muscular dystrophy, *Biochem. Cell Biol.*, 64, 1010, 1986.

29. **Davison, A., Tibbits, G., Shi, A., and Moon, J.**, Active oxygen in neuromuscular disorders, *Mol. Cell. Biochem.*, 84, 199, 1988.

30. **Komatsu, K., Inazuki, K., Hosoya, J., and Satoh, S.**, Beneficial effect of new thiol protease inhibitors, epoxide derivatives, on dystrophic mice, *Exp. Neurol.*, 91, 23, 1986.

31. **Tsuji, S. and Matsushita, H.**, Successful treatment of murine muscular dystrophy with the protease inhibitor bestatin, *J. Neurol. Sci.*, 72, 183, 1986.

32. **Gopalan, P., Dufresne, M. J., and Warner, A. H.**, Thiol protease and cathepsin D activities in selected tissues and cultured cells from normal and dystrophic mice, *Can. J. Physiol. Pharmacol.*, 65, 124, 1987.

33. **Turner, P. R., Westwood, T., Regen, C. M., and Steinhardt, R. A.**, Increased protein degradation results from elevated free calcium levels found in muscle from mdx mice, *Nature (London)*, 335, 735, 1988.

34. **Ashmore, C. P., Summers, P. J., and Lee, Y. B.**, Proteolytic enzyme activities and onset of muscular dystrophy in the chick, *Exp. Neurol.*, 94, 585, 1986.

35. **Murphy, M. E. and Kehrer, J. P.**, Oxidative stress and muscular dystrophy, *Chem.-Biol. Interact.*, 69, 101, 1989.

36. **Salminen, A. and Kilhström, M.**, Increased susceptibility to lipid peroxidation in skeletal muscles of dystrophic hamsters, *Experientia*, 45, 747, 1989.

37. **Ohta, K. and Muzuno, Y.**, Pathogenesis of progressive muscular dystrophy: studies on free radical metabolism in an animal model, *Acta Neurol. Scand*, 77, 108, 1988.

38. **Murphy, M. E.**, Oxidative stress in inherited muscular dystrophy in chickens, Ph.D. dissertation, University of Texas at Austin, TX, 1988.

39. **Murphy, M. E. and Kehrer, J. P.**, Oxidation state of tissue thiol groups and content of protein carbonyl groups in chickens with inherited muscular dystrophy, *Biochem. J.*, 260, 359, 1989.

40. **Hull, S. J. and Scott, M. L.**, Studies on the changes in reduced glutathione of chick tissues during onset and regression of nutritional muscular dystrophy, *J. Nutr.*, 106, 181, 1976.

41. **Shih, J. C. H., Jonas, R. H., and Scott, M. L.**, Oxidative deterioration of the muscle proteins during nutritional muscular dystrophy in chicks, *J. Nutr.*, 107, 1786, 1977.

42. **Hadlow, W. J.**, Myopathies of animals, in *The Striated Muscle*, Pearson, C. M. and Mostofi, F. K., Eds., Williams & Wilkins, Baltimore, MD, 1973, 364.

43. **Jackson, M. J., Jones, D. A., and Edwards, R. H. T.**, Vitamin E and muscle diseases, *J. Inher. Metab. Dis.*, 8, 84, 1985.

44. **Milhorat, A. T., Mackenzie, J. B., Ulick, S., Rosenkrantz, H., and Bartels, W. E.**, Observations on a biologically active vitamin E derivative present in hog gastric mucin and in hog stomach lining. The biologic activity of DL,alpha tocopherylhydroquinone, *Ann. N.Y. Acad. Sci.*, 52, 334, 1949.

45. **Murphy, M. E. and Kehrer, J. P.**, Simultaneous measurement of tocopherols and tocopheryl quinones in tissue fractions using high-performance liquid chromatography with redox-cycling electrochemical detection, *J. Chromatogr. Biomed. Appl.*, 421, 71, 1987.

46. **Murphy, M. E. and Kehrer, J. P.**, Altered contents of tocopherols in chickens with inherited muscular dystrophy, *Biochem. Med. Metab. Biol.*, 41, 234, 1989.

47. **Evarts, R. P. and Bieri, J. G.**, Ratios of polyunsaturated fatty acids to α-tocopherol in tissues of rats fed corn or soybean oils, *Lipids*, 9, 860, 1974.

48. **Haenen, G. R. M. M. and Bast, A.**, Protection against lipid peroxidation by a microsomal glutathione-dependent labile factor, *FEBS Lett.*, 159, 24, 1983.

49. **Ursini, F., Maiorino, M., and Gregolin, C.**, The selenoenzyme phospholipid hydroperoxide glutathione peroxidase, *Biochim. Biophys. Acta*, 839, 62, 1985.

50. **Reddy, C. C., Scholz, R. W., Thomas, C. E., and Massaro, E. J.**, Vitamin E dependent reduced glutathione inhibition of rat liver microsomal lipid peroxidation, *Life Sci.*, 31, 571, 1982.

51. **McCay, P. B., Gibson, D. D., Fong, K.-L., and Hornbrook K. R.**, Effect of glutathione peroxidase activity on lipid peroxidation in biological membranes, *Biochim. Biophys. Acta*, 431, 459, 1976.

52. **Burk, R. F., Trumble, M. J., and Lawrence, R. A.**, Rat hepatic cytosolic glutathione-dependent enzyme protection against lipid peroxidation in the NADPH-microsomal lipid peroxidation system, *Biochim. Biophys. Acta*, 618, 35, 1980.

53. **Murphy, M. E. and Kehrer, J. P.**, Lipid peroxidation inhibitor factors in liver and muscle of rat, mouse, and chicken, *Arch. Biochem. Biophys.*, 268, 585, 1989.

54. **Tirmenstein, M. and Reed, D. J.**, Effects of glutathione on the alpha-tocopherol-dependent inhibition of nuclear lipid peroxidation, *J. Lipid Res.*, 30, 959, 1989.

Chapter 11

PROTECTION AGAINST FREE RADICAL-MEDIATED TISSUE INJURY

Mary Treinen Moslen

TABLE OF CONTENTS

The goal of this volume was to critically discuss evidence for the participation of free radicals and reactive oxygen species (ROS) in tissue injury. Only a few ROS have an unpaired electron (e.g., superoxide and hydroxyl radical) and thus qualify as free radicals. Other ROS including peroxides, epoxides, and hypohalous acids, are logically considered under the topic of free radicals because ROS can form free radical species or can be produced by reactions involving free radicals. This concluding chapter will discuss the defense mechanisms that restrain and quench potentially injurious reactions involving free radicals and ROS.

I. TYPES OF EVIDENCE IMPLICATING FREE RADICALS IN TISSUE INJURY

Table 1 lists the major forms of tissue injury examined in this book where free radicals have been implicated in a "causal" role. Because most free radicals are very reactive and exist briefly, difficulties are frequently encountered documenting a role for free radicals or ROS in specific processes of cell injury. As Smith discussed in Chapter 1, three types of evidence can provide documentation of a role for free radicals in a given process: direct detection of radicals by spectra or spin traps, indirect detection of the characteristic products of radical-initiated reactions, and pharmacologic modulation of radical-mediated reactions by protection/potentiation studies.[1] These types of evidence support the radical-linked forms of tissue injury listed in Table 1.

Least frequently documented was the direct detection of free radical species by electron paramagnetic resonance (EPR) spectra or by spin-trap adducts within affected tissues. One example cited by Taylor and Shappell in Chapter 5 about heart injury due to ischemia-reflow[2] was the detection of EPR spectra for oxygen-centered radicals shortly after the onset of reflow and the attenuation of these radical spectra by infusion of superoxide dismutase (SOD) (see Section III.B, Chapter 5). Repeatedly noted as indicative of a free radical process was the detection of characteristic products of lipid peroxidation, since this reaction is well known to be initiated by radical addition or abstraction. For example, Table 1 in Chapter 9 shows a decrease in the peroxidation-vulnerable polyunsaturated fatty acid docosahexaenoate and an increase in the lipid peroxidation product malondialdehyde as evidence for hyperoxia-induced peroxidation in the retinas of rat neonates.[3]

Pharmacologic modulation studies involving vitamin E (α-tocopherol) were repeatedly cited because of the important role of vitamin E in the quenching of free radicals species formed during lipid peroxidation reactions (see Figure 8, Chapter 1).[1] Specifically, vitamin E deficiency enhanced the retinopathy of prematurity associated with hyperoxia;[3] vitamin E treatment had a beneficial effect on ischemia-reflow cardiac injury;[2] and ethanol administration led to an increase in hepatic concentrations of α-tocopherol quinone, which is a product of the reaction of ethanol with tocopherol.[4]

TABLE 1
Tissues Injured by Radicals or ROS

Tissue	Initiating Process	Ref.
Liver	CCl$_4$-derived radicals	1
	Diquat/ROS	1
	Ethanol (?)	4
Lung	Asbestos/Fe/ROS	6
	Hyperoxia/ROS	7
Retina	Hyperoxia/ROS	3
Central nervous system	Hemorrhage/Fe/ROS	8
Heart	Ischemia—reflow/ROS	2
Vascular system	Cu or Fe/lipid peroxides	9

An important point which several authors addressed is the difficulty of definitively distinguishing a causal role for free radicals and ROS in tissue injury processes involving chronic chemical exposure or a progressive disease. Chapter 3 reviews available evidence on liver injury by ethanol; as illustrated in Figure 3, ethanol could yield reactive free radical metabolites via its biotransformation by the microsomal ethanol-oxidizing system or via the metabolism of its major product acetaldehyde by xanthine oxidase. Yet, as Lauterburg and de Quay noted, correlations between radical formation after ethanol administration and the observed detrimental biologic effects remain unestablished.[4] Their work described another problem encountered in the ethanol studies, which was properly interpreting increases in some products of lipid peroxidation, since ethanol-induced slower metabolic clearance (e.g., inhibited metabolism) could account for the higher amounts of both malondialdehyde and ethane that have been found after ethanol administration.

Extensive studies may be necessary before a consensus can be reached that free radicals do *not* play a causal role in a disease process. A major factor in the negative conclusion reached in Kehrer and Murphy's review[5] about the role of free radicals in muscular dystrophy was the inability of vitamin E therapy to ameliorate the disease despite the critical role of this antioxidant in normal muscle function. These authors explained a confounding problem encountered due to the numerous alterations in cellular physiology and biochemistry that occur during tissue deterioration in chronic progressive disorders like muscular dystrophy. Specifically, it was often difficult to distinguish the effects of these secondary alterations from the primary changes that caused the injury.

II. MOLECULAR AND FUNCTIONAL ALTERATIONS DUE TO FREE RADICALS OR ROS

Injury to a tissue or organism is the end result of a series of molecular and functional alterations. Table 2 lists some of the altered structures and functions of biological entities which several authors attributed to free radicals

TABLE 2
Structural and Functional Alterations of Biological Entities Produced by Radicals or ROS

Structural alterations

Lipid peroxidation
Protein adducts
Hemoglobin adducts
Protein thiol oxidation

Functional alterations

Increased permeability of membranes
Impaired response of membranes to signals
Altered enzyme activity
Reduced affinity of binding proteins
Altered uptake of oxidized entities

or ROS. One unusual molecular alteration described by Corbett and Corbett[10] is the adduct between nucleic acids and arylamines that they observed after activating the respiratory burst of human granulocytes. In Figure 2 of Chapter 6, the ROS formed by the respiratory burst are linked with the conversion of arylamines to reactive species that are capable of forming such adducts. The vulnerability of protein thiols to radicals and ROS was a common theme. In Figure 6 of Chapter 5 about the role of ROS in ischemia-reflow, Taylor and Shappell postulated that ROS-mediated thiol modifications contribute to multiple functional changes.[2]

ROS were linked by Elliott and Schilling to an impaired ability of lung vascular cells to modify cell levels of the intracellular messenger Ca in response to a signal.[7] Alterations in leukocyte functions and adherence were described by Taylor and Shappell as consequences of ROS from ischemia-reflow.[2] Alterations of LDL (low-density lipoprotein) structure by oxidants were reported by Haberland and Smith to affect LDL uptake by its cellular receptors.[9]

III. SYSTEMS THAT PROTECT AGAINST FREE RADICALS OR ROS

Table 3 lists primary and auxiliary systems that protect intra- and extracellular spaces against free radicals and ROS. Endogenous antioxidants are specialized for function in lipid or aqueous environments. The major lipid-soluble species are vitamin E (α-tocopherol) and β-carotene, which has a particular role for the quenching of singlet oxygen.[11] The most important water-soluble antioxidants are the thiol-containing tripeptide glutathione (GSH) and vitamin C (ascorbate). Detoxification enzymes are widely distributed in cells, fluids, and tissues. Three enzymes detoxify ROS. Superoxide dismutase (SOD) converts O_2^- to H_2O_2 and catalase converts H_2O_2 to H_2O. Glutathione

TABLE 3
Primary and Auxiliary Protection Systems

Primary Systems	Auxiliary Systems
Antioxidants	**Antioxidant Regenerators**
Vitamin E	GSSG reductase
β-Carotene	Glucose-6-phosphate dehydrogenase
Glutathione	
Vitamin C	
Detoxification Enzymes	**Exporters**
Superoxide dismutase	GSSG transporter
Catalase	GS-conjugate transporter
Glutathione peroxidases	
Glutathione transferases	
Transition Metal Binders	
Transferrin	
Ferritin	
Ceruloplasmin	
Albumin	
Metallothionein	

peroxidases have a wider capacity for peroxide substrates. Several isozymes of glutathione transferases have peroxidase activity and can also detoxify epoxides and reactive alkenal products of lipid peroxidation.[12]

Transition metals, such as iron and copper, participate in the conversion of H_2O_2 to OH· and lipid peroxides to lipid oxide radicals. Therefore, plasma and intracellular proteins that restrict these conversions by binding transition metals serve an important role in protecting tissues against ROS.[13] Major extracellular metal binders are transferrin, with its high affinity for Fe, and ceruloplasmin, which more specifically binds Cu. Significant extracellular binding of transition metals is also provided by albumin, although it binds Fe and Cu with relatively less affinity because albumin is present in large amounts in plasma.[13] Intracellular proteins which bind transition metals are ferritin, which stores Fe, and metallothionein, which binds many metals including Cu.[14]

Three chapters in this book described experiments indicating a role for Fe in radical- or ROS-mediated injurious processes where the iron chelator desferrioxamine was used to attenuate injurious processes. Figure 1 in the chapter by Shull et al.[6] about asbestos-induced lung injury shows higher survival rates in lung fibroblasts exposed to asbestos when the cells were pretreated with desferrioxamine. Infusion of desferrioxamine with hemoglobin is shown in Figure 3 in the chapter by Sadrzadeh and Eaton[8] about hemoglobin-induced loss of brain Na^+/K^+-ATPase activity to provide excellent protection against this parameter of tissue injury. Diquat-induced hepatic injury via stimulation of ROS was found in experiments described by Smith[1] to be minimized by pretreatment with desferrioxamine and in consonant results the injury was enhanced by pretreatment with ferrous sulfate. Thus, desferriox-

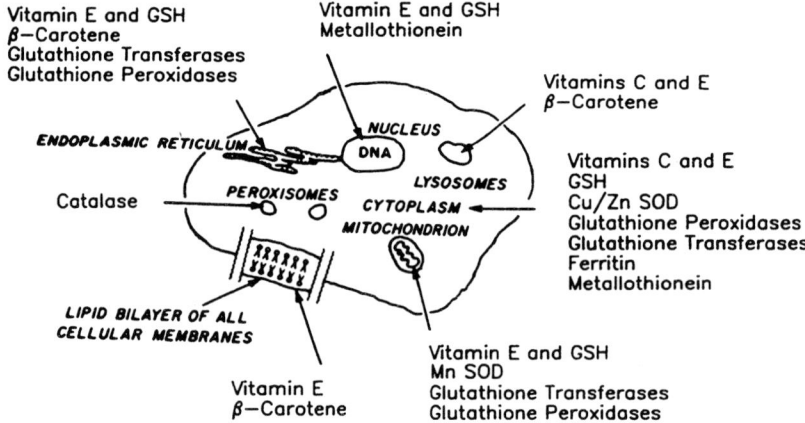

FIGURE 1. Distribution of antioxidants, detoxification enzymes, and transition metal binding proteins that comprise the intracellular defense system within cellular membranes and organelles. (Modified from Machlin, L .J. and Bendich, A., *Fed. Am. Soc. Exp. Biol.*, 1, 444, 1987.)

amine protection experiments can provide a "pharmacologic modulation" type of evidence for Fe-mediated injuries that is analogous to the type of evidence described previously for vitamin E modulation of free radical-mediated injury in other studies.

Protection systems are not uniformly distributed in extracellular fluids, tissues, or cellular organelles. Glutathione concentrations are about 100-fold higher in human alveolar lining fluid than in plasma, which may be very important in the protection of the lung against ROS.[15] Ascorbic acid is higher in the two intraocular fluids, the aqueous and vitreous humors, which surround the lens, than in other extracellular fluids of humans; the ascorbic acid in these intraocular fluids is thought important in the protection of the lens against UV light-induced forms of ROS.[16] Levels of α-tocopherol are 30 to 50 times higher in human red blood cells than mononuclear or polymorphonuclear cells.[17] An immunohistochemical study in hamster tissues revealed wide variability in the amounts of catalase and of mitochondrial and cytoplasmic forms of SOD among tissues, including higher activities of SOD in heart than in liver.[18] Heterogeneity within tissues was also observed, including higher activities of all the detoxification enzymes in kidney proximal tubular cells than in glomerular cells and in stomach surface mucous cells than in gland neck cells. These differences in distribution of detoxification enzymes likely have a basic pathophysiological relevance, such as a greater need for SOD detoxification of O_2^- in the heart with its high tissue O_2 than in the liver with its low tissue O_2.

Figure 1 indicates the distribution of protection systems within an idealized cell. Note the numerous protection systems in the cytoplasm where many reactions occur and metals are stored. The mitochondrion is defended against ROS by vitamin E, GSH, SOD, glutathione peroxidases, and transferases.

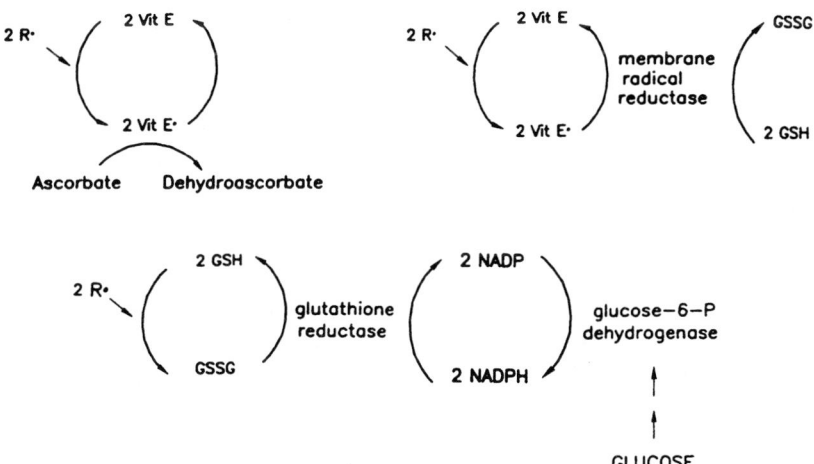

FIGURE 2. Oxidative damage to the polyunsaturated fatty acid component of a membrane phospholipid is illustrated from left to right. Sites where detoxification enzymes and antioxidants detoxify reactive ROS and lipid species are indicated from top to bottom.

Vital material of the nucleus is protected by water- and lipid-soluble antioxidants and by metallothionein.

IV. COMPLEMENTARY AND AUXILIARY PROTECTION SYSTEMS

The enzymes and antioxidants of the protection systems have some complementary functions related to their distribution in different spaces or their sequential reactions.[11,13] The schematic in Figure 2 illustrates the complementary involvement of multiple antioxidants and detoxification enzymes at various stages of lipid peroxidation. The oxidative damage is initiated by the reaction of O_2^- or H_2O_2 with the polyunsaturated fatty acid of a membrane phospholipid and leads to fragmentation of the oxidized fatty acid or its conversion to an unstable peroxide. SOD and/or catalase can intervene before the initial injurious reaction of the first ROS species. Vitamin E can donate a H· to the lipid peroxyl radical species and prevent it from abstracting a H· from another fatty acid, and thus propagating the radical reaction.

The phospholipid fatty acid peroxide (PLOOH), formed as a result of either a radical termination or radical propagation reaction, must be liberated (by a phospholipase) as a free fatty acid before its further detoxification to a fatty acid alcohol by classical GSH peroxidase or a GSH transferase with peroxidase activity.[19] Alternatively, as shown in Figure 2, instead of being converted intact to PLOOH, the fatty acid peroxide radical could fragment into reactive species including alkenals and dialdehydes. Alkenals, such as 4-hydroxynonenal, are able to attack and denature proteins.[20] Fortunately, certain GSH transferase isozymes very efficiently convert these alkenals to glutathione conjugates.[12]

Phospholipids are not the only lipid components of membranes that are converted to peroxides by lipid peroxidation reactions since cholesterol peroxides (ChOOH) are also formed. Membrane ChOOH is inert to GSH peroxidases but has been found to be reduced to the alcohol ChOH by a recently characterized special type of GSH peroxidase called "phospholipid hydroperoxide glutathione peroxidase".[19] This new enzyme differs from the classical GSH peroxidase because of its greater lipid solubility and ability to react directly with membrane-bound lipid peroxides. Interestingly, this new peroxidase can reduce the hydroperoxide of cholesterol ester (CEOOH), a major lipid in LDL, and thus can reduce all the major oxidized lipids in LDL including PLOOH, ChOOH, and CEOOH.[19] Given the importance of oxidized LDL in the development of atherosclerotic lesions, as reviewed by Haberland and Smith,[9] further studies are clearly needed to investigate any beneficial effects that phospholipid peroxide glutathione peroxidase may have detoxifying oxidized lipids in LDL.

The detoxification reactions schematized in Figure 2 transform the critical antioxidants, vitamin E and GSH, to inactive vitamin E products and glutathione disulfide (GSSG). The processes which generate these antioxidants are considered important auxiliary components of the protection system (Table 3). Figure 3 illustrates nonenzymatic regeneration of vitamin E from a radical product (ascorbate). One ascorbate molecule could reduce two vitamin E radicals by successive one-electron (i.e., H·) transfer reactions with the formation of an intermediate ascorbyl radical that disproportionates to the stable product dehydroascorbate.[21] Alternatively, as shown in Figure 3, vitamin E radicals can be regenerated enzymatically via a "membrane free radical reductase"-mediated reaction that is dependent upon GSH. Limited information is available about the distribution and function of the membrane free radical reductase, but its existence explains the ability of GSH to preserve vitamin E in the active α-tocopherol form that has been observed *in vitro* in experimental systems of membrane peroxidation.[22] In addition, a recent report indicates that the nonradical vitamin E product formed after vitamin E has sequentially donated two H· to two R· can also be regenerated in an ascorbate-dependent two-electron transfer process.[23]

Figure 3 illustrates the regeneration of GSSG. The GSSG reductase that catalyzes this reaction depends on the availability of NADPH generated from glucose by the glucose-6-phosphate dehydrogenase system.

A second type of auxiliary protection system (Table 3) is the plasma membrane exporter that removes GSSG and/or GS-conjugates, such as the alkenal GS-conjugate shown in Figure 2. These products of detoxification reactions can detrimentally interact with cell constituents if allowed to accumulate within cells. GSSG can interact with protein thiols (R—SH) to form R—S—S—G while glutathione conjugates can inhibit both GSH transferases and GSSG reductase.[24] GS-conjugate exporters have been identified in red blood cells, heart, and liver.[25-27]

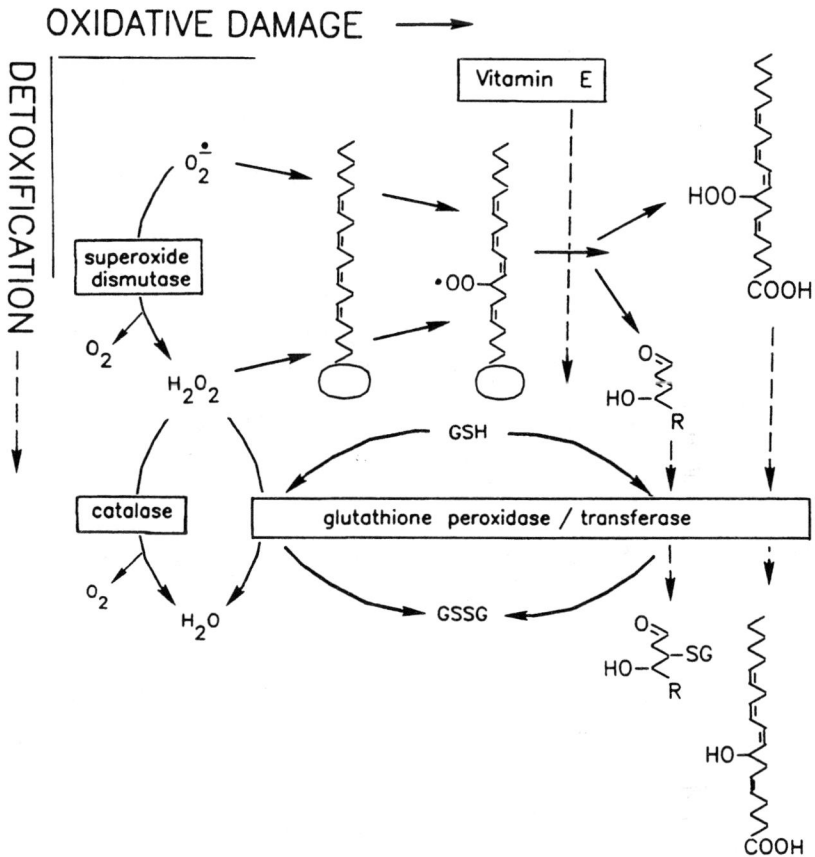

FIGURE 3.   Reactions involved in the regeneration of the lipid-soluble antioxidant vitamin E (top) and the water-soluble antioxidant glutathione (bottom).

The complementary multiplicity of enzymes able to detoxify H_2O_2 is vital to the protection of cellular organelles since catalase has a limited localization within cells compared to GSH peroxidase and GSH transferases with peroxidase functions (see Figure 1). The ability of the aqueous antioxidants ascorbate and GSH to regenerate the lipid-soluble α-tocopherol by direct or enzymatic reactions (see Figure 3) is a complementary linkage of the protection systems existing in different regions of the cell. The proteins listed in Table 3 that are able to bind transition metals have a complementary distribution in either extra- or intracellular spaces (e.g., transferrin vs ferritin, and ceruloplasmin vs metallothionein).

V. DEFICIENCY, INDUCTION, AND ACTIVATION OF PROTECTION SYSTEMS

A classical example of the severe consequences of a protection system deficiency is the hemolytic anemia which occurs in individuals with genetic glucose-6-phosphate dehydrogenase deficiency. Children with this problem have low plasma vitamin E levels and can respond to vitamin E therapy with improved erythrocyte survival.[28] As indicated by the interrelated reactions for antioxidant regeneration in Figure 3, deficiency of glucose-6-phosphate dehydrogenase should be associated with a primary impaired ability to supply the reducing equivalents needed to regenerate GSH and a secondary impaired ability to regenerate vitamin E by GSH-dependent reactions.

Cardiac injury associated with a deficiency of GSH peroxidase has been described in persons receiving inadequate amounts of selenium, which is an essential cofactor for GSH peroxidase. This problem has occurred endemically in regions of China with low levels of selenium in the soil[29] and in a few patients receiving protracted parenteral nutrition without selenium.[30] The endemic problem in China has been rectified by selenium supplements and for parenteral nutrition patients by addition of selenium to parenteral mixtures.

Specific inhibitors of protection enzymes provide valuable experimental tools to examine the role of ROS in tissue injury or functional changes associated with such injury. Bis(2-chloroethyl)-1-nitrosourea (BCNU) selectively inhibits GSSG reductase activity which leads to an impaired regeneration of the antioxidant GSH;[31] therefore, BCNU is a useful way to pharmacologically modulate injurious processes involving ROS (see Table 4 in Chapter 1).[1] In Table 2 of Chapter 8, BCNU enhanced the peroxide-induced loss of ability of the lung cells to regulate cytosolic calcium concentrations.

Induction of protection systems after exposure to free radicals or ROS may be an important adaptive response to nonlethal insults by these reactive species. Immunohistochemical studies in animals have revealed increased amounts of the metal-binding protein metallothionein in both the cytoplasm and nucleus of liver and kidney cells after treatment with metals that bind to metallothionein.[32] Activities of GSH peroxidases in various rat tissues were found markedly higher in adults than young or aged animals.[33] The investigators observing this age-dependent change suggested that the higher levels in adult rats could reflect an "inducible" response lost during aging. Interestingly, unlike the age-dependent changes with the classical GSH peroxidases, the phospholipid hydroperoxide glutathione peroxidase showed little change with age, which may reflect a "housekeeping" role for this membrane-localized enzyme.[33]

Figure 2 of Chapter 6 shows the nonuniform induction of the mRNAs for SOD and glutathione peroxidase in lung cells exposed to ROS.[6] Vitamin E-deprived rat pups exposed to hyperoxia were found in experiments described by Penn to have higher levels of GSH peroxidase in their retinas than rat pups exposed to hyperoxia without vitamin E manipulation.[3] Table 4 in the review about muscular dystrophy enumerates the antioxidants and detoxification enzymes reported to be elevated in affected muscles of several muscular dystrophy models.[5] The authors of this review note that the elevations of these protection systems with the development of muscular dystrophy may "reflect a normal compensatory response of a tissue to oxidative stress."

Activation of detoxification enzymes may be a faster adaptive response than enzyme induction. Exposure of hepatic microsomes to ROS activated the GSH transferase localized in this organelle; the activation was evident within 10 min after exposure to ROS and is thought to occur by the oxidation of a sulfhydryl group of the enzyme.[34] Such an activation could explain the 60% increase in GSH transferase activities of hepatic microsomes which my laboratory found within 2 h after *in vivo* treatment of rats with CCl_4,[35] a well known initiator of lipid peroxidation in hepatic microsomes. The biological relevance of the activation of microsomal GSH transferase needs to be further evaluated using, instead of xenobiotic substrates, endogenous substrates such as 4-hydroxynonenal and fatty acid peroxides.

VI. CONCLUSIONS

Knowledge about the role of radicals or ROS in specific diseases facilitates development of appropriate therapies to prevent or attenuate injurious processes. The value of this knowledge was most emphatically stated in Taylor and Shappell's chapter[2] about ischemia-reflow injury, which described the efficacy of many different types of interventive agents in clinical or experiment studies. Many questions remain about the way radicals and ROS alter biological molecules and larger entities. The types of questions remaining were repeatedly noted in the chapter by Haberland and Smith[9] about the role of oxidant-altered LDL in atherosclerosis. There is a great need to apply increasingly specific chemical methodologies and concepts to the study of LDL oxidation and other structure/function alterations involved in the tissue injuries that were discussed in this symposium. This need creates an exciting opportunity for formally trained chemists to advance our understanding of the chemical events involved in human diseases.

ACKNOWLEDGMENT

Supported by the John Sealy Memorial Endowment Fund.

REFERENCES

1. **Smith, C. V.**, Free Radical Mechanisms of Tissue Injury, in *Free Radical Mechanisms of Tissue Injury*, Moslen, M. T. and Smith, C. V., Eds., CRC Press, Boca Raton, FL, 1992, chap. 1.
2. **Taylor, A. A. and Shappell, S. B.**, Reactive oxygen species, neutrophil and endothelial adherence molecules, and lipid-derived inflammatory mediators in myocardial ischemia-reflow injury, in *Free Radical Mechanisms of Tissue Injury*, Moslen, M. T. and Smith, C. V., Eds., CRC Press, Boca Raton, FL, 1992, chap 5.
3. **Penn, J. S.**, Oxygen-induced retinopathy in the rat: a proposed role for peroxidation reactions in the pathogenesis, in *Free Radical Mechanisms of Tissue Injury*, Moslen, M. T. and Smith, C. V., Eds., CRC Press, Boca Raton, FL, 1992, chap 9.

4. **Lauterburg, B. H. and de Quay, B.,** Radicals and oxidants in ethanol-induced liver injury, in *Free Radical Mechanisms of Tissue Injury,* Moslen, M. T. and Smith, C. V., Eds., CRC Press, Boca Raton, FL, 1992, chap 3.
5. **Kehrer, J. P. and Murphy, M. E.,** Free radicals in muscular dystrophy, in *Free Radical Mechanisms of Tissue Injury,* Moslen, M. T. and Smith, C. V., Eds., CRC Press, Boca Raton, FL, 1992, chap 10.
6. **Shull, S., Monohar, M., Marsh, J. P., Janssen, Y. M. W., and Mossman, B. T.,** Role of iron and reactive oxygen species in asbestos-induced lung injury, in *Free Radical Mechanisms of Tissue Injury,* Moslen, M. T. and Smith, C. V., Eds., CRC Press, Boca Raton, FL, 1992, chap 7.
7. **Elliott, S. J. and Schilling, W. P.,** The vascular endothelium in oxidant-induced lung injury, in *Free Radical Mechanisms of Tissue Injury,* Moslen, M. T. and Smith, C. V., Eds., CRC Press, Boca Raton, FL, 1992, chap 8.
8. **Sadrzadeh, S. M. H. and Eaton, J. W.,** Hemoglobin-induced oxidant damage to the central nervous system, in *Free Radical Mechanisms of Tissue Injury,* Moslen, M. T. and Smith, C. V., Eds., CRC Press, Boca Raton, FL, 1992, chap 2.
9. **Haberland, M. E. and Smith, C. V.,** Lipid peroxide-dependent modifications of lipoproteins in atherosclerosis, in *Free Radical Mechanisms of Tissue Injury,* Moslen, M. T. and Smith, C. V., Eds., CRC Press, Boca Raton, FL, 1992, chap 4.
10. **Corbett, M. D. and Corbett, B. R.,** Bioactivation of xenobiotics by the respiratory burst of human granulocytes, in *Free Radical Mechanisms of Tissue Injury,* Moslen, M. T. and Smith, C. V., Eds., CRC Press, Boca Raton, FL, 1992, chap 6.
11. **Machlin, L. J. and Bendich, A.,** Free radical tissue damage: protective role of antioxidant nutrients, *Fed. Am. Soc. Exp. Biol. J.,* 1, 444, 1987.
12. **Alin, P., Danielson, U. H., and Mannervik, B.,** 4-Hydroxyalk-2-enals are substrates for glutathione transferase, *Proc. Meet. Fed. Eur. Biochem. Soc.,* 179, 267, 1985.
13. **Halliwell, B. and Gutteridge, J. M. C.,** Oxygen free radicals and iron in relation to biology and medicine: some problems and concepts, *Arch. Biochem. Biophys.,* 246, 501, 1986.
14. **Bremer, I.,** Involvement of metallothionein in the hepatic metabolism of copper, *J. Nutr.,* 117, 19, 1989.
15. **Cantin, A. M., North, S. L., Hubbard, R. C., and Crystal, R. C.,** Normal alveolar epithelial lining fluid contains high levels of glutathione, *J. Appl. Physiol.,* 63, 152, 1987.
16. **Varma, S. D.,** Ascorbic acid and the eye with special reference to the lens, *Ann. N.Y. Acad. Sci.,* 498, 280, 1987.
17. **Ogihara, T., Miyake, M., Kawamura, N., Tamai, H., Kitagawa, M., and Mino, M.,** Tocopherol concentrations of leukocytes in neonates, *Ann. N.Y. Acad. Sci.,* 570, 487, 1989.
18. **Oberley, T. D., Oberley, L. W., Slattery, A. F., Lauchner, L. J., and Elwell, J. H.,** Immunohistochemical localization of antioxidant enzymes in adult syrian hamster tissues and during kidney development, *Am. J. Pathol.,* 137, 199, 1990.
19. **Thomas, J. P., Geiger, P. G., Maiorino, M., Ursini, R., and Girotti, A. W.,** Enzymatic reduction of phospholipid and cholesterol hydroperoxides in artificial bilayers and lipoproteins, *Biochim. Biophys. Acta,* 1045, 252, 1990.
20. **Benedetti, A., Comporti, M., and Esterbauer, H.,** Identification of 4-hydroxynonenal as a cytotoxic product originating from the peroxidation of liver microsomal lipids, *Biochim. Biophys. Acta,* 620, 281, 1980.
21. **Bendich, A., Machline, L. J., and Scandurra, O.,** The antioxidant role of vitamin C, *Adv. Free Radical Biol. Med.,* 2, 419, 1986.
22. **McCay, P. B. and Powell, S. R.,** Relationship between glutathione and chemically induced lipid peroxidation, in *Glutathione: Chemical, Biochemical, and Medical Aspects,* Part B, David, D., Poulson, R., and Avramovic, O., Eds., 1989, 111.

23. **Liebler, D. C., Kaysen, K. L., and Kennedy, T. A.,** Redox cycles of vitamin E: hydrolysis and ascorbic acid dependent reduction of 8a-(alkyldioxy)tocopherones, *Biochemistry*, 28, 9772, 1989.
24. **Bilzer, M., Krauth-Siegel, R. L., Schirmer, R. H., Akerboom, T. P. M., Sies, H., and Schulz, G. E.,** Interaction of a glutathione S-conjugate with glutathione reductase: kinetic and X-ray crystallography studies, *Eur. J. Biochem.*, 138, 373, 1984.
25. **LaBelle, E. F., Singh, S. V., Srivastava, S. R., and Awasthi, Y. C.,** Evidence for different transport systems for oxidized glutathione and S-nitrophenyl glutathione in human erythrocytes, *Biochem. Biophys. Res. Commun.*, 139, 538, 1986.
26. **Ishikawa, T., Esterbauer, H., and Sies, H.,** Role of cardiac glutathione transferase and of the glutathione S-conjugate export system in biotransformation of 4-hydroxynonenal in the heart, *J. Biol. Chem.*, 261, 1576, 1986.
27. **Lindwall, G. and Boyer, T. D.,** Excretion of glutathione conjugates by primary cultured rat hepatocytes, *J. Biol. Chem.*, 262, 5151, 1987.
28. **Hafez, M., Amar, E.-S., Zedan, M., Hammad, H., Sorour, A. H., El-Desouky, E.-S.A., and Gamil, N.,** Improved erythrocyte survival with combined vitamin E and selenium therapy in children with glucose-6-phosphate dehydrogenase deficiency and mild chronic hemolysis, *J. Pediatr.*, 108, 558, 1986.
29. **Ge, K., Xue, A., Bai, J., and Wang, S.,** Keshan disease: an endemic cardiomyopathy in China, *Virchows Arch. Pathol. Anat. Physiol.*, 401, 1, 1983.
30. **Fleming, C. R., Lie, J. T., McCall, J. T., O'Brien, J. F., Baillie, E. E., and Thistle, J. L.,** Selenium deficiency and fatal cardiomyopathy in a patient on home parenteral nutrition, *Gastroenterology*, 83, 689, 1982.
31. **Kehrer, J. P.,** The effect of BCNU on tissue glutathione reductase activity, *Toxicol. Lett.*, 17, 63, 1983.
32. **Banerjee, D., Onosaka, S., and Cherian, M. G.,** Immunohistochemical localization of metallothionein in cell nucleus and cytoplasm of rat liver and kidney, *Toxicology*, 24, 95, 1982.
33. **Zhang, L., Maiorino, M., Roveri, A., and Ursini, F.,** Phospholipid hydroperoxide glutathione peroxidase: specific activity in tissues of rats of different age and comparison with other glutathione peroxidases, *Biochem. Biophys. Acta*, 1006, 140, 1989.
34. **Aniya, L. and Anders, M. W.,** Activation of rat liver microsomal glutathione S-transferase by reduced oxygen species, *J. Biol. Chem.*, 264, 1998, 1989.
35. **Botti, B., Moslen, M. T., Trieff, N. M., and Reynolds, E. S.,** Transient decrease of liver cytosolic glutathione S-transferase activities in rats given 1,2-dibromoethane or CCl$_4$, *Chem. Biol. Interaction*, 42, 259, 1982.

INDEX

I

K

L